Molecular Modelling
for Beginners

Molecular Modelling for Beginners

Alan Hinchliffe
UMIST, Manchester, UK

WILEY

Other Wiley Editorial Offices

John Wiley & Sons Inc., 111 River Street, Hoboken, NJ 07030, USA

Jossey-Bass, 989 Market Street, San Francisco, CA 94103-1741, USA

Wiley-VCH Verlag GmbH, Boschstr. 12, D-69469 Weinheim, Germany

John Wiley & Sons Australia Ltd, 33 Park Road, Milton, Queensland 4064, Australia

John Wiley & Sons (Asia) Pte Ltd, 2 Clementi Loop #02-01, Jin Xing Distripark, Singapore 129809

John Wiley & Sons Canada Ltd, 22 Worcester Road, Etobicoke, Ontario, Canada M9W 1L1

Wiley also publishes its books in a variety of electronic formats. Some content that appears in print may not be
available in electronic books.

Library of Congress Cataloging-in-Publication Data

(to follow)

British Library Cataloguing in Publication Data
A catalogue record for this book is available from the British Library
ISBN 0 470 84309 8 (Hardback)
     0 470 84310 1 (Paperback)

Typeset in 10.5/13pt Times by Thomson Press (India) Ltd., Chennai
Printed and bound in Great Britain by TJ International Ltd., Padstow, Cornwall
This book is printed on acid-free paper responsibly manufactured from sustainable forestry
in which at least two trees are planted for each one used for paper production.

Contents

Preface

There is nothing radically new about the techniques we use in modern molecular modelling. Classical mechanics hasn't changed since the time of Newton, Hamilton and Lagrange, the great ideas of statistical mechanics and thermodynamics were discovered by Ludwig Boltzmann and J. Willard Gibbs amongst others and the basic concepts of quantum mechanics appeared in the 1920s, by which time J. C. Maxwell's famous electromagnetic equations had long since been published.

The chemically inspired idea that molecules can profitably be treated as a collection of balls joined together with springs can be traced back to the work of D. H. Andrews in 1930. The first serious molecular Monte Carlo simulation appeared in 1953, closely followed by B. J. Alder and T. E. Wainwright's classic molecular dynamics study of hard disks in 1957.

The Hartrees' 1927 work on atomic structure is the concrete reality of our everyday concept of atomic orbitals, whilst C. C. J. Roothaan's 1951 formulation of the HF–LCAO model arguably gave us the basis for much of modern molecular quantum theory.

If we move on a little, most of my colleagues would agree that the two recent major advances in molecular quantum theory have been density functional theory, and the elegant treatment of solvents using ONIOM. Ancient civilizations believed in the cyclical nature of time and they might have had a point for, as usual, nothing is new. Workers in solid-state physics and biology actually proposed these models many years ago. It took the chemists a while to catch up.

Scientists and engineers first got their hands on computers in the late 1960s. We have passed the point on the computer history curve where every 10 years gave us an order of magnitude increase in computer power, but it is no coincidence that the growth in our understanding and application of molecular modelling has run in parallel with growth in computer power. Perhaps the two greatest driving forces in recent years have been the PC and the graphical user interface. I am humbled by the fact that my lowly 1.2 GHz AMD Athlon office PC is far more powerful than the world-beating mainframes that I used as a graduate student all those years ago, and that I can build a molecule on screen and run a B3LYP/6-311++G(3d, 2p) calculation before my eyes (of which more in Chapter 20).

We have also reached a stage where tremendously powerful molecular modelling computer packages are commercially available, and the subject is routinely taught as part of undergraduate science degrees. I have made use of several such packages to produce the screenshots; obviously they look better in colour than the greyscale of this text.

There are a number of classic (and hard) texts in the field; if I'm stuck with a basic molecular quantum mechanics problem, I usually reach for Eyring, Walter and Kimball's *Quantum Chemistry*, but the going is rarely easy. I make frequent mention of this volume throughout the book.

Equally, there are a number of beautifully produced elementary texts and software reference manuals that can apparently transform you into an expert overnight. It's a two-edged sword, and we are victims of our own success. One often meets self-appointed experts in the field who have picked up much of the jargon with little of the deep understanding. It's no use (in my humble opinion) trying to hold a conversation about gradients, hessians and density functional theory with a colleague who has just run a molecule through one package or another but hasn't the slightest clue what the phrases or the output mean.

It therefore seemed to me (and to the Reviewers who read my New Book Proposal) that the time was right for a middle course. I assume that you are a 'Beginner' in the sense of *Chambers Dictionary*–'someone who begins; a person who is in the early stages of learning or doing anything . . .' – and I want to tell you how we go about modern molecular modelling, why we do it, and most important of all, explain much of the basic theory behind the mouse clicks. This involves mathematics and physics, and the book neither pulls punches nor aims at instant enlightenment. Many of the concepts and ideas are difficult ones, and you will have to think long and hard about them; if it's any consolation, so did the pioneers in our subject. I have given many of the derivations in full, and tried to avoid the dreaded phrase 'it can be shown that'.

There are various strands to our studies, all of which eventually intertwine. We start off with molecular mechanics, a classical treatment widely used to predict molecular geometries. In Chapter 8 I give a quick guide to statistical thermodynamics (if such a thing is possible), because we need to make use of the concepts when trying to model arrays of particles at non-zero temperatures. Armed with this knowledge, we are ready for an assault on Monte Carlo and Molecular Dynamics.

Just as we have to bite the bullet of statistical mechanics, so we have to bite the equally difficult one of quantum mechanics, which occupies Chapters 11 and 12. We then turn to the quantum treatment of atoms, where many of the sums can be done on a postcard if armed with knowledge of angular momentum.

The Hartree–Fock and HF–LCAO models dominate much of the next few chapters, as they should. The Hartree–Fock model is great for predicting many molecular properties, but it can't usually cope with bond-breaking and bond-making. Chapter 19 treats electron correlation and Chapter 20 deals with the very topical density functional theory (DFT). You won't be taken seriously if you have not done a DFT calculation on your molecule.

Quantum mechanics, statistical mechanics and electromagnetism all have a certain well-deserved reputation amongst science students; they are hard subjects. Unfortunately all three feature in this new text. In electromagnetism it is mostly a matter of getting to grips with the mathematical notation (although I have spared you

Maxwell's equations), whilst in the other two subjects it is more a question of mastering hard concepts. In the case of quantum mechanics, the concepts are often in direct contradiction to everyday experience and common sense. I expect from you a certain level of mathematical competence; I have made extensive use of vectors and matrices not because I am perverse, but because such mathematical notation brings out the inherent simplicity and beauty of many of the equations. I have tried to help by giving a mathematical Appendix, which should also make the text self-contained.

I have tried to put the text into historical perspective, and in particular I have quoted directly from a number of what I call *keynote papers*. It is interesting to read at first hand how the pioneers put their ideas across, and in any case they do it far better than me. For example, I am not the only author to quote Paul Dirac's famous statement

> *The underlying Physical Laws necessary for the mathematical theory of a large part of physics and the whole of chemistry are thus completely known, and the difficulty is only that exact application of these laws leads to equations much too complicated to be soluble.*

I hope you have a profitable time in your studies, and at the very least begin to appreciate what all those options mean next time you run a modelling package!

Alan Hinchliffe

alan.hinchliffe@umist.ac.uk

Manchester 2003

List of Symbols

$\langle \cdots \rangle$	Mean value/time average
a_0	Atomic unit of length (the bohr)
A	Thermodynamic Helmholtz energy
α	GTO orbital exponent; exchange parameter in $X\alpha$ DFT
α_A	Hückel π-electron Coulomb integral for atom A
α_e	Vibration–rotation coupling constant
$\boldsymbol{\alpha}$	Electric polarizability matrix
$\mathbf{B}$	Wilson B matrix
β^0_{AB}	Bonding parameter in semi-empirical theories (e.g. CNDO)
β_{AB}	Hückel π-electron resonance integral for bonded pairs A, B
χ	Electronegativity; basis function in LCAO theories
C_6, C_{12}	Lennard-Jones parameters
C_v, C_p	Heat capacities at constant volume and pressure.
d	Contraction coefficient in, for example, STO-nG expansion
$D(\varepsilon)$	Density of states
D_0	Spectroscopic dissociation energy
D_e	Thermodynamic dissociation energy
$d\tau$	Volume element
E	Electron affinity
E_h	Atomic unit of energy (the hartree)
$\mathbf{E(r)}$	Electric field vector ($\mathbf{r}$ = field point)
ε	Particle energy
$\mathbf{F}$	Force (a vector quantity)
Φ	Total mutual potential energy
$\phi(\mathbf{r})$	Electrostatic potential ($\mathbf{r}$ = field point)
$\mathbf{g}$	Gradient vector
G	Thermodynamic gibbs energy
$\mathbf{H}$	Hessian matrix
H	Thermodynamic enthalpy; classical hamiltonian
$\mathbf{h}_1$	Matrix of one-electron integrals in LCAO models
$H_v(\xi)$	Hermite polynomial of degree v
I	Ionization energy
ϵ_0	Permittivity of free space
ϵ_r	Relative permittivity

j	Square root of -1
$\mathbf{J}$, $\mathbf{K}$ and $\mathbf{G}$	Coulomb, exchange and G matrices from LCAO models
k_s	Force constant
$\mathbf{l}$, $\mathbf{L}$	Angular momentum vectors
L-J	Lennard-Jones (potential)
μ	Reduced mass
n	Amount of substance
p	Pressure
$\mathbf{P}(\mathbf{r})$	Dielectric polarization ($\mathbf{r}$ = field point)
$\mathbf{p}_e$	Electric dipole moment
q	Normal coordinate; atomic charge; molecular partition function
$\mathbf{q}$	Quaternion
Q	Partition function
Q_A	Point charge
$\mathbf{q}_e$	Electric second moment tensor
$\mathbf{\Theta}_e$	Electric quadrupole moment tensor
R	Gas constant
$\mathbf{R}$	Rotation matrix
$\rho(\mathbf{r})$	Electrostatic charge distribution ($\mathbf{r}$ = field point)
$\mathbf{r}$, $\mathbf{R}$	Field point vectors
R_∞	Rydberg constant for one-electron atom with infinite nuclear mass.
$\rho_1(\mathbf{x}_1)$	One-electron density function
$\rho_2(\mathbf{x}_1, \mathbf{x}_2)$	Two-electron density function
$\mathbf{R}_A$	Position vector
R_e	Equilibrium bond length
R_H	Rydberg constant for hydrogen
S	Thermodynamic entropy
U	Mutual potential energy
U, U_{th}	Thermodynamic internal energy
V	Volume
ω	Angular vibration frequency
$\omega_e x_e$	Anharmonicity constant
$\psi(\mathbf{r})$	Orbital (i.e. single-particle wavefunction)
$\Psi(\mathbf{R}, t)$	Time-dependent wavefunction
$\Psi(\mathbf{R}_1, \mathbf{R}_2, \ldots)$	Many-particle wavefunction
Z	Atomic number
ζ	STO orbital exponent

1 Introduction

1.1 Chemical Drawing

A vast number of organic molecules are known. In order to distinguish one from another, chemists give them names. There are two kinds of names: *trivial* and *systematic*. Trivial names are often brand names (such as aspirin, and the amino acid phenylanine shown in Figure 1.1). Trivial names don't give any real clue as to the structure of a molecule, unless you are the recipient of divine inspiration. The IUPAC systematic name for phenylanine is 2-amino-3-phenyl-propionic acid. Any professional scientist with a training in chemistry would be able to translate the systematic name into Figure 1.1 or write down the systematic name, given Figure 1.1. When chemists meet to talk about their work, they draw structures. If I wanted to discuss the structure and reactivity of phenylanine with you over a cup of coffee, I would draw a sketch, such as those shown in Figure 1.1, on a piece of paper. There are various conventions that we can follow when drawing chemical structures, but the conventions are well understood amongst professionals. First of all, I haven't shown the hydrogen atoms attached to the benzene ring (or indeed the carbon atoms within), and I have taken for granted that you understand that the normal valence of carbon is four. Everyone understands that hydrogens are present, and so we needn't clutter up an already complicated drawing.

The right-hand sketch is completely equivalent to the left-hand one; it's just that I have been less explicit with the CH_2 and the CH groups. Again, everyone knows what the symbols mean.

I have drawn the benzene ring as alternate single and double bonds, yet we understand that the C—C bonds in benzene are all the same. This may not be the case in the molecule shown; some of the bonds may well have more double bond character than others and so have different lengths, but once again it is a well-understood convention. Sometimes a benzene ring is given its own symbol Ph or ϕ. Then again, I have drawn the NH_2 and the OH groups as 'composites' rather than showing the individual O—H and N—H bonds, and so on. I have followed to some extent the convention that all atoms are carbon atoms unless otherwise stated.

Much of this is personal preference, but the important point is that no one with a professional qualification in chemistry would mistake my drawing for another molecule. Equally, given the systematic name, no one could possibly write down an incorrect molecule.

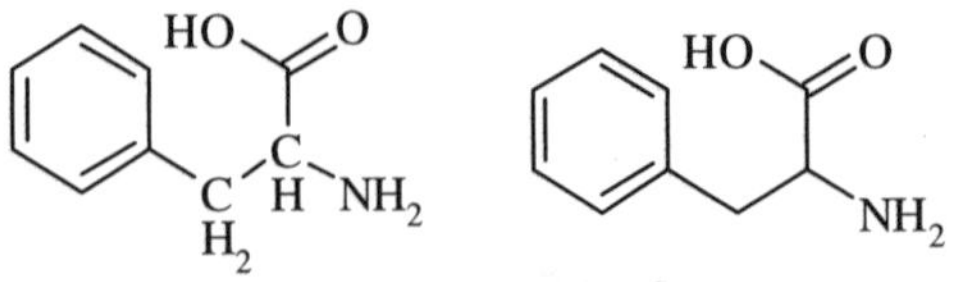

Figure 1.1   Two-dimensional drawings of phenylanine

You might like to know that phenylanine is not just another dull amino acid. A search through the Internet reveals that it is a molecule of great commercial and (alleged) pharmacological importance. One particular World Wide Web (www) site gives the following information.

Phenylanine

• Relates to the action of the central nervous system

• Can elevate mood, decrease pain, aid in memory and learning, and suppress appetite

• Can be used to treat schizophrenia, Parkinson's disease, obesity, migraines, menstrual cramps, depression

and you can order a pack of tablets in exchange for your credit card number.

The aim of Chapter 1 is to tell you that chemistry is a well-structured science, with a vast literature. There are a number of important databases that contain information about syntheses, crystal structures, physical properties and so on. Many databases use a molecular structure drawing as the key to their information, rather than the systematic name. Structure drawing is therefore a key chemical skill.

1.2 Three-Dimensional Effects

Chemical drawings are inherently two-dimensional objects; they give information about what is bonded to what. Occasionally, the lengths of the lines joining the atoms are scaled to reflect accurate bond lengths.

Molecules are three-dimensional entities, and that's where the fun and excitement of chemistry begins. In order to indicate three-dimensional effects on a piece of paper, I might draw the molecule CBrClFH (which is roughly tetrahedral) as in Figure 1.2. The top left-hand drawing is a two-dimensional one, with no great attempt to show the arrangement of atoms in space. The next three versions of the same molecule show the use of 'up', 'down' and 'either' arrows to show the relative

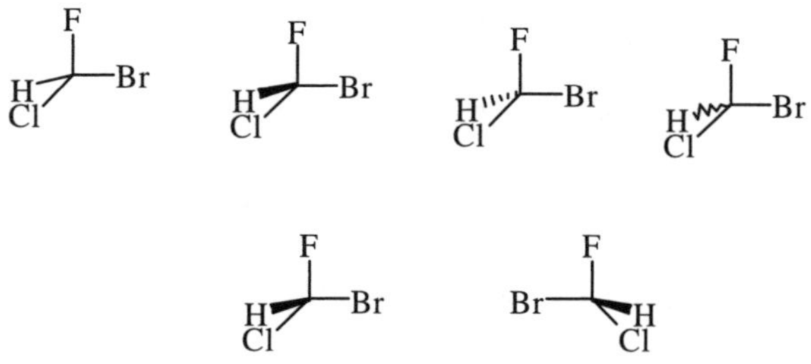

Figure 1.2 Two-dimensional drawings

dispositions of the bonds in space more explicitly. The bottom two drawings are two-dimensional attempts at the three-dimensional structure of the molecule and its mirror image. Note that the molecule cannot be superimposed on its mirror image. The central carbon atom is a *chiral* centre, and the two structures are *enantiomers*.

This *chirality* may be important in certain contexts, and it is often necessary to be aware of the correct spatial arrangement of atoms around each chiral centre in a molecule. Such information has to be obtained experimentally. A given molecule might have a number of chiral centres, not just one. Except in situations where there is opposed chirality on adjacent carbon atoms, chiral molecules exhibit the property of optical activity, considered below.

1.3 Optical Activity

Perhaps at this stage I should remind you about the two ways that chemists label optically active molecules.

The first method is 'operational', and relates to how a beam of polarized light is rotated as it passes through a solution of the molecule. If the plane of polarization is rotated to the right (i.e. clockwise when viewed against the light), then the molecule is said to be *dextrorotatory*, and given a symbol D (or +). If the plane of polarization is rotated to the left, then the molecule is said to be *laevorotatory* and is given the symbol L (or −).

Note that this method gives no information about the actual spatial arrangement of atoms around a chiral centre, nor about the number of chiral centres. The only way to be certain of the configuration of a compound is by deducing the molecular structure from, for example, X-ray and neutron diffraction studies, which brings me to the second way to label optically active compounds.

Once the correct structure is known, the chiral centre is labelled according to a standard IUPAC method, often referred to as the Cahn–Ingold–Prelog system (named after its originators). Atoms around a chiral centre are given a priority in order of decreasing atomic number. When two or more atoms connected to the asymmetric carbon are the same, the highest atomic number for the group second

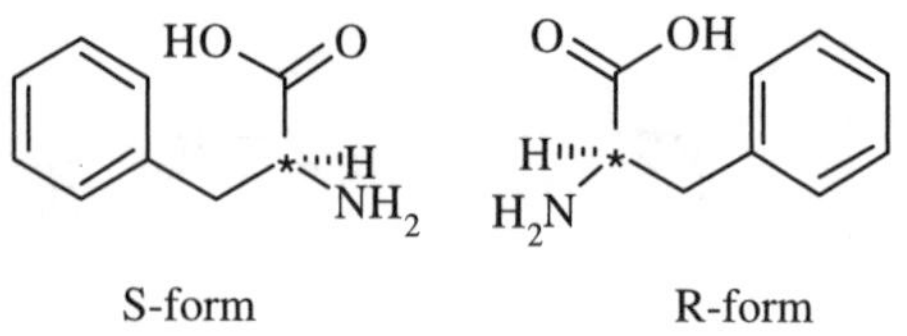

Figure 1.3   R and S forms of phenylanine

outer atoms determines the order, and so on. The molecule is then oriented so that the atom of lowest priority is towards the rear. The centre is then R (from the latin *rectus*, right) or S (from the latin *sinister*, left) according to the rotation from highest to lowest priority group; the rotation is clockwise for R and anticlockwise for S.

There is no connection between the D and L, and the R and S nomenclatures. A molecule labelled D could be either R or S, and a molecule labelled L could also be R or S.

I'm going to use phenylanine to exemplify many of the molecular modelling procedures we will meet throughout this text. The molecule has a single chiral centre, labelled * in Figure 1.3, and it is found that solutions of the naturally occurring form rotate the plane of polarized light to the left (and so it is the L-form). There are two possibilities for the absolute molecular structure (see Figure 1.3) and it turns out that the L form has the stereochemistry shown on the left-hand side of Figure 1.3. It is therefore S in the Cahn–Ingold–Prelog system.

1.4 Computer Packages

Over the years, several chemical drawing computer packages have appeared in the marketplace. They are all very professional and all perform much the same function. Which one you choose is a matter of personal preference; I am going to use MDL ISIS/Draw for my chemical drawing illustrations. At the time of writing, it is possible to obtain a free download from the Internet. Set your web browser to locate ISIS/Draw (http://www.mdli.com/) and follow the instructions; be sure to download the Help file, and any other add-ins that are on offer.

To make sure you have correctly followed instructions, use your copy to reproduce the screen shown in Figure 1.4. As the name suggests, 'AutoNom Name' is a facility to translate the structure into the IUPAC name.

1.5 Modelling

The title of this and many other texts includes the word 'modelling', which begs the question as to the meaning of the word 'model'. My 1977 edition of *Chambers*

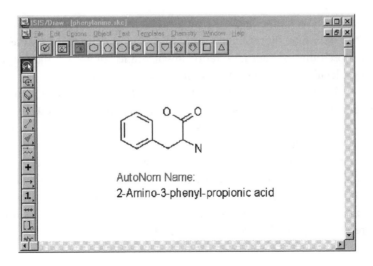

Figure 1.4 ISIS/Draw screen grab for phenylanine

Dictionary gives the following definition:

model, *mod'l*, *n.* plan, design (*obs*): a preliminary solid representation, generally small, or in plastic material, to be followed in construction: something to be copied: a pattern: an imitation of something on a smaller scale: a person or thing closely resembling another: ...

This definition captures the status of modelling in the 1970s, and Figure 1.5 shows a photograph of a plastic model of L-phenylanine. Such plastic models were fine in their day, but they took a considerable time to build, they tended to be unstable and,

Figure 1.5 Plastic model of L-phenylanine

more importantly, you had to know the molecular structure before you could actually build one. Not only that, they gave no sense of temperature in that they didn't vibrate or show the possibility of internal rotation about single bonds, and they referred only to isolated molecules at infinite separation in the gas phase.

As we will shortly see, a given molecule may well have very many plausible stable conformations. Plastic models gave clues as to which conformations were unlikely on the grounds of steric repulsions, but by and large they didn't help us identify the 'true' molecular geometry.

We have come a long way since then. Computer simulation has long taken over from mechanical model building, and by the end of this book you should at the very least know how to construct both quantum mechanical and classical computer models of a molecular system, how to predict the molecular geometry, how to simulate the temperature and how to allow for solvent effects.

1.6 Molecular Structure Databases

Molecular geometries can be determined for gas-phase molecules by microwave spectroscopy and by electron diffraction. In the solid state, the field of structure determination is dominated by X-ray and neutron diffraction and very many crystal structures are known. Nuclear magnetic resonance (NMR) also has a role to play, especially for proteins. All of these topics are well discussed in every university-level general chemistry text.

Over the years, a vast number of molecular structures have been determined and there are several well-known structural databases. One is the Cambridge Structural Database (CSD) (http://ccdc.cam.ac.uk/), which is supported by the Cambridge Crystallographic Data Centre (CCDC). The CCDC was established in 1965 to undertake the compilation of a computerized database containing comprehensive data for organic and metal–organic compounds studied by X-ray and neutron diffraction. It was originally funded as part of the UK contribution to international data compilation. According to its mission statement, the CCDC serves the scientific community through the acquisition, evaluation, dissemination and use of the world's output of small molecule crystal structures. At the time of writing, there are some 272 000 structures in the database.

For each entry in the CSD, three types of information are stored. First, the *bibliographic information*: who reported the crystal structure, where they reported it and so on. Next comes the *connectivity* data; this is a list showing which atom is bonded to which in the molecule. Finally, the *molecular geometry* and the crystal structure. The molecular geometry consists of cartesian coordinates. The database can be easily reached through the Internet, but individual records can only be accessed on a fee-paying basis.

The Brookhaven Protein Data Bank (PDB) is the single worldwide repository for the processing and distribution of three-dimensional biological macromolecular structural data. It is operated by the Research Collaboratory for Structural Bioinformatics. At the time of writing, there were 19 749 structures in the databank, relating to proteins, nucleic acids, protein–nucleic acid complexes and viruses. The databank is available free of charge to anyone who can navigate to their site http://www.rcsb.org/. Information can be retrieved from the main website. A four-character alphanumeric identifier, such as 1PCN, represents each structure. The PDB database can be searched using a number of techniques, all of which are described in detail at the homepage.

1.7 File Formats

The Brookhaven PDB (.pdb) file format is widely used to report and distribute molecular structure data. A typical .pdb file for phenylanine would start with bibliographic data, then move on to the cartesian coordinates (expressed in ångstroms and relative to an arbitrary reference frame) and connectivity data as shown below. The only parts that need concern us are the atom numbers and symbols, the geometry and the connectivity.

HETATM	1	N	PHE	1	− 0.177	1.144	0.013
HETATM	2	H	PHE	1	0.820	1.162	− 0.078
HETATM	3	CA	PHE	1	− 0.618	1.924	1.149
HETATM	4	HA	PHE	1	− 1.742	1.814	1.211
HETATM	5	C	PHE	1	− 0.290	3.407	0.988
HETATM	6	O	PHE	1	0.802	3.927	0.741
HETATM	7	CB	PHE	1	0.019	1.429	2.459
HETATM	8	1HB	PHE	1	0.025	0.302	2.442
HETATM	9	2HB	PHE	1	1.092	1.769	2.487
HETATM	10	CG	PHE	1	− 0.656	1.857	3.714
HETATM	11	CD1	PHE	1	− 0.068	1.448	4.923
HETATM	12	HD1	PHE	1	0.860	0.857	4.900
HETATM	13	CD2	PHE	1	− 1.829	2.615	3.757
HETATM	14	HD2	PHE	1	− 2.301	2.975	2.829
HETATM	15	CE1	PHE	1	− 0.647	1.783	6.142
HETATM	16	HE1	PHE	1	− 0.176	1.457	7.081
HETATM	17	CE2	PHE	1	− 2.411	2.946	4.982
HETATM	18	HE2	PHE	1	− 3.338	3.538	4.999
HETATM	19	CZ	PHE	1	− 1.826	2.531	6.175
HETATM	20	HZ	PHE	1	− 2.287	2.792	7.139
HETATM	21	O	PHE	1	− 1.363	4.237	1.089
HETATM	22	H		22	− 0.601	1.472	− 0.831
HETATM	23	H		23	− 1.077	5.160	0.993
CONECT	1		2	3	22		
CONECT	2		1				

CONECT	3	1	4	5	7
CONECT	4	3			
CONECT	5	3	6	21	
CONECT	6	5			
CONECT	7	3	10	8	9
CONECT	8	7			
CONECT	9	7			
CONECT	10	7	11	13	
CONECT	11	10	15	12	
CONECT	12	11			
CONECT	13	10	17	14	
CONECT	14	13			
CONECT	15	11	19	16	
CONECT	16	15			
CONECT	17	13	19	18	
CONECT	18	17			
CONECT	19	15	17	20	
CONECT	20	19			
CONECT	21	5	23		
CONECT	22	1			
CONECT	23	21			
END					

Records 1–23 identify the atoms so for example, atom 1 is a nitrogen with cartesian coordinates $x = -0.177\,\text{Å}$, $y = 1.144\,\text{Å}$ and $z = 0.013\,\text{Å}$. The PHE identifies the aminoacid residue, of which there is just one in this simple case. The record

CONECT 1 2 3 22

tells us that atom 1 (N) is joined to atoms 2, 3 and 22.

1.8 Three-Dimensional Displays

We are going to need to display three-dimensional structures as we progress through the text. There are many suitable packages and I am going to use WebLabViewer throughout this chapter. The 'Lite' version of this package can be downloaded free from the publisher's Internet site http://www.accelrys.com/. It displays molecular structures; there is also a 'Pro' version, which will additionally perform limited molecular modelling. (The packages have recently been renamed Discovery Studio ViewerPro and Discovery Studio ViewerLite.)

There are several different ways to represent a molecular model; these are often referred to as the *rendering*. First, we have the *line representation* of phenylanine,

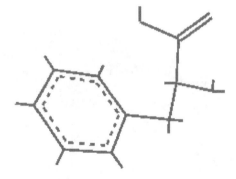

Figure 1.6 Line representation

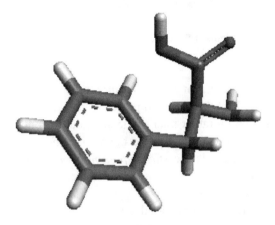

Figure 1.7 Stick representation

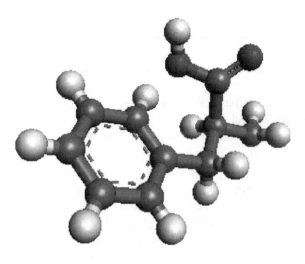

Figure 1.8 Ball-and-stick representation

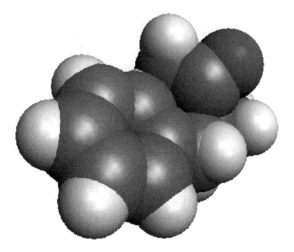

Figure 1.9   CPK space-filling rendering

Figure 1.6. Line models are drawn using thin lines. Each representation has a number of options. In this case, I set the option to perceive aromaticity so the benzene ring is shown as an aromatic entity rather than single and double bonds. The groups can be coloured, although this doesn't show in the figure. The molecule can be rotated and otherwise manipulated. Next is the *stick representation*, Figure 1.7. Once again, I selected the option to perceive aromaticity. Another representation is the *ball-and-stick representation*, Figure 1.8.

There are other representations, based on space filling. For example, the CPK (Corey–Pauling–Koltun) rendering shown in Figure 1.9 refers to a popular set of atomic radii used for depicting space-filling pictures or building plastic models.

1.9 Proteins

Figure 1.10 shows procolipase, as extracted from the Brookhaven PDB. The PDB serial number is 1PCN, and the molecule contains some 700 atoms.

Hydrogen atoms were not determined in the experimental studies, but most packages have an option to add hydrogen atoms. Protein structures can be complex and difficult to interpret. For that reason, workers in the field have developed alternative methods of rendering these molecules. The idea is to identify the structural backbone of the molecule. To do this, we identify the backbone of each amino acid, and these are linked together as Figure 1.11 shows.

This particular rendering is the *tube representation*. Again, the use of colour makes for easier visualization, although this is lost in the monochrome illustrations given here.

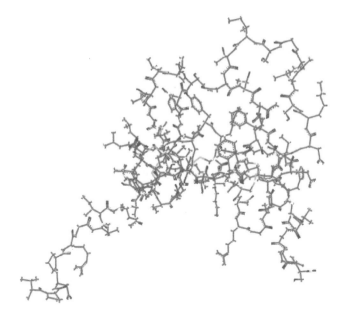

Figure 1.10 Procolipase

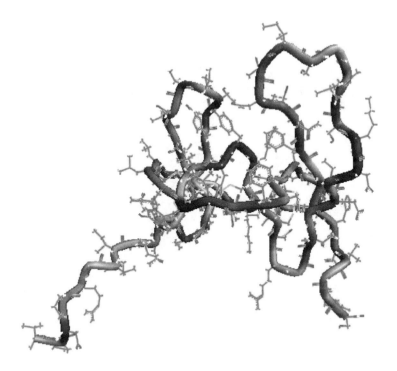

Figure 1.11 Procolipase tube representation

2 Electric Charges and Their Properties

As far as we can tell, there are four fundamental types of interactions between physical objects. There is the *weak nuclear interaction* that governs the decay of beta particles, and the *strong nuclear interaction* that is responsible for binding together the particles in a nucleus. The familiar *gravitational* interaction holds the earth very firmly in its orbit round the sun, and finally we know that there is an *electromagnetic* interaction that is responsible for binding atomic electrons to nuclei and for holding atoms together when they combine to form molecules.

Of the four, the gravitational interaction is the only one we would normally come across in our everyday world. This is because gravitational interactions between bodies always add. The gravitational interaction between two atoms is negligible, but when large numbers of fundamental particles such as atoms are aggregated together, the gravitational interaction becomes significant.

You may think it bizarre that there are four types of interaction, yet on the other hand you might wonder why there should be just four. Why not one, three or five? Should there not be a unifying theory to explain why there are four, and whether they are related? As I write, there is no such unifying theory despite tremendous research activity.

2.1 Point Charges

In this chapter I am going to concentrate on electric charges and their properties. It turns out that there are two types of electric charge in nature, which we might choose to call type X and type Y (or Red and Blue for that matter, but X and Y will do for now). Experimental evidence shows the existence of an electrostatic force between electric charges; the force between two X-type charges is always repulsive, as is the force between two Y-type charges. The force between an X and a Y-type is always attractive. For this reason, the early experimenters decided to classify charges as positive or negative, because a positive quantity times a positive quantity gives a positive quantity, a negative quantity times a negative quantity gives a positive quantity, whilst a negative quantity times a positive quantity gives a negative quantity.

I'm sure you know that the best known fundamental particles responsible for these charges are electrons and protons, and you are probably expecting me to tell you that the electrons are the negatively charged particles whilst protons are positively charged. It's actually just a convention that we take, we could just as well have called electrons positive.

Whilst on the subject, it is fascinating to note that the charge on the electron is exactly equal and opposite of that on a proton. Atoms and molecules generally contain exactly the same number of electrons and protons, and so the net charge on a molecule is almost always zero. Ions certainly exist in solutions of electrolytes, but the number of Na^+ ions in a solution of sodium chloride is exactly equal to the number of Cl^- ions and once again we are rarely aware of any imbalance of charge.

A thunderstorm results when nature separates out positive and negative charges on a macroscopic scale. It is thought that friction between moving masses of air and water vapour detaches electrons from some molecules and attaches them to others. This results in parts of clouds being left with an excess of charge, often with spectacular results. It was investigations into such atmospheric phenomena that gave the first clues about the nature of the electrostatic force.

We normally start any study of charges at rest (*electrostatics*) by considering the force between two point charges, as shown in Figure 2.1.

The term 'point charge' is a mathematical abstraction; obviously electrons and protons have a finite size. Just bear with me for a few pages, and accept that a point charge is one whose dimensions are small compared with the distance between them. An electron is large if you happen to be a nearby electron, but can normally be treated as a point charge if you happen to be a human being a metre away.

The concept of a point charge may strike you as an odd one, but once we have established the magnitude of the force between two such charges, we can deduce the force between any arbitrary charge distributions on the grounds that they are composed of a large number of point charges.

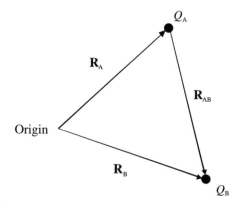

Figure 2.1   Point charges

In Figure 2.1 we have point charge Q_A at position vector $\mathbf{R}_A$ and Q_B at point $\mathbf{R}_B$. From the laws of vector analysis, the vector

$$\mathbf{R}_{AB} = \mathbf{R}_B - \mathbf{R}_A$$

joins Q_A to Q_B, and points from Q_A to Q_B as shown. I have indicated the direction of the vectors with arrows.

2.2 Coulomb's Law

In 1785, Charles Augustin de Coulomb became the first person to give a mathematical form to the force between point charges. He measured the force directly between two very small charged bodies, and was able to show that the force exerted by Q_A on Q_B was

- proportional to the inverse square of the distance between Q_A and Q_B when both charges were fixed;

- proportional to Q_A when Q_B and R_{AB} were fixed; and

- proportional to Q_B when Q_A and R_{AB} were fixed.

He also noticed that the force acted along the line joining the centres of the two charges, and that the force was either attractive or repulsive depending on whether the charges were different or of the same type. The sign of the product of the charges therefore determines the direction of the force.

A mathematical result of these observations can be written in scalar form as

$$F_{\text{A on B}} \propto \frac{Q_A Q_B}{R_{AB}^2} \tag{2.1}$$

Forces are vector quantities, and Equation (2.1) is better written in vector form as

$$\mathbf{F}_{\text{A on B}} \propto \frac{Q_A Q_B}{R_{AB}^3} \mathbf{R}_{AB}$$

When Coulomb first established his law, he had no way to quantify charge and so could not identify the proportionality constant. He took it to be unity, and thereby defined charge in terms of the force between charges. Modern practice is to regard charge and force as independent quantities and because of this a dimensioned proportionality constant is necessary. For a reason that need not concern us, this is taken as $1/(4\pi\epsilon_0)$, where the permittivity of free space ϵ_0 is an experimentally determined quantity with the approximate value $\epsilon_0 = 8.854 \times 10^{-12}\,\mathrm{C^2\,N^{-1}\,m^{-2}}$. Coulomb's law is therefore

$$\mathbf{F}_{\mathrm{A\ on\ B}} = \frac{1}{4\pi\epsilon_0}\frac{Q_A Q_B}{R_{AB}^3}\mathbf{R}_{AB} \tag{2.2}$$

and it applies to measurements done in free space. If we repeat Coulomb's experiments with the charges immersed in different media, then we find that the law still holds but with a different proportionality constant. We modify the proportionality constant using a quantity ϵ_r called the *relative permittivity*. In older texts, ϵ_r is called the *dielectric constant*. Our final statement of Coulomb's law is therefore

$$\mathbf{F}_{\mathrm{A\ on\ B}} = \frac{1}{4\pi\epsilon_r\epsilon_0}\frac{Q_A Q_B}{R_{AB}^3}\mathbf{R}_{AB} \tag{2.3}$$

According to Newton's Third Law, we know that if Q_A exerts a force $\mathbf{F}_{\mathrm{A\ on\ B}}$ on Q_B, then Q_B should exert an equal and opposite force on Q_A. Coulomb's law satisfies this requirement, since

$$\mathbf{F}_{\mathrm{B\ on\ A}} = \frac{1}{4\pi\epsilon_r\epsilon_0}\frac{Q_A Q_B}{R_{BA}^3}\mathbf{R}_{BA}$$

(the vector $\mathbf{R}_{BA}$ points in the opposite direction to $\mathbf{R}_{AB}$ and so one force is exactly the negative of the other, as it should be).

2.3 Pairwise Additivity

Suppose we now add a third point charge Q_C with position vector $\mathbf{R}_C$, as shown in Figure 2.2. Since Q_A and Q_B are point charges, the addition of Q_C cannot alter the force between Q_A and Q_B.

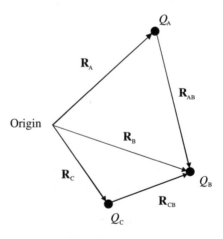

Figure 2.2   Third charge added

The total force on Q_B now comprises two terms, namely the force due to point charge Q_A and the force due to point charge Q_C. This total force is given by

$$\mathbf{F}_B = \frac{Q_B}{4\pi\epsilon_0}\left(Q_A\frac{\mathbf{R}_{AB}}{R_{AB}^3} + Q_C\frac{\mathbf{R}_{CB}}{R_{CB}^3}\right) \tag{2.4}$$

This may seem at first sight to be a trivial statement; surely all forces act this way. Not necessarily, for I have assumed that the addition of Q_C did not have any effect on Q_A and Q_B (and so did not influence the force between them).

The generic term *pairwise additive* describes things such as forces that add as above. Forces between point electric charges are certainly pairwise additive, and so you might imagine that forces between atoms and molecules must therefore be pairwise additive, because atoms and molecules consist of (essentially) point charges. I'm afraid that nature is not so kind, and we will shortly meet situations where forces between the composites of electrons and protons that go to make up atoms and molecules are far from being pairwise additive.

2.4 The Electric Field

Suppose now we have a point charge Q at the coordinate origin, and we place another point charge q at point P that has position vector $\mathbf{r}$ (Figure 2.3). The force exerted by Q on q is

$$\mathbf{F} = \frac{1}{4\pi\epsilon_0}\frac{Qq}{r^3}\mathbf{r}$$

which I can rewrite trivially as

$$\mathbf{F} = \left(\frac{1}{4\pi\epsilon_0}\frac{Q}{r^3}\mathbf{r}\right)q$$

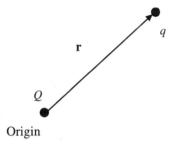

Figure 2.3 Field concept

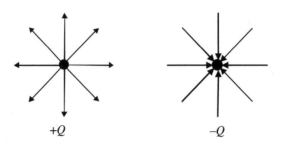

$+Q$ $-Q$

Figure 2.4 Field lines for point charges

The point is that the term in brackets is to do with Q and the vector $\mathbf{r}$, and contains no mention of q. If we want to find the force on any arbitrary q at $\mathbf{r}$, then we calculate the quantity in brackets once and then multiply by q. One way of thinking about this is to imagine that the charge Q creates a certain field at point $\mathbf{r}$, which determines the force on any other q when placed at position $\mathbf{r}$.

This property is called the *electric field* $\mathbf{E}$ at that point. It is a vector quantity, like force, and the relationship is that

$$\mathbf{F}(\text{on } q \text{ at } \mathbf{r}) = q\mathbf{E}(\text{at } \mathbf{r})$$

Comparison with Coulomb's law, Equation (2.3), shows that the electric field at point $\mathbf{r}$ due to a point charge Q at the coordinate origin is

$$\mathbf{E} = \frac{1}{4\pi\epsilon_0}\frac{Q\mathbf{r}}{r^3} \tag{2.5}$$

$\mathbf{E}$ is sometimes written $\mathbf{E}(\mathbf{r})$ to emphasize that the electric field depends on the position vector $\mathbf{r}$.

Electric fields are vector fields and they are often visualized as *field lines*. These are drawn such that their spacing is inversely proportional to the strength of the field, and their tangent is in the direction of the field. They start at positive charges and end at negative charges, and two simple examples are shown in Figure 2.4. Here the choice of eight lines is quite arbitrary. Electric fields that don't vary with time are called *electrostatic* fields.

2.5 Work

Look again at Figure 2.3, and suppose we move point charge q whilst keeping Q fixed in position. When a force acts to make something move, energy is transferred. There is a useful word in physical science that is to do with the energy transferred, and it is *work*. Work measures the energy transferred in any change, and can be calculated from the change in energy of a body when it moves through a distance under the influence of a force.

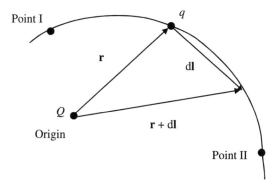

Figure 2.5   Electrostatic work

We have to be careful to take account of the energy balance. If a body gains energy, then this energy has to come from somewhere, and that somewhere must lose energy. What we do is to divide the universe into two parts: the bits we are interested in called the *system* and the rest of the universe that we call the *surroundings*.

Some texts focus on the work done *by* the system, and some concern themselves with the work done *on* the system. According to the Law of Conservation of Energy, one is exactly the equal and opposite of the other, but we have to be clear which is being discussed. I am going to write w_{on} for the work done *on* our system. If the system gains energy, then w_{on} will be positive. If the system loses energy, then w_{on} will negative.

We also have to be careful about the phrase 'through a distance'. The phrase means 'through a distance that is the projection of the force vector on the displacement vector', and you should instantly recognize a vector scalar product (see the Appendix).

A useful formula that relates to the energy gained by a system (i.e. w_{on}) when a constant force $\mathbf{F}$ moves its point of application through $\mathbf{l}$ is

$$w_{on} = -\mathbf{F} \cdot \mathbf{l} \tag{2.6}$$

In the case where the force is not constant, we have to divide up the motion into differential elements $d\mathbf{l}$, as illustrated in Figure 2.5. The energy transferred is then given by the sum of all the corresponding differential elements dw_{on}. The corresponding formulae are

$$dw_{on} = -\mathbf{F} \cdot d\mathbf{l}$$

$$w_{on} = -\int \mathbf{F} \cdot d\mathbf{l} \tag{2.7}$$

We now move q by an infinitesimal vector displacement $d\mathbf{l}$ as shown, so that it ends up at point $\mathbf{r} + d\mathbf{l}$. The work done on the system in that differential change is

$$dw_{on} = -\mathbf{F} \cdot d\mathbf{l}$$

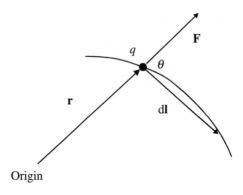

Figure 2.6   Relationship between vectors

If the angle between the vectors $\mathbf{r}_I$ and $\mathbf{dl}$ is θ, then we have

$$dw_{on} = -Fdl\cos\theta$$

and examination of Figure 2.6 shows that $dl\cos\theta$ is the radial distance moved by charge q, which we will write as dr. Hence

$$dw_{on} = -\frac{1}{4\pi\epsilon_0}\frac{Qq}{r^2}dr$$

The total work done moving from position I to position II is therefore found by integrating

$$w_{on} = -\frac{1}{4\pi\epsilon_0}\int_I^{II}\frac{Qq}{r^2}dr$$

$$= \frac{1}{4\pi\epsilon_0}Qq\left(\frac{1}{r_{II}}-\frac{1}{r_I}\right) \tag{2.8}$$

The work done depends only on the initial and final positions of the charge q; it is independent of the way we make the change.

Another way to think about the problem is as follows. The force is radial, and we can divide the movement from position I to position II into infinitesimal steps, some of which are parallel to $\mathbf{F}$ and some of which are perpendicular to $\mathbf{F}$. The perpendicular steps count zero towards w_{on}, the parallel steps only depend on the change in the (scalar) radial distance.

2.6 Charge Distributions

So far I have concentrated on point charges, and carefully skirted round the question as to how we deal with continuous distributions of charge. Figure 2.7 shows a charge

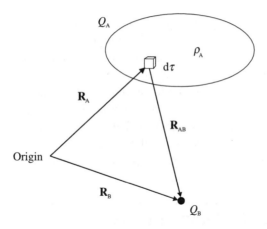

Figure 2.7 Charge distribution

distribution Q_A. The density of charge need not be constant through space, and we normally write $\rho(\mathbf{r})$ for the density at the point whose position vector is $\mathbf{r}$. The charge contained within the volume element $d\tau$ at $\mathbf{r}$ is therefore $\rho(\mathbf{r})\,d\tau$ and the relationship between $\rho(\mathbf{r})$ and Q_A is discussed in the Appendix. It is

$$Q_A = \int \rho(\mathbf{r})\,d\tau \tag{2.9}$$

In order to find the force between the charge distribution and the point charge Q_B we simply extend our ideas about the force between two point charges; one of the point charges being $\rho(\mathbf{r})\,d\tau$ and the other Q_B.

The total force is given by the sum of all possible contributions from the elements of the continuous charge distribution Q_A with point charge Q_B. The practical calculation of such a force can be a nightmare, even for simple charge distributions. One of the reasons for the nightmare is that forces are vector quantities; we need to know about both their magnitude and their direction.

In the next section I am going to tell you about a very useful scalar field called the mutual potential energy U. This field has the great advantage that it is a scalar field, and so we don't need to worry about direction in our calculations.

2.7 The Mutual Potential Energy *U*

Suppose now we start with charge q at infinity, and move it up to a point with vector position $\mathbf{r}$, as shown in Figure 2.3. The work done is

$$w_{on} = \frac{1}{4\pi\epsilon_0}\frac{Qq}{r} \tag{2.10}$$

and this represents the energy change on building up the charge distribution, with the charges initially at infinite separation. It turns out that this energy change is an important property, and we give it a special name (the *mutual potential energy*) and a special symbol U (occasionally Φ).

Comparison of the equations for force, work and mutual potential energy given above suggests that there might be a link between the force and the mutual potential energy; at first sight, one expression looks like the derivative of the other. I am going to derive a relationship between force and mutual potential energy. The relationship is perfectly general; it applies to all forces provided that they are constant in time.

2.8 Relationship Between Force and Mutual Potential Energy

Consider a body of mass m that moves in (say) the x-direction under the influence of a constant force. Suppose that at some instant its speed is v. The kinetic energy is $\frac{1}{2}mv^2$. You are probably aware of the law of conservation of energy, and know that when I add the potential energy U to the kinetic energy, I will get a constant energy ε

$$\varepsilon = \tfrac{1}{2}mv^2 + U \tag{2.11}$$

I want to show you how to relate U to the force $\mathbf{F}$. If the energy ε is constant in time, then $\mathrm{d}\varepsilon/\mathrm{d}t = 0$. Differentiation of Equation (2.11) with respect to time gives

$$\frac{\mathrm{d}\varepsilon}{\mathrm{d}t} = mv\frac{\mathrm{d}v}{\mathrm{d}t} + \frac{\mathrm{d}U}{\mathrm{d}t}$$

and so, by the chain rule

$$\frac{\mathrm{d}\varepsilon}{\mathrm{d}t} = mv\frac{\mathrm{d}v}{\mathrm{d}t} + \frac{\mathrm{d}U}{\mathrm{d}x}\frac{\mathrm{d}x}{\mathrm{d}t}$$

If the energy ε is constant, then its first differential with respect to time is zero, and v is just $\mathrm{d}x/\mathrm{d}t$. Likewise, $\mathrm{d}v/\mathrm{d}t$ is the acceleration and so

$$0 = \left(m\frac{\mathrm{d}^2x}{\mathrm{d}t^2} + \frac{\mathrm{d}U}{\mathrm{d}x}\right)\frac{\mathrm{d}x}{\mathrm{d}t} \tag{2.12}$$

Equation (2.12) is true if the speed is zero, or if the term in brackets is zero. According to Newton's second law of motion, mass times acceleration is force and so

$$\mathbf{F} = -\frac{\mathrm{d}U}{\mathrm{d}x} \tag{2.13}$$

which gives us the link between force and mutual potential energy.

When working in three dimensions, we have to be careful to distinguish between vectors and scalars. We treat a body of mass m whose position vector is $\mathbf{r}$. The velocity is $\mathbf{v} = d\mathbf{r}/dt$ and the kinetic energy is $\frac{1}{2}m(d\mathbf{r}/dt) \cdot (d\mathbf{r}/dt)$. Analysis along the lines given above shows that the force $\mathbf{F}$ and U are related by

$$\mathbf{F} = -\operatorname{grad} U \tag{2.14}$$

where the gradient of U is discussed in the Appendix and is given in Cartesian coordinates by

$$\operatorname{grad} U = \frac{\partial U}{\partial x}\mathbf{e}_x + \frac{\partial U}{\partial y}\mathbf{e}_y + \frac{\partial U}{\partial z}\mathbf{e}_z \tag{2.15}$$

Here $\mathbf{e}_x$, $\mathbf{e}_y$ and $\mathbf{e}_z$ are unit vectors pointing along the Cartesian axes.

2.9 Electric Multipoles

We can define exactly an array of point charges by listing the magnitudes of the charges, together with their position vectors. If we then wish to calculate (say) the force between one array of charges and another, we simply apply Coulomb's law [Equation (2.3)] repeatedly to each pair of charges. Equation (2.3) is exact, and can be easily extended to cover the case of continuous charge distributions.

For many purposes it proves more profitable to describe a charge distribution in terms of certain quantities called the *electric moments*. We can then discuss the interaction of one charge distribution with another in terms of the interactions between the electric moments.

Consider first a pair of equal and opposite point charges, $+Q$ and $-Q$, separated by distance R (Figure 2.8). This pair of charges usually is said to form an *electric dipole* of magnitude QR. In fact, electric dipoles are vector quantities and a more rigorous definition is

$$\mathbf{p}_{\mathrm{e}} = Q\mathbf{R} \tag{2.16}$$

where the vector $\mathbf{R}$ points from the negative charge to the positive charge.

We sometimes have to concern ourselves with a more general definition, one relating to an arbitrary array of charges such as that shown in Figure 2.9. Here we have four point charges: Q_1 whose position vector is $\mathbf{R}_1$; Q_2 whose position vector is $\mathbf{R}_2$, Q_3 whose position vector is $\mathbf{R}_3$, and Q_4 whose position vector is $\mathbf{R}_4$. We define the *electric dipole moment* $\mathbf{p}_{\mathrm{e}}$ of these four charges as

$$\mathbf{p}_{\mathrm{e}} = \sum_{i=1}^{4} Q_i\mathbf{R}_i$$

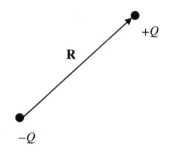

Figure 2.8 Simple electric dipole

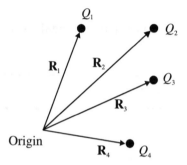

Figure 2.9 Generalized electric dipole

It is a vector quantity with x, y and z Cartesian components $\sum_{i=1}^{4} Q_i X_i$, $\sum_{i=1}^{4} Q_i Y_i$ and $\sum_{i=1}^{4} Q_i Z_i$. This is consistent with the elementary definition given above; in the case of two equal and opposite charges, $+Q$ and $-Q$, a distance d apart, the electric dipole moment has magnitude Qd and points from the negative charge to the positive. The generalization to n charges is obvious; we substitute n for 4 in the above definition.

There are several general rules about electric moments of charge distributions, and we can learn a couple of the ones that apply to dipole moments by considering the simple arrays shown in Figure 2.10 and keeping the definitions in mind.

I have joined up the charges with lines in order to focus attention on the charge systems involved; there is no implication of a 'bond'. We don't normally discuss the electric dipole due to a point charge (1). Examination of the charge distributions (2)–(6) and calculation of their electric dipole moment for different coordinate origins suggests the general result; neutral arrays of point charges have a unique electric dipole moment that does not depend on where we take the coordinate origin. Otherwise, we have to state the coordinate origin when we discuss the electric dipole moment.

I can prove this from Equation (2.16), generalized to n point charges

$$\mathbf{p}_e = \sum_{i=1}^{n} Q_i \mathbf{R}_i \tag{2.17}$$

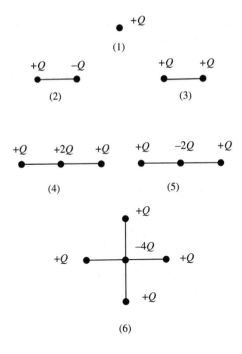

Figure 2.10   Simple arrays of point charges

Suppose that we move the coordinate origin so that each point charge Q_i has a position vector $\mathbf{R}'_i$, where

$$\mathbf{R}_i = \mathbf{R}'_i + \Delta$$

with Δ a constant vector. From the definition of the electric dipole moment we have

$$\mathbf{p}_e = \sum_{i=1}^{n} Q_i \mathbf{R}_i$$

and so, with respect to the new coordinate origin

$$\mathbf{p}'_e = \sum_{i=1}^{n} Q_i \mathbf{R}'_i$$

$$= \sum_{i=1}^{n} Q_i (\mathbf{R_i} - \Delta)$$

$$= \mathbf{p}_e - \Delta \sum_{i=1}^{n} Q_i$$

The two definitions only give the same vector if the sum of charges is zero. We often use the phrase *gauge invariant* to describe quantities that don't depend on the choice of coordinate origin.

Arrays (5) and (6) each have a centre of symmetry. There is a general result that any charge distribution having no overall charge but a centre of symmetry must have a zero dipole moment, and similar results follow for other highly symmetrical arrays of charges.

2.9.1 Continuous charge distributions

In order to extend the definition of an electric dipole to a continuous charge distribution, such as that shown in Figure 2.7, we first divide the region of space into differential elements $d\tau$. If $\rho(\mathbf{r})$ is the charge density then the change in volume element $d\tau$ is $\rho(\mathbf{r})d\tau$. We then treat each of these volume elements as point charges and add (i.e. integrate). The electric dipole moment becomes

$$\mathbf{p}_e = \int \mathbf{r}\rho(\mathbf{r})\,d\tau \tag{2.18}$$

2.9.2 The electric second moment

The electric dipole moment of an array of point charges is defined by the following three sums

$$\sum_{i=1}^{n} Q_i X_i, \qquad \sum_{i=1}^{n} Q_i Y_i \quad \text{and} \quad \sum_{i=1}^{n} Q_i Z_i$$

and we can collect them into a column vector in an obvious way

$$\mathbf{p}_e = \begin{pmatrix} \sum_{i=1}^{n} Q_i X_i \\ \sum_{i=1}^{n} Q_i Y_i \\ \sum_{i=1}^{n} Q_i Z_i \end{pmatrix} \tag{2.19}$$

The six independent quantities $\sum_{i=1}^{n} Q_i X_i^2, \sum_{i=1}^{n} Q_i X_i Y_i, \sum_{i=1}^{n} Q_i X_i Z_i, \ldots, \sum_{i=1}^{n} Q_i Z_i^2$ are said to define the *electric second moment* of the charge distribution.

We usually collect them into a real symmetric 3×3 matrix $\mathbf{q}_e$

$$
\mathbf{q}_e = \begin{pmatrix}
\sum_{i=1}^{n} Q_i X_i^2 & \sum_{i=1}^{n} Q_i X_i Y_i & \sum_{i=1}^{n} Q_i X_i Z_i \\
\sum_{i=1}^{n} Q_i Y_i X_i & \sum_{i=1}^{n} Q_i Y_i^2 & \sum_{i=1}^{n} Q_i Y_i Z_i \\
\sum_{i=1}^{n} Q_i Z_i X_i & \sum_{i=1}^{n} Q_i Z_i Y_i & \sum_{i=1}^{n} Q_i Z_i^2
\end{pmatrix} \tag{2.20}
$$

The matrix is symmetric because of the obvious equalities of the off-diagonal sums such as

$$
\sum_{i=1}^{n} Q_i X_i Y_i \quad \text{and} \quad \sum_{i=1}^{n} Q_i Y_i X_i
$$

There are, unfortunately, many different definitions related to the second (and higher) moments in the literature. There is little uniformity of usage, and it is necessary to be crystal clear about the definition and choice of origin when dealing with these quantities.

Most authors prefer to work with a quantity called the *electric quadrupole moment* rather than the second moment, but even then there are several different conventions. A common choice is to use the symbol $\mathbf{\Theta}_e$ and the definition

$$
\mathbf{\Theta}_e = \frac{1}{2} \begin{pmatrix}
\sum_{i=1}^{n} Q_i(3X_i^2 - R_i^2) & 3\sum_{i=1}^{n} Q_i X_i Y_i & 3\sum_{i=1}^{n} Q_i X_i Z_i \\
3\sum_{i=1}^{n} Q_i Y_i X_i & \sum_{i=1}^{n} Q_i(3Y_i^2 - R_i^2) & 3\sum_{i=1}^{n} Q_i Y_i Z_i \\
3\sum_{i=1}^{n} Q_i Z_i X_i & 3\sum_{i=1}^{n} Q_i Z_i Y_i & \sum_{i=1}^{n} Q_i(3Z_i^2 - R_i^2)
\end{pmatrix} \tag{2.21}
$$

Note that the diagonal elements of this matrix sum to zero and so the matrix has zero trace (the *trace* being the sum of the diagonal elements, see the Appendix). Some authors don't use a factor of $\frac{1}{2}$ in their definition. Quadrupole moments are gauge invariant provided the electric dipole moment and the charge are both zero.

Figure 2.11 shows an octahedrally symmetrical array of point charges. Each point charge has magnitude Q, apart from the central charge that has magnitude $-6Q$ in order to make the system neutral. The distance between each axial point charge and the central one is a.

If I choose to direct the Cartesian axes along the symmetry axes, then the second moment matrix is

$$
\mathbf{q}_e = Qa^2 \begin{pmatrix}
2 & 0 & 0 \\
0 & 2 & 0 \\
0 & 0 & 2
\end{pmatrix}
$$

whilst the quadrupole moment matrix is zero.

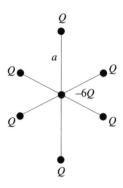

Figure 2.11   Octahedral charge distribution

If I now reduce the symmetry of the charge distribution by placing charges $2Q$ along the vertical axis (taken for the sake of argument as the x-axis) and $-8Q$ at the centre (to keep the electrical balance), the second moment matrix becomes

$$\mathbf{q}_e = Qa^2 \begin{pmatrix} 4 & 0 & 0 \\ 0 & 2 & 0 \\ 0 & 0 & 2 \end{pmatrix}$$

whilst the quadrupole moment matrix is now

$$\mathbf{\Theta}_e = Qa^2 \begin{pmatrix} 2 & 0 & 0 \\ 0 & -1 & 0 \\ 0 & 0 & -1 \end{pmatrix}$$

The electric quadrupole moment measures deviations from spherical symmetry. It is zero when the charge distribution has spherical symmetry. It always has zero trace (because of the definition), but it isn't always diagonal. Nevertheless, it can always be made diagonal by a rotation of the coordinate axes.

Finally, consider a linear array formed by the top $(+Q)$, central $(+2Q)$ and lower charges $(-3Q)$. We find

$$\mathbf{q}_e = Qa^2 \begin{pmatrix} -2 & 0 & 0 \\ 0 & 0 & 0 \\ 0 & 0 & 0 \end{pmatrix}; \quad \mathbf{\Theta}_e = Qa^2 \begin{pmatrix} -2 & 0 & 0 \\ 0 & 1 & 0 \\ 0 & 0 & 1 \end{pmatrix}$$

In cases where the symmetry of the problem determines that the second moment tensor only has one non-zero component, we speak colloquially of *the* second moment (which in this case is $-2Qa^2$).

2.9.3 Higher electric moments

The set of 10 independent quantities $\sum_{i=1}^{n} Q_i X_i^3$, $\sum_{i=1}^{n} Q_i X_i^2 Y_i$ through $\sum_{i=1}^{n} Q_i Z_i^3$ defines the electric third moment of the charge distribution, and so on. We rarely encounter such higher moments of electric charge in chemistry.

2.10 The Electrostatic Potential

Electrostatic forces are vector quantities, and we have to worry about their magnitude and direction. I explained earlier that it is more usual to work with the mutual potential energy U rather than the force $\mathbf{F}$, if only because U is a scalar quantity. In any case we can recover one from the other by the formula

$$\mathbf{F} = -\operatorname{grad} U$$

Similar considerations apply when dealing with electrostatic fields. They are vector fields with all the inherent problems of having to deal with both a magnitude and a direction. It is usual to work with a scalar field called the *electrostatic potential* ϕ. This is related to the electrostatic field $\mathbf{E}$ in the same way that U is related to $\mathbf{F}$

$$\mathbf{E} = -\operatorname{grad} \phi$$

We will hear more about the electrostatic potential in later sections. In the meantime, I will tell you that the electrostatic potential at field point $\mathbf{r}$ due to a point charge Q at the coordinate origin is

$$\phi(\mathbf{r}) = \frac{Q}{4\pi\epsilon_0} \frac{1}{r} \tag{2.22}$$

The electric field and the electrostatic potential due to an electric dipole, quadrupole and higher electric moments are discussed in all elementary electromagnetism texts. The expressions can be written exactly in terms of the various distances and charges involved. For many applications, including our own, it is worthwhile examining the mathematical form of these fields for points in space that are far away from the charge distribution. We then refer to (for example) the 'small' electric dipole and so on.

The electrostatic potential at field point $\mathbf{r}$ due to a small electric dipole $\mathbf{p}_e$ at the coordinate origin turns out to be

$$\phi(\mathbf{r}) = \frac{1}{4\pi\epsilon_0} \frac{\mathbf{p}_e \cdot \mathbf{r}}{r^3} \tag{2.23}$$

which falls off as $1/r^2$. It falls off faster with r than the potential due to a point charge because of the cancellation due to plus and minus charges. This is in fact a general rule, the electrostatic potential for a small electric multipole of order l falls off as $r^{-(l+1)}$ so dipole moment potentials fall off faster than those due to point charges, and so on.

2.11 Polarization and Polarizability

In electrical circuits, charges are stored in capacitors, which at their simplest consist of a pair of conductors carrying equal and opposite charges. Michael Faraday (1837) made a great discovery when he observed that filling the space between the plates of a parallel plate capacitor with substances such as mica increased their ability to store charge. The multiplicative factor is called the relative permittivity and is given a symbol ϵ_r, as discussed above. I also told you that the older name is the dielectric constant.

Materials such as glass and mica differ from substances such as copper wire in that they have few conduction electrons and so make poor conductors of electric current. We call materials such as glass and mica *dielectrics*, to distinguish them from metallic conductors.

Figure 2.12 shows a two-dimensional picture of a dielectric material, illustrated as positively charged nuclei each surrounded by a localized electron cloud. We now apply an electrostatic field, directed from left to right. There is a force on each charge, and the positive charges are displaced to the right whilst the negative charges move a corresponding distance to the left, as shown in Figure 2.13.

The macroscopic theory of this phenomenon is referred to as *dielectric polarization,* and we focus on the induced dipole moment $d\mathbf{p}_e$ per differential volume $d\tau$. Because it is a macroscopic theory, no attention is paid to atomic details; we assume that there are a large number of atoms or molecules within the volume element $d\tau$ (or that the effects caused by the discrete particles has somehow been averaged out).

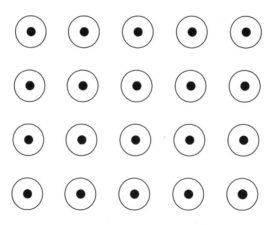

Figure 2.12   Dielectric slab

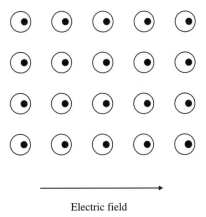

Electric field

Figure 2.13   Dielectric with applied field

We relate the induced electric dipole to the volume of a differential element by

$$d\mathbf{p}_e = \mathbf{P}d\tau \tag{2.24}$$

where the dielectric polarization $\mathbf{P}$ is an experimentally determined quantity. $\mathbf{P}$ can depend on the applied field in all manner of complicated ways, but for very simple media and for low field strengths, it turns out that $\mathbf{P}$ is directly proportional to $\mathbf{E}$. We write

$$\mathbf{P} = (\epsilon_r - 1)\epsilon_0 \, \mathbf{E} \tag{2.25}$$

where ϵ_r is the relative permittivity of the dielectric. The polarization acts so as to reduce the field inside a dielectric.

2.12 Dipole Polarizability

At the microscopic level, we concern ourselves with the various electric moments that are induced in each atom or molecule. Consider the simple case shown in Figure 2.14, where we apply a weak electric field E in a direction parallel to a molecule's electric dipole moment. This causes charge redistribution and we can write

$$p_e(\text{induced}) = p_e(\text{permanent}) + \alpha E \tag{2.26}$$

I have distinguished between the permanent electric dipole, the one a molecule has in free space with no fields present, from the induced dipole. I have also used the symbol α for the *dipole polarizability*.

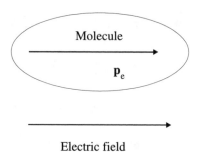

Figure 2.14 Induced molecular dipole

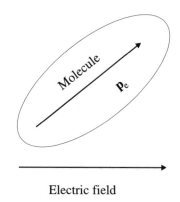

Figure 2.15 The more general case

In the general case, the field need not be weak and the induced dipole need not be in the same direction as either the applied field or the permanent dipole moment. This is shown in Figure 2.15. α cannot be a scalar, since the directions of the applied field and the induced dipole need not be the same.

The dipole polarizability $\boldsymbol{\alpha}$ is a special type of physical property called a tensor, just like the electric second moment. We can represent $\boldsymbol{\alpha}$ as a 3×3 real symmetric matrix

$$\boldsymbol{\alpha} = \begin{pmatrix} \alpha_{xx} & \alpha_{xy} & \alpha_{xz} \\ \alpha_{yx} & \alpha_{yy} & \alpha_{yz} \\ \alpha_{zx} & \alpha_{zy} & \alpha_{zz} \end{pmatrix} \tag{2.27}$$

and I will write the more general expression as

$$\mathbf{p}_e(\text{induced}) = \mathbf{p}_e(\text{permanent}) + \boldsymbol{\alpha} \cdot \mathbf{E} + \text{higher order terms} \tag{2.28}$$

The higher order terms are the *hyperpolarizabilities*; they feature in advanced texts of this kind. We are not going to meet them again. We interpret Equation (2.28) as a matrix equation; the **p**s are column vectors.

2.12.1 Properties of polarizabilities

The matrix $\boldsymbol{\alpha}$ can always be written in diagonal form by a suitable rotation of the cartesian axes to give

$$\boldsymbol{\alpha} = \begin{pmatrix} \alpha_{aa} & 0 & 0 \\ 0 & \alpha_{bb} & 0 \\ 0 & 0 & \alpha_{cc} \end{pmatrix}$$

The quantities α_{aa}, α_{bb} and α_{cc} are called the *principal values* of the polarizability tensor. For molecules with symmetry, the principal axes of polarizability correspond to the molecular symmetry axes. For a linear molecule the components that refer to perpendicular axes are equal and usually different from the parallel component, and the matrix is usually written

$$\boldsymbol{\alpha} = \begin{pmatrix} \alpha_{aa} & 0 & 0 \\ 0 & \alpha_{bb} & 0 \\ 0 & 0 & \alpha_{bb} \end{pmatrix}$$

2.13 Many-Body Forces

It is instructive to calculate the work done in building up an array of charges such as that shown in Figure 2.9. We will assume that all the charges are point charges and so cannot be polarized. We start with all the charges (n in total) at infinity, and move them in turn from infinity to their position as shown.

Moving Q_1 from infinity to position vector $\mathbf{R}_1$ takes no energy, because no other charges are present. Moving Q_2 from infinity to position vector $\mathbf{R}_2$ involves an energy change,

$$U_{12} = \frac{1}{4\pi\epsilon_0} \frac{Q_1 Q_2}{R_{12}}$$

where R_{12} is the length of the vector joining the charges. We now move Q_3 from infinity to position vector $\mathbf{R}_3$. This charge moves in the field due to Q_1 and Q_2, involving an energy cost

$$U_{13} + U_{23} = \frac{1}{4\pi\epsilon_0} \frac{Q_1 Q_3}{R_{13}} + \frac{1}{4\pi\epsilon_0} \frac{Q_2 Q_3}{R_{23}}$$

The total work done U_{tot} is seen to be

$$U_{\text{tot}} = \sum_{i=1}^{n-1} \sum_{j=i+1}^{n} U_{ij} \tag{2.29}$$

This expression holds because of the pairwise additivity of the forces between point charges. The expression would not necessarily hold if the charges were themselves charge distributions because the addition of further charges could polarize the existing ones and so alter the forces already calculated.

Whatever the case, U_{tot} will depend on the coordinates of all the charges present

$$U_{tot} = U_{tot}(\mathbf{R}_1, \mathbf{R}_2, \ldots, \mathbf{R}_n) \tag{2.30}$$

and it is always possible to write a formal expansion

$$U_{tot}(\mathbf{R}_1, \mathbf{R}_2, \ldots, \mathbf{R}_n) = \sum_{\text{pairs}} U^{(2)}(\mathbf{R}_i, \mathbf{R}_j)$$

$$+ \sum_{\text{triples}} U^{(3)}(\mathbf{R}_i, \mathbf{R}_j, \mathbf{R}_k) + \cdots + U^{(n)}(\mathbf{R}_1, \mathbf{R}_2, \ldots, \mathbf{R}_n) \tag{2.31}$$

involving distinct pairs, triples, etc. of particles. The $U^{(2)}$ terms are referred to as the pair contributions, the $U^{(3)}$'s are the three body terms and so on. Terms higher than the second are identically zero for the interaction between point charges.

3 The Forces Between Molecules

When molecules are near enough to influence one another, we need to concern ourselves with the balance between the forces of attraction and repulsion. We know that such forces exist, for otherwise there would be nothing to bring molecules together into the solid and liquid states, and all matter would be gaseous. A study of the forces between atomic or molecular species constitutes the subject of *intermolecular forces*.

People have speculated about the nature of intermolecular forces ever since the ideas of atoms and molecules first existed. Our present ideas, that molecules attract at long range but repel strongly at short range, began to emerge in the nineteenth century due to the experimental work of Rudolf J. E. Clausius and Thomas Andrews. It was not until the early twentieth century, when the principles of quantum mechanics became established, that we could truly say that we understood the detailed mechanism of intermolecular forces.

Although we talk about intermolecular *forces*, it is more usual and convenient to focus on the *mutual potential energy*, discussed in Chapter 2. If we start with two argon atoms at infinite separation, then their mutual potential energy at separation R tells us the energy change on bringing the atoms together to that distance from infinity.

Even for the simplest pair of molecules, the intermolecular mutual potential energy will depend on their relative orientations in addition to their separation. Perhaps you can now see why the study of intermolecular forces has taken so much effort by so many brilliant scientists, over very many years.

3.1 The Pair Potential

So, to start with, we concern ourselves with two atomic or molecular species, A and B, and ask how they interact. No chemical reaction is implied, and I should say straightaway that I am not going to be concerned with bond making and bond breaking in this chapter. That is the subject of *valence theory*. In the (unusual) case that the two species A and B concerned are ions, you may think that the problem is more or less solved. We simply calculate their mutual Coulomb potential energy as discussed in Chapter 2

$$U_{AB} = \frac{1}{4\pi\epsilon_0} \frac{Q_A Q_B}{R_{AB}} \qquad (3.1)$$

You would certainly be on the right lines in this approach except that ions aren't point charges and they can be polarized just like any other continuous charge distribution. But, as I explained in Chapter 2, we rarely have to concern ourselves with ions and to get started we will consider the very simple case of a pair of like atoms (such as two argon atoms). We know from experiment that the general form of their mutual potential energy must be that shown in Figure 3.1.

This curve is meant to be schematic, and I have been careful not to quantify the axes. It has all the right features but we have yet to discover its precise form. We very often speak about the *pair potential*, and write it $U(R)$, where R is the separation between the two atoms. The zero of $U(R)$, denoted by the horizontal line, is commonly taken to refer to the two atoms at infinity. We often characterize the curve in terms of a small number of parameters; for example, the collision diameter σ being the distance at which $U(R) = 0$, the minimum R_{min}, and minus the value of $U(R)$ at R_{min} (often written ε, and as defined is a positive quantity).

We now need to investigate more closely the precise form of this pair potential. The potential comprises a repulsive part (important for small R) and an attractive part

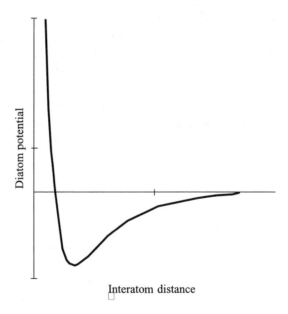

Figure 3.1   Schematic Ar–Ar interaction

(important for large R). It turns out that there are three major contributions to the attractive part, as we will see below.

3.2 The Multipole Expansion

Suppose that we have two molecules with centres a distance R apart (Figure 3.2). The distance R is taken to be large compared with a molecular dimension. Each molecule consists of a number of charged particles, and in principle we can write down an expression for the mutual potential energy of these two molecules in terms of the pair potentials between the various point charges. The basic physical idea of the *multipole expansion* is to make use of the fact that several of these particles go to form molecule A, and the remainder to form molecule B, each of which has a distinct chemical identity. We therefore seek to write the mutual potential energy of A and B in terms of the properties of the two molecular charge distributions and their separation.

3.3 The Charge–Dipole Interaction

I can illustrate the ideas by considering an elementary textbook problem, namely the mutual potential energy of a simple small electric dipole and a point charge. Suppose that we have a simple dipole consisting of a pair of charges, Q_A and Q_B, aligned along the horizontal axis and equally separated from the coordinate origin by distance d. We introduce a third charge Q as shown in Figure 3.3, with the scalar distance R

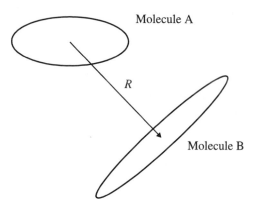

Figure 3.2 Two interacting molecules

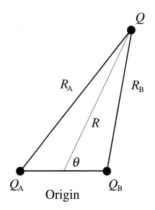

Figure 3.3 Charge–dipole interaction

from the origin. This point charge makes an angle θ with the electric dipole, as shown. The two point charges Q_A and Q_B have a mutual potential energy of

$$\frac{1}{4\pi\epsilon_0} \frac{Q_A Q_B}{2d} \tag{3.2}$$

but we are going to investigate what happens to the mutual potential energy of the system as we change the position vector of Q, and so we ignore this term since it remains constant.

The mutual potential energy U(charge–dipole) of the point charge and the electric dipole is given exactly by

$$U(\text{charge–dipole}) = \frac{1}{4\pi\epsilon_0} Q\left(\frac{Q_A}{R_A} + \frac{Q_B}{R_B}\right) \tag{3.3}$$

This can also be written in terms of R and θ as

$$U(\text{charge–dipole}) = \frac{1}{4\pi\epsilon_0} Q\left(\frac{Q_A}{\sqrt{(R^2 + d^2 + 2dR\cos\theta)}} + \frac{Q_B}{\sqrt{(R^2 + d^2 - 2dR\cos\theta)}}\right) \tag{3.4}$$

and once again, this is an exact expression.

In the case where the point charge Q gets progressively far away from the coordinate origin, we can usefully expand the two denominators using the binomial theorem to give

$$U(\text{charge–dipole}) = \frac{1}{4\pi\epsilon_0} Q\left(\frac{(Q_A + Q_B)}{R} + \frac{(Q_B - Q_A)d}{R^2}\cos\theta\right.$$
$$\left. + \frac{(Q_A + Q_B)d^2}{2R^3}(3\cos^2\theta - 1) + \cdots\right) \tag{3.5}$$

The first term on the right-hand side contains the sum of the two charges making up the dipole. Very often, we deal with simple dipoles that carry no overall charge, and this term is zero because $Q_A = -Q_B$. The second term on the right-hand side obviously involves the electric dipole moment, whose magnitude is $(Q_B - Q_A)d$. The third term involves the electric second moment whose magnitude is $(Q_B + Q_A)d^2$ and so on. The mutual potential energy is therefore seen to be a sum of terms; each is a product of a moment of the electric charge, and a function of the inverse distance. Hopefully, as R increases, the magnitude of the successive terms will become less and eventually the mutual potential energy will be dominated by the first few terms in the expansion.

In the more general case where we replace Q_A and Q_B with an arbitrary array of point charges $Q_1, Q_2, \ldots, Q_n$, whose position vectors are $\mathbf{R}_1, \mathbf{R}_2, \ldots, \mathbf{R}_n$ (or for that matter a continuous charge distribution), it turns out that we can always write the mutual interaction potential with Q as

$$U = \frac{Q}{4\pi\epsilon_0}\left\{ \left(\sum_{i=1}^{n} Q_i\right)\frac{1}{R} - \left(\sum_{i=1}^{n} Q_i \mathbf{R}_i\right) \cdot \operatorname{grad}\left(\frac{1}{R}\right) + \text{higher order terms}\right\} \quad (3.6)$$

The first summation on the right-hand side gives the overall charge of the charge distribution. The second term involves the electric dipole moment; the third term involves the electric quadrupole moment and so on.

3.4 The Dipole–Dipole Interaction

Consider now a slightly more realistic model for the interaction of two simple (diatomic) molecules, Figure 3.4. Molecule A consists of two point charges, Q_{1A} and

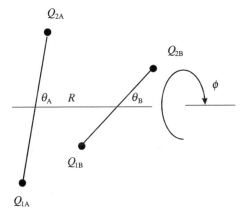

Figure 3.4   Multipole expansion for a pair of diatomics

Q_{2A}, and molecule B consists of two point charges, Q_{1B} and Q_{2B}. The overall charge on molecule A is therefore $Q_A = Q_{1A} + Q_{2A}$ with a similar expression for molecule B. The electric dipole moments of A and B are written $\mathbf{p}_A$ and $\mathbf{p}_B$ in an obvious notation, and their scalar magnitudes are written p_A and p_B. The second moments of the two molecules are each determined by a scalar value q_A and q_B, simply because they are linear.

Molecule A is centred at the origin, whilst molecule B has its centre a distance R away along the horizontal axis. The inclinations to the axis are θ_A and θ_B, and ϕ gives the relative orientation of the two molecules. The sizes of the two molecules are much less than their separation, so we can make the same approximations as for the small dipole. After some standard analysis we find that the mutual potential energy of A and B is

$$
\begin{aligned}
(4\pi\epsilon_0)U_{AB} = {} & \frac{Q_A Q_B}{R} + \frac{1}{R^2}(Q_B p_A \cos\theta_A - Q_A p_B \cos\theta_B) \\
& - \frac{p_A p_B}{R^3}(2\cos\theta_A \cos\theta_B - \sin\theta_A \sin\theta_B \cos\phi) \\
& + \frac{1}{2R^3}(Q_A q_B(3\cos^2\theta_B - 1) + Q_B q_A(3\cos^2\theta_A - 1)) \\
& + \cdots
\end{aligned}
\tag{3.7}
$$

The physical interpretation is as follows. The first term on the right-hand side gives the mutual potential energy of the two charged molecules A and B. The second term gives a contribution due to each charged molecule with the other dipole. The third term is a dipole–dipole contribution and so on.

If A and B correspond to uncharged molecules, then the leading term is seen to be the dipole–dipole interaction

$$
(4\pi\epsilon_0)U_{AB,\text{dip–dip}} = -\frac{p_A p_B}{R^3}(2\cos\theta_A \cos\theta_B - \sin\theta_A \sin\theta_B \cos\phi)
\tag{3.8}
$$

The sign and magnitude of this term depends critically on the relative orientation of the two molecules. Table 3.1 shows three possible examples, all of which have $\phi = 0$.

Table 3.1 Representative dipole–dipole terms for two diatomics

θ_A	θ_B	Relative orientations	Expression for dipole–dipole U
0	0	Parallel	$-2p_A p_B/4\pi\epsilon_0 R^3$
0	π	Antiparallel	$+2p_A p_B/4\pi\epsilon_0 R^3$
0	$\pi/2$	Perpendicular	0

3.5 Taking Account of the Temperature

We now imagine that the two molecules undergo thermal motion; we keep their separation R constant but allow the angles to vary. The aim is to calculate the average dipole–dipole interaction. Some orientations of the two dipoles will be more energetically favoured than others and we allow for this by including a Boltzmann factor $\exp(-U/k_BT)$, where k_B is the Boltzmann constant and T the thermodynamic temperature. It is conventional to denote mean values by 'carets' $\langle\cdots\rangle$ and the mean value of the dipole–dipole interaction is given formally by

$$\langle U_{AB}\rangle_{dip-dip} = \frac{\int U_{AB}\exp(-\frac{U_{AB}}{k_BT})d\tau}{\int \exp(-\frac{U_{AB}}{k_BT})d\tau} \tag{3.9}$$

The integral has to be done over all possible values of the angles, keeping R fixed. After some standard integration, we find

$$\langle U_{AB}\rangle_{dip-dip} = -\frac{2p_A^2 p_B^2}{3k_BT(4\pi\epsilon_0)^2}\frac{1}{R^6} \tag{3.10}$$

The overall value is therefore negative, and the term is inversely dependent on the temperature. It also falls off as $1/R^6$.

3.6 The Induction Energy

The next step is the case of two interacting molecules, one of which has a permanent dipole moment and one of which is polarizable but does not have a permanent electric dipole moment.

Figure 3.5 shows molecule A with a permanent dipole moment $\mathbf{p}_A$. I have indicated the direction of $\mathbf{p}_A$ in the diagram, and an arbitrary point P in molecule B. The dipole $\mathbf{p}_A$ is a distance R from point P, and makes an angle θ as shown. The molecules are sufficiently far apart for the precise location of the point P inside the second molecule to be irrelevant.

The basic physical idea is that the electric dipole $\mathbf{p}_A$ induces a dipole in molecule B, since B is polarizable. We evaluate the potential energy involved and finally average over all possible geometrical arrangements, for a fixed value of the intermolecular separation. The steps involved are as follows. The electrostatic potential due to the small dipole $\mathbf{p}_A$ is

$$\phi_A(\mathbf{R}) = \frac{1}{4\pi\epsilon_0}\frac{\mathbf{p}_A\cdot\mathbf{R}}{R^3}$$

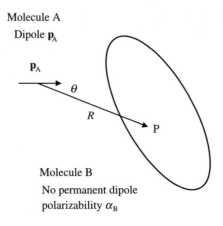

Molecule A
Dipole $\mathbf{p}_A$

$\mathbf{p}_A$

θ

R

P

Molecule B
No permanent dipole
polarizability α_B

Figure 3.5   Dipole-induced dipole

This is related to the electrostatic field by the general formula

$$\mathbf{E}(\mathbf{R}) = -\mathrm{grad}\,\phi(\mathbf{R})$$

and direct differentiation gives the following formula

$$\mathbf{E}_A(\mathbf{R}) = -\frac{1}{4\pi\epsilon_0}\left\{\frac{\mathbf{p}_A}{R^3} - 3\frac{\mathbf{p}_A \cdot \mathbf{R}}{R^5}\mathbf{R}\right\} \tag{3.11}$$

Molecule A therefore generates an electrostatic field in the region of molecule B, according to the above vector equation. The modulus of this vector at point P is

$$E_A = \frac{1}{4\pi\epsilon_0}\frac{p_A}{R^3}\sqrt{(1 + 3\cos^2\theta)} \tag{3.12}$$

This electrostatic field induces a dipole in molecule B. For the sake of argument, I will assume that the induced dipole is in the direction of the applied field (and so we need not worry about the fact that the polarizability is a tensor property). Calculation of the resulting mutual potential energy U_{AB} gives

$$U_{AB} = -\frac{1}{(4\pi\epsilon_0)^2}\frac{\alpha_B p_A^2}{R^6}\frac{1}{2}(3\cos^2\theta + 1) \tag{3.13}$$

Polarizabilities are positive quantities and so U_{AB} is negative for all values of θ at a given intermolecular separation This is quite different to the dipole–dipole interaction, where some alignments of the dipoles gave a positive contribution to the mutual potential energy and some gave a negative one.

Finally, we have to average over all possible alignments keeping the internuclear separation fixed. This averaging again has to be done using the Boltzmann weightings, and we find eventually an expression for the *induction* contribution to the mutual potential energy of A and B

$$\langle U_{AB}\rangle_{ind} = -\frac{1}{(4\pi\epsilon_0)^2}\frac{p_A^2\alpha_B}{R^6} \qquad (3.14)$$

Note that the interaction falls off as $1/R^6$ just as for the dipole–dipole interaction, but this time there is no temperature dependence. For two identical A molecules each with permanent electric dipole $\mathbf{p}_A$ and polarizability $\boldsymbol{\alpha}_A$ the expression becomes

$$\langle U_{AA}\rangle_{ind} = -\frac{2}{(4\pi\epsilon_0)^2}\frac{p_A^2\alpha_A}{R^6} \qquad (3.15)$$

This of course has to be added to the dipole–dipole expression of the previous section.

3.7 Dispersion Energy

It is an experimental fact that inert gases can be liquefied. Atoms don't have permanent electric moments, so the dipole–dipole and induction contributions to the mutual potential energy of an array of inert gas atoms must both be zero. There is clearly a third interaction mechanism (referred to as *dispersion*), and this was first identified by Fritz W. London in 1930.

The two contributions to the mutual potential energy discussed in previous sections can be described by classical electromagnetism. There is no need to invoke the concepts of quantum mechanics. Dispersion interactions can only be correctly described using the language of quantum mechanics. Nevertheless, the following qualitative discussion is to be found in all elementary texts.

The electrons in an atom or molecule are in continual motion, even in the ground state. So, although on average the dipole moment of a spherically-symmetrical system is zero, at any instant a temporary dipole moment can occur. This temporary dipole can induce a further temporary dipole in a neighbouring atom or molecule and, as in the case of the inductive interaction; the net effect will be attractive.

Paul K. L. Drude gave a simple quantum mechanical description of the effect, and his theory suggests that the dispersion contribution can be written

$$\langle U\rangle_{disp} = -\left(\frac{D_6}{R^6} + \frac{D_8}{R^8} + \frac{D_{10}}{R^{10}} + \cdots\right) \qquad (3.16)$$

The first term (which I have written D_6) is to be identified with the instantaneous dipole-induced dipole mechanism. The higher terms are caused by instantaneous quadrupole-induced quadrupoles, etc. According to Drude's theory

$$D_6 = -\frac{3\alpha^2\varepsilon_1}{4(4\pi\epsilon_0)^2} \qquad (3.17)$$

In this expression, ε_1 is the first excitation energy of the atomic or molecular species concerned. The dispersion energy is again seen to be attractive and to fall off as $1/R^6$.

3.8 Repulsive Contributions

When two molecular species approach so closely that their electron clouds overlap, the positively charged nuclei become less well shielded by the negative electrons and so the two species repel each other. The repulsive term is sometimes written

$$U_{rep} = A\exp(-BR) \qquad (3.18)$$

where A and B are specific to the particular molecular pair and have to be determined from experiment. The precise form of the repulsive term is not well understood; all that is certain is that it must fall off quickly with distance, and the exponential function is therefore a possible suitable candidate.

The total interaction is $U = U_{rep} + U_{dip-dip} + U_{ind} + U_{disp}$, which we can write

$$U = A\exp(-BR) - \frac{C}{R^6} \qquad (3.19)$$

since all the attractive forces fall off as $1/R^6$. This is known as the exp-6 potential. In the *Lennard-Jones* (L-J) *12–6* potential, we take a repulsive term proportional to $1/R^{12}$ and so

$$U_{L\text{-}J} = \frac{C_{12}}{R^{12}} - \frac{C_6}{R^6} \qquad (3.20)$$

Once again the coefficients C_{12} and C_6 have to be determined from experiments on the species under study. The L-J potential usually is written in terms of the well depth ε and the distance of closest approach σ as follows

$$U_{L\text{-}J} = 4\varepsilon\left(\left(\frac{\sigma}{R}\right)^{12} - \left(\frac{\sigma}{R}\right)^6\right) \qquad (3.21)$$

The two L-J parameters σ and ε have been deduced for a range of atoms. The quantity ε/k_B (which has dimensions of temperature) is usually recorded in the literature rather than ε. Sample atomic parameters are shown in Table 3.2.

Table 3.2 Representative L-J atomic parameters

	$(\varepsilon/k_B)/K$	σ/pm
He	10.22	258
Ne	35.7	279
Ar	124	342
Xe	229	406

Table 3.3 L-J parameters for simple molecules

	$(\varepsilon/k_B)/K$	σ/pm
H_2	33.3	297
O_2	113	343
N_2	91.5	368
Cl_2	357	412
Br_2	520	427
CO_2	190	400
CH_4	137	382
CCl_4	327	588
C_2H_4	205	423
C_6H_6	440	527

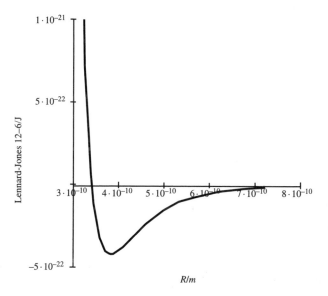

Figure 3.6 Lennard-Jones 12–6 potential for argon–argon

Over the years, people have extended these ideas to the interaction of simple molecules. Some caution is needed: the interaction between two molecules will generally depend on the precise details of their orientation, and the values given in Table 3.3 must be interpreted as some kind of geometrical average. These values were taken from the classic text *Molecular Theory of Gases and Liquids* [1]. Figure 3.6 shows a L-J 12–6 plot for argon–argon.

3.9 Combination Rules

A large number of L-J parameters have been deduced over the years, but they relate to pairs of like atoms. Rather than try to deduce corresponding parameters for unlike pairs, it is usual to use so-called *combination rules*, which enable us to relate the C_{12} and the C_6 parameters for an unlike-atom pair A–B to those of A–A and B–B. The use of such combination rules is common in subjects such as chemical engineering, and is widely applied to many physical properties.

There are three common combination rules in the literature, as follows

$$C_{12,ij} = \left(\frac{R_i^*}{2} + \frac{R_j^*}{2} \right)^{12} \sqrt{\varepsilon_i \varepsilon_j}$$

$$C_{6,ij} = 2 \left(\frac{R_i^*}{2} + \frac{R_j^*}{2} \right)^6 \sqrt{\varepsilon_i \varepsilon_j} \tag{3.22}$$

where R_i^* is the minimum energy separation for two atoms of type i and ε_i the well depth;

$$C_{12,ij} = 4(\sigma_i \sigma_j)^6 \sqrt{\varepsilon_i \varepsilon_j}$$

$$C_{6,ij} = 4(\sigma_i \sigma_j)^3 \sqrt{\varepsilon_i \varepsilon_j} \tag{3.23}$$

and finally

$$C_{6,ij} = C \frac{\alpha_i \alpha_j}{\sqrt{\frac{\alpha_i}{N_i}} + \sqrt{\frac{\alpha_j}{N_j}}}$$

$$C_{12,ij} = \tfrac{1}{2} C_{6,ij} (R_i + R_j)^6 \tag{3.24}$$

where α_i is the dipole polarizability of atom i, N_i the number of valence electrons and R_i the van der Waals radius.

3.10 Comparison with Experiment

You will have got the idea by now that we have to determine the parameters in any pair potential by appeal to experiment. There are two kinds of experiment to consider. First, there are those that are essentially in the gas phase, where pairs of atoms genuinely interact with each other unencumbered by other species. This means that the total mutual potential energy is given by the sum of the interacting pairs.

Second, there are experiments that essentially relate to condensed phases, where the interacting particles are sufficiently close to raise doubts about the credibility of the pairwise additivity assumption.

3.10.1 Gas imperfections

The deviation of gases from perfect behaviour can be expressed in the form of a virial equation of state

$$\frac{pV}{nRT} = 1 + \frac{nB(T)}{V} + \frac{n^2 C(T)}{V^2} + \cdots \tag{3.25}$$

where the virial coefficients $B(T), C(T), \ldots$ depend on the temperature and on the characteristics of the species under study. Here, n is the amount of substance, p the pressure, V the volume, R the gas constant and T the thermodynamic temperature. $B(T)$ is called the *second virial coefficient* whilst $C(T)$ is called the *third virial coefficient* and so on. They have to be determined experimentally by fitting the pVT data of the gas under study.

The virial equation of state has a special significance in that the virial coefficients can be related directly to the molecular properties. $B(T)$ depends on the pair potential $U(R)$ in the following way

$$B(T) = 2\pi \int_0^\infty \left(1 - \exp\left(-\frac{U(R)}{k_B T}\right)\right) R^2 \, dR \tag{3.26}$$

3.10.2 Molecular beams

In a molecular beam experiment, a beam of mono-energetic molecules is produced and allowed to collide either with other molecules in a scattering chamber, or with a similar beam travelling at right angles to the original beam. Measurements of the amount by which the incident beam is reduced in intensity, or the number of molecules scattered in a particular direction, allow determination of the parameters in the pair potential.

3.11 Improved Pair Potentials

The L-J 12−6 potential for a pair of interacting atoms

$$U_{\text{L-J}}(R) = \frac{C_{12}}{R^{12}} - \frac{C_6}{R^6}$$

$$= 4\varepsilon \left(\left(\frac{\sigma}{R}\right)^{12} - \left(\frac{\sigma}{R}\right)^6\right)$$

contains two parameters (ε and σ, or C_{12} and C_6,) that have to be determined by experimental observation. The exp-6 model

$$U(R) = A \exp(-BR) - \frac{C}{R^6}$$

contains three parameters, which allows for a little more flexibility.

The Born–Mayer–Huggins potential

$$U(R) = A \exp(-BR) - \frac{C}{R^6} - \frac{D}{R^8} \tag{3.27}$$

contains four parameters.

More recent investigations have concentrated on pair potentials having many more disposable parameters, for example

$$U(r) = \exp\left(A\left(1 - \frac{R}{\sigma}\right)\right) \sum_{i=0}^{n} B_i \left(\frac{R}{\sigma} - 1\right)^i$$
$$+ \frac{C_6}{\left(D + \left(\frac{R}{\sigma}\right)^6\right)} + \frac{C_8}{\left(D + \left(\frac{R}{\sigma}\right)^8\right)} + \frac{C_{10}}{\left(D + \left(\frac{R}{\sigma}\right)^{10}\right)} \tag{3.28}$$

There is no obvious relationship between the various parameters in these different models; they all have to be determined by fitting experimental data. Roughly speaking, the more parameters the better.

3.12 Site–Site Potentials

The L-J potential plays an important role in the history of molecular modelling. Early work focused on atoms, but as I explained there were many ambitious attempts to model simple molecules as if they were in some way L-J atoms, and the parameters have to be interpreted as applying to some kind of average over molecular rotations (and presumably vibrations).

Suppose now that we want to try to understand the interaction between two dinitrogen molecules in more depth. In view of our discussion above, the instantaneous interaction energy clearly will depend on the separation of the two diatoms, together with their mutual angular arrangement in space.

Figure 3.7 shows two such dinitrogens, oriented arbitrarily in space with respect to each other. Nitrogen A and nitrogen B make up a stable diatom, as do atoms C and D. We ignore the fact that the molecules have vibrational energy, and the two diatoms are taken to be rigid. As a first approximation, the mutual potential energy of the pair of diatoms could be calculated by adding together the appropriate L-J parameters.

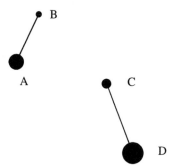

Figure 3.7 Two dinitrogen molecules, arbitrarily oriented

If I write as shorthand the L-J interaction between atoms A and C as

$$U_{\text{L-J}}(\text{A}, \text{C}) = 4\varepsilon \left(\left(\frac{\sigma}{R_{\text{AC}}} \right)^{12} - \left(\frac{\sigma}{R_{\text{AC}}} \right)^{6} \right)$$

then the total interaction between the two diatomics is taken as

$$U_{\text{L-J}} = U_{\text{L-J}}(\text{A}, \text{C}) + U_{\text{L-J}}(\text{A}, \text{D}) + U_{\text{L-J}}(\text{B}, \text{C}) + U_{\text{L-J}}(\text{B}, \text{D}) \qquad (3.29)$$

Such a potential is called a *site–site potential*. We will meet such potentials later in the book.

4 Balls on Springs

The theory of intermolecular forces relates to atomic and/or molecular species that are some distance apart (say, a few bond lengths). We saw in Chapter 3 that progress can be made in such a theory without normally invoking the concepts of quantum mechanics. If we truly want to understand why two atoms combine to give a chemical bond, and how bonds get broken and reformed in chemical reactions, then we enter the realms of *valence theory*. Quantum mechanics plays a dominant part in such discussions.

These are simple-minded comments and my arbitrary division of molecular interactions is subjective. At first sight, the stability of an NaCl ion pair can be explained in terms of elementary electrostatics, and we can usefully model argon liquid without recourse to quantum mechanics (apart from the London dispersion potential, which is a 'pure' quantum mechanical effect). A C—C bond in ethane is at first sight a quantum mechanical animal, and we will certainly have to invoke quantum mechanical ideas to explain the reaction of ethene with dichlorine. But there are grey areas that I can bring to your attention by considering the phenomenon of hydrogen bonding. The hydrogen bond is an attractive interaction between a proton donor X—H and a proton acceptor Y in the same or a different molecule

$$X—H \cdots Y$$

The bond usually is symbolized by three dots, as shown above, in order to reconcile the existence of compounds such as

$$NH_3 \cdots HCl$$

with the trivalence of nitrogen, the divalence of oxygen in oxonium salts and other compounds that apparently break the classical valence rules. Hydrogen bonds typically have strengths of $10-100 \, kJ \, mol^{-1}$. The lone pairs of oxygen and nitrogen and the partially charged character of the proton were eventually recognized as the sources of this bond. The first reference to this 'weak bond' were made by W. M. Latimer and W. H. Rodebush in 1920 [2].

The individual monomers X—H and Y retain their chemical identity to a large extent on hydrogen bond formation. In other words, no new covalent bond gets made. A great deal of evidence suggests that simple electrostatic models of the H bond give perfectly acceptable quantitative descriptions of the structure, vibrations and electric dipole moments of such hydrogen-bonded species. The hydrogen-bonded species

$$FHF^-$$

is well known and has been well studied, but it cannot be written

$$F-H\cdots F^-$$

because the proton is equally shared between the two fluorine atoms. Such a species is best thought of as covalently bound, and has to be treated by the methods of molecular quantum theory.

Having warned about bond breaking and bond making, I should tell you that a great deal of molecular modelling is concerned with the prediction and rationalization of molecular bond lengths and bond angles. Here we usually deal with isolated molecules in the gas phase and the theoretical treatments often refer to $0\,K$. A surprising amount of progress can be made by treating molecules as structureless balls (atoms) held together with springs (bonds). The array of balls and springs is then treated according to the laws of classical mechanics. Such calculations are remarkably accurate, and are taken very seriously.

4.1 Vibrational Motion

To get started, consider a particle of mass m lying on a frictionless horizontal table, and attached to the wall by a spring, as shown in Figure 4.1. The particle is initially at rest, when the length of the spring is R_e (where the subscript 'e' stands for equilibrium). If we stretch the spring, it exerts a restoring force on the particle, whilst if we compress the spring there is also a force that acts to restore the particle to its equilibrium position. If R denotes the length of the spring, then the extension is

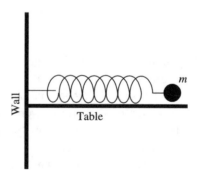

Figure 4.1   Ball attached to the wall by a spring

$R - R_e$, and if F_s is the restoring force due to the spring, then it is often observed experimentally that the force is directly proportional to the extension

$$F_s = -k_s(R - R_e) \qquad (4.1)$$

The constant of proportionality k_s is called the force constant and it tells us the strength of the spring. This law is known as *Hooke's Law* and it applies to very many springs made from many different materials. It invariably fails for large values of the extension, but is good for small deviations from equilibrium.

Suppose that we now set the particle in motion, so that it oscillates about R_e. According to Newton's second law we have

$$m\frac{d^2R}{dt^2} = -k_s(R - R_e) \qquad (4.2)$$

This second-order differential equation has the general solution

$$R = R_e + A \sin\left(\sqrt{\frac{k_s}{m}}t\right) + B \cos\left(\sqrt{\frac{k_s}{m}}t\right) \qquad (4.3)$$

where A and B are constants of integration. These constants have to be fixed by taking account of the *boundary conditions*. For example, if the particle starts its motion at time $t = 0$ from $R = R_e$, then we have

$$R_e = R_e + A \sin\left(\sqrt{\frac{k_s}{m}}0\right) + B \cos\left(\sqrt{\frac{k_s}{m}}0\right)$$

from which we deduce that $B = 0$ for this particular case. Normally we have to find A and B by a similar procedure.

The trigonometric functions sine and cosine repeat every 2π and a little manipulation shows that the general solution of Equation (4.3) can also be written

$$R = R_e + A \sin\left(\sqrt{\frac{k_s}{m}}\left(t + 2\pi\sqrt{\frac{m}{k_s}}\right)\right) + B \cos\left(\sqrt{\frac{k_s}{m}}\left(t + 2\pi\sqrt{\frac{m}{k_s}}\right)\right)$$

The quantity $\sqrt{k_s/m}$ has the dimension of inverse time and obviously it is an important quantity. We therefore give it a special symbol (ω) and name (the *angular vibration frequency*). We often write the general solution as

$$R = R_e + A \sin(\omega t) + B \cos(\omega t) \qquad (4.4)$$

A typical solution is shown as Figure 4.2 (for which I took $A = 1$ m, $B = 0$, $m = 1$ kg and $k_s = 1$ N m^{-1}). Such motions are called *simple harmonic*. At any given time, the displacement of the particle from its equilibrium position may be non-zero, but it should be clear from Figure 4.2 that the average value of the displacement $R - R_e$ is

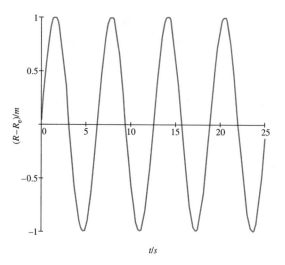

Figure 4.2   Simple harmonic motion

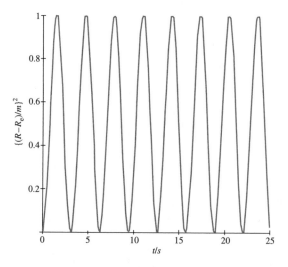

Figure 4.3   Variation of $(R - R_e)^2$ with time

zero. As noted in Chapter 3, it is usual to denote average values by $\langle \cdots \rangle$ and so we write

$$\langle R - R_e \rangle = 0$$

It should also be clear from Figure 4.3 that the average value of $(R - R_e)^2$ is not zero. A direct calculation using $\langle \sin^2(\omega t) \rangle = \frac{1}{2}$ gives

$$\langle (R - R_e)^2 \rangle = \frac{1}{2}(A^2 + B^2) \tag{4.5}$$

4.2 The Force Law

If we use the general one-dimensional result that links force and mutual potential energy

$$U(x) = -\int F(x)\mathrm{d}x$$

we find

$$U(R) = U(R_\mathrm{e}) + \tfrac{1}{2}k_\mathrm{s}(R - R_\mathrm{e})^2 \qquad (4.6)$$

where $U(R_\mathrm{e})$ is the value of $U(R)$ at the equilibrium position. As mentioned earlier, we often set this constant of integration arbitrarily to zero and so

$$U(R) = \tfrac{1}{2}k_\mathrm{s}(R - R_\mathrm{e})^2 \qquad (4.7)$$

Because the motion is simple harmonic, we refer to this potential as a *harmonic potential*. The potential energy varies over each cycle, in the same way as shown in Figure 4.3. The average value of the potential energy over a cycle is

$$\langle U(R)\rangle = \tfrac{1}{2}k_\mathrm{s}\langle (R - R_\mathrm{e})^2\rangle$$
$$= \tfrac{1}{4}(A^2 + B^2)$$

Finally, the kinetic energy T is given by

$$T(R) = \tfrac{1}{2}m\left(\frac{\mathrm{d}R}{\mathrm{d}t}\right)^2$$

The average value is

$$\langle T(R)\rangle = \tfrac{1}{4}m\omega^2(A^2 + B^2)$$
$$= \tfrac{1}{4}k_\mathrm{s}(A^2 + B^2)$$

The kinetic and potential energies vary with time, but the total energy $U(R) + T(R)$ is constant; it does not vary with time. The average value of the kinetic energy over a cycle is equal to the average value of the potential energy, each of which is one half of the total energy. I am going to use the symbol ε for energy when referring to a single atom or molecule, throughout the text. The total energy ε can take any value with no restrictions.

4.3 A Simple Diatomic

Consider next the Hooke's Law model of a diatomic molecule, Figure 4.4. The atoms

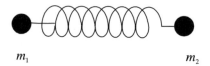

m_1 m_2

Figure 4.4   Diatomic molecule

have masses m_1 and m_2, and the spring has a harmonic force constant of k_s. I am going to consider the motion along the horizontal axis, which I will refer to as the x-axis. The x coordinates of the two atoms are x_1 and x_2 relative to an arbitrary axis, the equilibrium length of the spring is R_e and the length of the extended spring at some given time is

$$R = x_2 - x_1$$

The spring extension is therefore

$$x_2 - x_1 - R_e$$

Considering atom 1, the spring exerts a force of $k_s(x_2 - x_1 - R_e)$ and so, according to Newton's second law

$$m_1 \frac{d^2 x_1}{dt^2} = k_s(x_2 - x_1 - R_e)$$
$$= k_s(R - R_e) \tag{4.8}$$

As far as atom 2 is concerned, the extended spring exerts a force of magnitude $k_s(x_2 - x_1 - R_e)$ in the direction of decreasing x_2 and so

$$m_2 \frac{d^2 x_2}{dt^2} = -k_s(x_2 - x_1 - R_e)$$
$$= -k_s(R - R_e) \tag{4.9}$$

After a little rearrangement we find

$$\frac{d^2 R}{dt^2} = -\frac{k_s}{m_2}(R - R_e) - \frac{k_s}{m_1}(R - R_e)$$
$$= -k_s \left(\frac{1}{m_1} + \frac{1}{m_2} \right)(R - R_e) \tag{4.10}$$

We now define a quantity μ called the *reduced mass* by

$$\frac{1}{\mu} = \frac{1}{m_1} + \frac{1}{m_2}$$

and so we have

$$\mu \frac{d^2 R}{dt^2} = -k_s(R - R_e) \tag{4.11}$$

which is identical to Equation (4.2) already derived for a single particle of mass μ on a spring. The general solution is therefore

$$R = R_e + A \sin \left(\sqrt{\frac{k_s}{\mu}} t \right) + B \cos \left(\sqrt{\frac{k_s}{\mu}} t \right) \tag{4.12}$$

and the angular frequency is

$$\omega = \sqrt{\frac{k_s}{\mu}}$$

It is easy to demonstrate that the potential energy is

$$U = \tfrac{1}{2} k_s (x_2 - x_1 - R_e)^2$$

and the total energy ε_{vib} of the harmonically vibrating diatomic is therefore

$$\varepsilon_{vib} = \frac{1}{2} m_1 \left(\frac{dx_1}{dt} \right)^2 + \frac{1}{2} m_2 \left(\frac{dx_2}{dt} \right)^2 + \tfrac{1}{2} k_s (x_2 - x_1 - R_e)^2 \tag{4.13}$$

4.4 Three Problems

This simple treatment suggests three problems. First, how do we determine the spring constant for a simple molecule such as $^1H^{35}Cl$ or $^{12}C^{16}O$? Second, how good is the harmonic approximation? And third, have we missed anything by trying to treat a molecular species as if it obeyed the laws of classical mechanics rather than quantum mechanics?

The three questions are interlinked, but let me start with the third one. The experimental evidence suggests that we have made a serious error in neglecting the quantum mechanical details. If we irradiate a gaseous sample of $^1H^{35}Cl$ with infrared radiation, it is observed that the molecules strongly absorb radiation of wavenumber $2886\,\text{cm}^{-1}$. With hindsight we would of course explain the observation by saying that the molecular vibrational energies are *quantized*. A major flaw of the classical treatment is that the total vibrational energy is completely unrestricted and quantization does not arise.

The quantum mechanical treatment of a harmonically vibrating diatomic molecule is given in all the elementary chemistry texts. The results are quite different from the classical ones, in that

1. the vibrational energy cannot take arbitrary values, it is *quantized*;

2. there is a single quantum number v, which takes values $0, 1, 2, \ldots$, called the *vibrational quantum number*; and

3. vibrational energies ε_{vib} are given by

$$\varepsilon_{\text{vib}} = \frac{h}{2\pi}\sqrt{\frac{k_s}{\mu}}\left(v + \frac{1}{2}\right)$$

where h is Planck's constant.

The results are usually summarized on an energy level diagram, such as Figure 4.5.

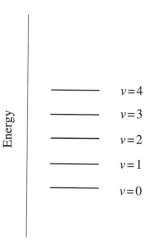

Figure 4.5 Vibrational energy levels

I have just drawn the first four vibrational energy levels, but there are an infinite number of them. According to the harmonic model, the spacing between the levels is constant.

A careful investigation into the mechanism by which electromagnetic radiation interacts with matter suggests that transitions between these vibrational energy levels are allowed, provided the vibrational quantum number changes by just 1 unit. So molecules with $v = 0$ can absorb radiation of exactly the right energy for promotion to $v = 1$. Molecules with $v = 1$ can either absorb radiation with exactly the right energy for promotion to $v = 2$ or they can emit radiation and fall to $v = 0$ and so on.

According to the quantum model, then, molecules can only have certain vibrational energies and this behaviour is totally at variance with the classical treatment. Also, the quantum treatment differs from the classical treatment in that the lowest energy is that with $v = 0$, where the energy is non-zero. This is called the *zero-point* energy. According to the classical treatment, a molecule can be completely at rest. According to the quantum treatment, the lowest vibrational energy allowed is the zero-point energy.

How do we measure the spring constant? According to the harmonic quantum model, the energy difference between any consecutive pair of energy levels is given by

$$\Delta \varepsilon = \frac{h}{2\pi} \sqrt{\frac{k_s}{\mu}}$$

so all we need to do is measure this energy difference experimentally. The reduced mass μ of the $^1\text{H}^{35}\text{Cl}$ isotopic species is 1.6267×10^{-27} kg and substitution of the experimental value ($2886\,\text{cm}^{-1}$) into the energy difference gives the harmonic force constant as $480.7\,\text{N}\,\text{m}^{-1}$.

In fact, there is more to the experiment than I have told you. Spectroscopic experiments are done at finite temperatures and a given sample of N molecules may have many energy levels populated. Relative populations N_v are given by the Boltzmann formula

$$N_v \propto \exp\left(-\frac{\varepsilon_v}{k_B T}\right)$$

Substitution of values into the formula shows that for many everyday diatomic molecules at everyday temperatures, the only vibrational level populated is that with $v = 0$. So an infrared absorption spectrum should just show a single absorption, corresponding to the transition $v = 0$ to $v = 1$.

A closer examination of the $^1\text{H}^{35}\text{Cl}$ spectrum shows weak absorptions at 5668, 8347, ... cm^{-1}, which are nearly (but not exactly) two and three times the fundamental vibration frequency. The existence of these lines in the spectrum shows that our assumption of Hooke's Law is not completely correct. Figure 4.6 shows the 'experimental' energy level diagram compared with the harmonic one.

Our conclusion from the experimental data is that vibrational energy levels get progressively closer together as the quantum number increases. This suggests that whilst the harmonic model is a reasonable one, we need to look more carefully at the form of the potential in order to get better agreement with experiment.

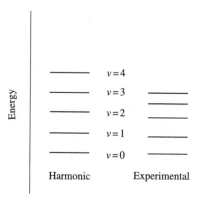

Figure 4.6   Harmonic vs. experimental energy levels

4.5 The Morse Potential

Professional spectroscopists would be unhappy with the idea of using Hooke's Law as a model for the vibrational motion. They would be more concerned with matching their experimental energy levels to a more accurate potential. Many such potentials have been used over the years, with that due to Morse being widely quoted in elementary chemistry texts. The Morse potential is as follows

$$U = D_e(1 - \exp(-\beta(R - R_e)))^2 \tag{4.14}$$

where D_e is the depth of the potential well, i.e. the thermodynamic dissociation energy, and

$$\beta = \sqrt{\frac{\left(\frac{d^2 U}{dR^2}\right)}{2D_e}}$$
$$= \frac{\omega_e}{2}\sqrt{\frac{2\mu}{D_e}} \tag{4.15}$$

This potential contains three parameters, D_e, ω_e and R_e, and so should be capable of giving a better representation to the potential energy curve than the simple harmonic, which contains just the two parameters, k_s and R_e.

In the case of $^1H^{35}Cl$, a simple calculation shows that the dissociation energy

$$D_e = D_0 + \tfrac{1}{2}h(2\pi\omega_e)$$

is $4.430 + 0.186\,eV = 4.616\,eV$. The Morse potential for $^1H^{35}Cl$ is shown in Figure 4.7 compared with the simple harmonic model. The full curve is the simple harmonic potential, the dashed curve the Morse potential.

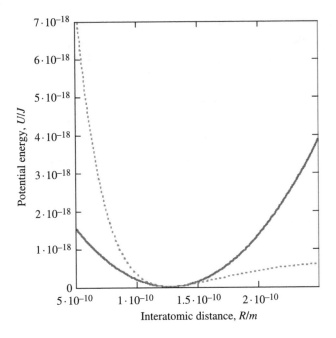

Figure 4.7 Simple harmonic and Morse curves for HCl

4.6 More Advanced Potentials

More often than not the following spectroscopic constants are available for a diatomic molecule:

R_e the equilibrium internuclear separation

D_e the dissociation energy

k_s the force constant

$\omega_e x_e$ the anharmonicity constant (sometimes written x_e only)

α_e the vibration–rotation coupling constant

Usually these five constants can be found to good experimental accuracy.

There are a number of three- to five-parameter potential functions for bonded pairs in the literature, of which the Morse potential is the most popular. Jack Linnett [3] made a careful study of many such functions, for example the four-parameter potential

$$U(R) = \frac{a}{R^m} - b\exp(-nR)$$

The four parameters a, m, b and n in this reciprocal–exponential function are deduced by fitting spectroscopic data. At this point I should explain how we recover the force constant from such a complicated expression, and to do this I'll use a Taylor expansion of the potential about the equilibrium bond length

$$U(R) = U(R_e) + (R - R_e)\left(\frac{dU}{dR}\right)_{R=R_e} + \tfrac{1}{2}(R - R_e)^2 \left(\frac{d^2U}{dR^2}\right)_{R=R_e} + \cdots \quad (4.16)$$

$U(R)$ is obviously equal to $U(R_e)$ when $R = R_e$, and this fixes the constant of integration. The equation is sometimes written as

$$U(R) - U(R_e) = (R - R_e)\left(\frac{dU}{dR}\right)_{R=R_e} + \tfrac{1}{2}(R - R_e)^2 \left(\frac{d^2U}{dR^2}\right)_{R=R_e} + \cdots$$

or even

$$U(R) = (R - R_e)\left(\frac{dU}{dR}\right)_{R=R_e} + \tfrac{1}{2}(R - R_e)^2 \left(\frac{d^2U}{dR^2}\right)_{R=R_e} + \cdots$$

where it is understood that $U(R)$ is measured relative to the potential energy minimum (that is to say, we take the zero as $U(R_e)$).

The quantity dU/dR is of course the gradient of U. The second derivative evaluated at the minimum where $R = R_e$ is called the (harmonic) force constant. To find R_e we solve the equation $dU/dR = 0$ and substitute this value into the second derivative to evaluate the force constant. In the special case of a harmonic potential, the second derivative is a constant and is equal to the force constant.

5 Molecular Mechanics

In Chapter 4 I showed you how to use classical mechanics to model the vibrational motion of a diatomic molecule. I also explained the shortcomings of this treatment, and hinted at applications where a quantum mechanical model would be more appropriate. We will deal specifically with quantum mechanical models in later chapters.

5.1 More About Balls on Springs

It is time to move on to more complicated molecules, and I want to start the discussion by considering the arrangement of balls on springs shown in Figure 5.1.

We assume that the springs each satisfy Hooke's Law. I will call the spring constant of the left-hand spring k_1 and the spring constant of the right-hand spring k_2. The equilibrium position corresponds to the two masses having x coordinates $R_{1,e}$ and $R_{2,e}$, and we constrain the motion so that the springs can only move along the x-axis. The particle masses are shown in Figure 5.1.

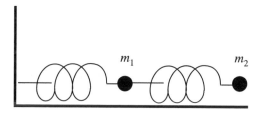

Figure 5.1   Two balls, two springs

We then stretch the system, so extending the two springs, and I will call the instantaneous positions of the two masses, x_1 and x_2. The extensions of the springs from their equilibrium positions are

$$\xi_1 = x_1 - R_{1,e} \quad \text{and} \quad \xi_2 = x_2 - R_{2,e}$$

Consider the left-hand spring; it exerts a restoring force on particle 1 of $-k_1\xi_1$. Now consider the right-hand spring. This spring is stretched by an amount $(\xi_2 - \xi_1)$, and so it exerts a force of $k_2(\xi_2 - \xi_1)$; this force acts to the left on particle 2 and to the right on particle 1. Application of Newton's second law gives

$$k_2(\xi_2 - \xi_1) - k_1\xi_1 = m_1 \frac{d^2\xi_1}{dt^2}$$

$$-k_2(\xi_2 - \xi_1) = m_2 \frac{d^2\xi_2}{dt^2}$$

(5.1)

There are many different solutions to these simultaneous differential equations, but it proves possible to find two particularly simple ones called *normal modes of vibration*. These have the property that both particles execute simple harmonic motion at the same angular frequency. Not only that, every possible vibrational motion of the two particles can be described as linear combinations of the normal modes.

Having said that it proves possible to find such solutions where both particles vibrate with the same angular frequency ω, let me assume that there exist such solutions to the equations of motion such that

$$\xi_1(t) = A \sin(\omega t + \phi_1)$$
$$\xi_2(t) = B \sin(\omega t + \phi_2)$$

where A, B, ϕ_1 and ϕ_2 are constants that have to be determined from the boundary conditions.

Differentiating these two equations with respect to time gives

$$\frac{d^2\xi_1(t)}{dt^2} = -\omega^2 A \sin(\omega t + \phi_1)$$

$$\frac{d^2\xi_2(t)}{dt^2} = -\omega^2 B \sin(\omega t + \phi_2)$$

and substituting these expressions into the equations of motion gives

$$-\frac{(k_1 + k_2)}{m_1}\xi_1 + \frac{k_2}{m_1}\xi_2 = -\omega^2\xi_1$$

$$\frac{k_2}{m_2}\xi_1 - \frac{k_2}{m_2}\xi_2 = -\omega^2\xi_2$$

(5.2)

These two equations are simultaneously valid only when ω has one of two possible values called the *normal mode angular frequencies*. In either case, both particles oscillate with the same angular frequency.

In order to investigate these normal modes of vibration, I write the above equations in matrix form, and then find the eigenvalues and eigenvectors as follows

$$
\begin{pmatrix}
-\dfrac{(k_1 + k_2)}{m_1} & \dfrac{k_2}{m_1} \\
\dfrac{k_2}{m_2} & -\dfrac{k_2}{m_2}
\end{pmatrix}
\begin{pmatrix} \xi_1 \\ \xi_2 \end{pmatrix}
= -\omega^2 \begin{pmatrix} \xi_1 \\ \xi_2 \end{pmatrix}
\tag{5.3}
$$

Matrix diagonalization gives the allowed values of ω^2 (the eigenvalues), and for each value of $-\omega^2$ we calculate the relevant combinations of the ξs (the eigenvectors). The eigenvectors of the matrix are called the *normal coordinates*.

5.2 Larger Systems of Balls on Springs

For a molecule comprising N atoms, there are $3N$ Cartesian coordinates. Of these, three can be associated with the position of the centre of mass of the whole molecule and three for the orientation of the molecule at the centre of mass (two for linear molecules). This leaves $3N - 6$ vibrational degrees of freedom ($3N - 5$ if the molecule is linear), and it is appropriate to generalize some concepts at this point. I am going to use matrix notation in order to make the equations look friendlier.

The molecular potential energy U will depend on $p = 3N - 6$ (independent) variables. For the minute, let me call them $q_1, q_2, \ldots, q_p$, and let me also write $q_{1,e}, q_{2,e}, \ldots, q_{p,e}$ for their 'equilibrium' values. These coordinates are often referred to as *internal coordinates*, and they will be linear combinations of the Cartesian coordinates.

First of all, for the sake of neatness, I will collect all the qs into a column matrix $\mathbf{q}$. I will also collect together the 'equilibrium' values into a column matrix $\mathbf{q}_e$ and the extensions into a column ξ

$$
\mathbf{q} = \begin{pmatrix} q_1 \\ q_2 \\ \ldots \\ q_p \end{pmatrix}; \qquad
\mathbf{q}_e = \begin{pmatrix} q_{1,e} \\ q_{2,e} \\ \ldots \\ q_{p,e} \end{pmatrix}; \qquad
\xi = \begin{pmatrix} q_1 - q_{1,e} \\ q_2 - q_{2,e} \\ \ldots \\ q_p - q_{p,e} \end{pmatrix}
\tag{5.4}
$$

I will now write $U(\mathbf{q})$ to indicate the dependence of U on these variables. If I use Taylor's Theorem to expand $U(\mathbf{q})$ about the point $\mathbf{q}_e$, then the one-dimensional equation

$$
U(R) - U(R_e) = (R - R_e)\left(\frac{dU}{dR}\right)_{R=R_e} + \tfrac{1}{2}(R - R_e)^2 \left(\frac{d^2 U}{dR^2}\right)_{R=R_e} + \cdots
$$

(given previously in Chapter 4) has to be modified to take account of the larger number of variables. First derivatives become partial first derivatives, and we have to take account of the 'mixed' second-order derivatives

$$U(\mathbf{q}) - U(\mathbf{q}_e) = \sum_{i=1}^{p} \xi_i \left(\frac{\partial U}{\partial q_i}\right)_{\xi_i=0} + \frac{1}{2}\sum_{i=1}^{p}\sum_{j=1}^{p} \xi_i\xi_j \left(\frac{\partial^2 U}{\partial q_i \partial q_j}\right)_{\xi_i=0, \xi_j=0} + \cdots \quad (5.5)$$

In ordinary vector differentiation, we meet the gradient of a scalar field f, defined in Cartesian coordinates as

$$\mathrm{grad} f = \frac{\partial f}{\partial x}\mathbf{e}_x + \frac{\partial f}{\partial y}\mathbf{e}_y + \frac{\partial f}{\partial z}\mathbf{e}_z$$

where $\mathbf{e}_x$, $\mathbf{e}_y$ and $\mathbf{e}_z$ are Cartesian unit vectors. When dealing with functions of many variables it proves useful to make a generalization and write the gradient of U, for example, as

$$\mathrm{grad}\, U = \begin{pmatrix} \dfrac{\partial U}{\partial q_1} \\[2mm] \dfrac{\partial U}{\partial q_2} \\[1mm] \cdots \\[1mm] \dfrac{\partial U}{\partial q_p} \end{pmatrix} \quad (5.6)$$

so grad U is a column matrix that stores all the partial derivatives. This 'vector' will occur many times through the text, and I am going to give it the symbol $\mathbf{g}$ (for gradient).

The second derivatives can be collected into a symmetric $p \times p$ matrix that is called the *hessian* of U and I will give this the symbol $\mathbf{H}$. In the case where $p = 3$, we have

$$\mathbf{H} = \begin{pmatrix} \dfrac{\partial^2 U}{\partial q_1^2} & \dfrac{\partial^2 U}{\partial q_1 \partial q_2} & \dfrac{\partial^2 U}{\partial q_1 \partial q_3} \\[3mm] \dfrac{\partial^2 U}{\partial q_2 \partial q_1} & \dfrac{\partial^2 U}{\partial q_2^2} & \dfrac{\partial^2 U}{\partial q_2 \partial q_3} \\[3mm] \dfrac{\partial^2 U}{\partial q_3 \partial q_1} & \dfrac{\partial^2 U}{\partial q_3 \partial q_2} & \dfrac{\partial^2 U}{\partial q_3^2} \end{pmatrix} \quad (5.7)$$

The Taylor expansion then becomes

$$U(\mathbf{q}) - U(\mathbf{q}_e) = \xi^{\mathrm{T}}\mathbf{g} + \tfrac{1}{2}\xi^{\mathrm{T}}\mathbf{H}\xi + \cdots \quad (5.8)$$

Both the gradient and the hessian have to be evaluated at the point $\mathbf{q}_e$, and so you will sometimes see the equation written with an 'e' subscript

$$U(\mathbf{q}) - U(\mathbf{q}_e) = \xi^{\mathrm{T}}\mathbf{g}_e + \tfrac{1}{2}\xi^{\mathrm{T}}\mathbf{H}_e\xi + \cdots$$

The superscript T, as in ξ^T, indicates the transpose of a matrix; the transpose of a column matrix is a row matrix. The hessian is often referred to as the force constant matrix.

Finally, if I denote the $3N$ Cartesian coordinates $X_1, X_2, \ldots, X_{3N}$, we usually write the transformation from Cartesian coordinates to internal coordinates as

$$\mathbf{q} = \mathbf{BX} \qquad (5.9)$$

where the rectangular matrix $\mathbf{B}$ is called the Wilson B matrix. The $\mathbf{B}$ matrix has p rows and $3N$ columns.

5.3 Force Fields

I have been vague so far about which variables are the 'correct' ones to take. Chemists visualize molecules in terms of bond lengths, bond angles and dihedral angles, yet this information is also contained in the set of Cartesian coordinates for the constituent atoms. Both are therefore 'correct'; it is largely a matter of personal choice and professional training. I should mention that there are only $3N - 6$ vibrational coordinates, and so we have to treat the $3N$ Cartesian coordinates with a little care; they contain three translational and three rotational degrees of freedom. I will return to this technical point later.

Spectroscopists usually are interested in finding a set of equilibrium geometric parameters and force constants that give an exact fit with their experimental data. This is harder than it sounds, because for a molecule comprising N atoms and hence $p = 3N - 6$ vibrational degrees of freedom, there are $\frac{1}{2}p(p - 1)$ force constants (diagonal and off-diagonal). In order to measure the individual force constants, the spectroscopist usually has to make experimental measurements on all possible isotopically labelled species. It turns out that there are many more unknowns than pieces of experimental information. Spectroscopists usually want a *force field* (comprising force constants, equilibrium quantities and every other included parameter) that is specific for a given molecule. They want to match up 'theory' with their incredibly accurate measurements.

Many of the 'off-diagonal' force constants turn out to be small, and spectroscopists have developed systematic simplifications to the force fields in order to make as many as possible of the small terms vanish. If the force field contains only 'chemical' terms such as bond lengths, bond angles and dihedral angles, then it is referred to as a *valence force field* (VFF). There are other types of force field in the literature, intermediate between the VFF and the general force field discussed above.

5.4 Molecular Mechanics

Molecular modellers usually have a quite different objective; they want a force field that can be transferred from molecule to molecule, in order to predict (for example) the geometry of a new molecule by using data derived from other related molecules. They make use of the bond concept, and appeal to traditional chemists' ideas that a molecule comprises a sum of bonded atoms; a large molecule consists of the same features we know about in small molecules, but combined in different ways.

The term *molecular mechanics* was coined in the 1970s to describe the application of classical mechanics to determinations of molecular equilibrium structures. The method was previously known by at least two different names, the *Westheimer* method and the *force-field* method. The name and acronym, MM, are now firmly established quantities.

The idea of treating molecules as balls joined by springs can be traced back to the 1930 work of D. H. Andrews [4]. A key study to the development of MM was that by R. G. Snyder and J. H. Schachtschneider [5] who showed that transferable force constants could be obtained for alkanes provided that a few off-diagonal terms were retained. These authors found that off-diagonal terms are usually largest when neighbouring atoms are involved, and so we have to take account of non-bonded interactions, but only between next-nearest neighbours.

A final point for consideration is that we must also take account of the chemical environment of a given atom. An sp carbon atom is different from an sp^2 carbon atom and so on. It is traditional to speak of *atom types* in molecular mechanics.

Our idea is to treat the force field as a set of constants that have to be fixed by appeal to experiment or more rigorous calculation. In molecular mechanics we take account of non-bonded interactions, and also the chemical sense of each atom. A valence force field that contains non-bonded interactions is often referred to as a *Urey–Bradley force field*.

5.4.1 Bond-stretching

If we consider phenylanine (see Figure 5.2) we can identify a variety of bond types

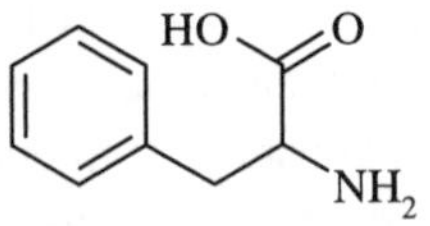

Figure 5.2   Phenylanine

including $C(sp^2)$—$C(sp^2)$, $C(sp^2)$—$C(sp^3)$, O—H, C=O and so on. If we assume that Hooke's Law is adequate, then each bond stretch between atom types A and B makes a contribution to the total molecular potential energy of

$$U_{AB} = \tfrac{1}{2}k_{AB}(R_{AB} - R_{e,AB})^2 \qquad (5.10)$$

in an obvious notation. Here k_{AB} is the force constant, R_{AB} the instantaneous bond length and $R_{e,AB}$ the equilibrium bond length.

Other scientists recommend the Morse potential

$$U_{AB} = D(1 - \exp(-\alpha(R_{AB} - R_{AB,e})))^2$$

whilst some recommend the addition of extra terms to the simple Harmonic expression

$$U_{AB} = k_1(R_{AB} - R_{e,AB})^2 + k_2(R_{AB} - R_{e,AB})^4 \qquad (5.11)$$

5.4.2 Bond-bending

Next we have to consider the bond-bending vibrations. It is usual to write these as harmonic ones, typically for the connected atoms A—B—C

$$U_{ABC} = \tfrac{1}{2}k_{ABC}(\theta_{ABC} - \theta_{e,ABC})^2 \qquad (5.12)$$

k is the force constant, and the subscript 'e' refers to the equilibrium value where the molecule is at rest. A variation on the theme is given by

$$U_{ABC} = \frac{k_{ABC}}{2\sin^2\theta_{ABC,e}}(\cos\theta_{ABC} - \cos\theta_{ABC,e})^2 \qquad (5.13)$$

5.4.3 Dihedral motions

Next we must consider the dihedral angle ABCD between the four bonded atoms A, B, C and D (see Figure 5.3). Some authors divide these into *proper* dihedrals, where we might expect full rotation about the connecting bond B—C, and *improper*

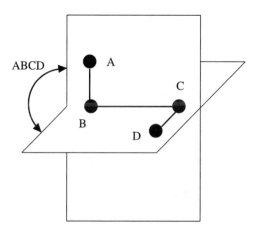

Figure 5.3   Dihedral angle

dihedrals where the rotation is limited. For example, if C—D were a C—H fragment of a methyl group, then we would be expect full rotation about B—C and a three-fold symmetry in the potential energy term. A—CH—CH— linkage in a benzene ring would only show a moderate flexing from its planar value (angle zero).

If we use χ to denote the ABCD angle, then a popular dihedral potential is given by

$$U = \frac{U_0}{2}(1 - \cos(n(\chi - \chi_e))) \qquad (5.14)$$

Here n is the periodicity parameter, which would be 3 for a methyl group. χ_e is the equilibrium torsional angle. A more complicated example is given by

$$U = \frac{V_1}{N_d}(1 + \cos(n_1\chi - g_1)) + \frac{V_2}{N_d}(1 + \cos(n_2\chi - g_2)) + \frac{V_3}{N_d}(1 + \cos(n_3\chi - g_3))$$

The Vs are energy terms, the ns are periodicity parameters, the gs are phase parameters and N_d is a constant that depends on the number of bonds.

Some authors treat improper dihedrals in the same way as bond-bending, and take a contribution to the molecular potential energy as

$$U_{ABCD} = \tfrac{1}{2}k_{ABCD}(\chi_{ABCD} - \chi_{e,ABCD})^2 \qquad (5.15)$$

where χ is the dihedral angle, as above.

5.4.4 Out-of-plane angle potential (inversion)

Next we consider the out-of-plane potential terms. Imagine molecule ABCD in Figure 5.4 to be ammonia, a molecule with a very low barrier to inversion; as ψ changes from positive to negative, the molecule inverts. We can write the inversion

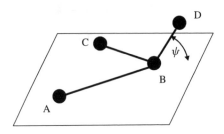

Figure 5.4   Out-of-plane (inversion) potential

potential either in terms of the angle indicated (ψ) or in terms of the height (h) of atom D above the plane of atoms A, B and C. Four examples follow:

$$U = \frac{k_1}{2 \sin^2 \psi_e} (\cos \psi - \cos \psi_e)^2$$

$$U = k_2 h^2 \tag{5.16}$$

$$U = k_3 (1 + k_4 \cos (n\psi))$$

where n is a periodicity parameter, and finally

$$U = k_5 (1 + \cos (n\psi - k_6))$$

The ks are constants that have to be fixed against experiment, and ψ_e is the equilibrium value.

5.4.5 Non-bonded interactions

I mentioned earlier that molecular mechanics force fields have to be transferable from molecule to molecule, and explained the necessity for non-bonded interactions. These are usually taken to be Lennard-Jones 12–6 type, and they are included between all non-bonded pairs of atoms.

$$U_{L\text{-}J} = \frac{C_{12}}{R^{12}} - \frac{C_6}{R^6}$$

The Born–Mayer–Huggins potential

$$U_{BMH} = A \exp(-BR) - \frac{C_6'}{R^6} - \frac{C_8}{R^8}$$

is sometimes used when dealing with polar species. B is a parameter determined by the size and 'softness' of an ion, C_6' (not the same as C_6 in $U_{L\text{-}J}$) has to do with dipole–dipole interactions whilst C_8 is determined by dipole–quadrupole interactions.

Some force fields make special provision for hydrogen-bonded atoms; they treat them as non-bonded interactions but soften the Lennard-Jones 12–6 potential for A—H $\cdots$ B to a 12–10 version

$$U_{HB} = \frac{C_{12}}{R_{HB}^{12}} - \frac{C_{10}}{R_{HB}^{10}} \tag{5.17}$$

Other authors take the view that hydrogen bonds are perfectly respectable chemical bonds that should be treated just like any other bond. They are therefore given a force constant and so on.

5.4.6 Coulomb interactions

Many force fields take account of electronegativity differences between atoms and add electrostatic terms. Atomic charges Q_A and Q_B are assigned to atoms A and B according to the rules of the particular force field, and we write

$$U_{AB} = \frac{1}{4\pi\epsilon_0} \frac{Q_A Q_B}{R_{AB}}$$

5.5 Modelling the Solvent

I should remind you that the electrostatic expression above relates only to point charges in free space. In the presence of a dielectric material (such as water), the force between point charges is reduced by a factor ϵ_r called the relative permittivity. Many force fields were developed at a time when it was not feasible to include a solvent explicitly in such calculations.

Various attempts were made to allow for the effect of a solvent; the most obvious thing to do is to alter the relative permittivity even though no solvent molecules are actually taken into account. There is no agreement between authors as to the correct value of ϵ_r and values ranging between 1 and 80 have been used for water. Some force fields take ϵ_r proportional to the distance between the point charges. I will explain some more up-to-date ways of modelling the solvent in later chapters.

5.6 Time-and-Money-Saving Tricks

All the contributions to the molecular potential energy U given above can be done on a pocket calculator. The larger the molecular system, the larger the number of individual contributions to U and the relationship between molecular size and computational effort is roughly dependent on the square of the number of atoms. Over the years, people have tried to reduce the computational time for a given problem by the use of various tricks of the trade. Two such methods are as follows.

5.6.1 United atoms

Some professional force fields use the so-called *united atom* approach. Here, we regard (for example) a CH_3 group as a pseudo-atom, X, and develop parameters for a $C(sp^2)$—X stretch, and so on. It is customary to treat methyl, methylene and methane groups as united atoms, especially when dealing with large biological systems.

5.6.2 Cut-offs

For a large molecule, there are many more non-bonded interactions than bonded interactions. Molecular mechanics force fields very often cut these off to zero at some finite
distance, in order to save computer time. This can sometimes lead to mathematical
difficulties because of the discontinuity, and various ingenious methods have been proposed to circumvent the problem (other than actually retaining the terms). I will show
you in a later chapter that there are other problems associated with this cut-off procedure;
it's a real problem, not just one that I have mentioned out of historical interest. Figure 5.5

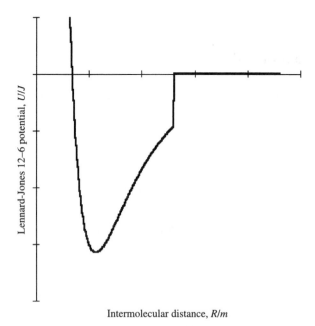

Figure 5.5   Schematic cut-off of L-J potential

shows a Lennard-Jones 12–6 potential with a cut-off (after a certain value of R, the
potential is set to zero).

5.7 Modern Force Fields

A 'standard' modern molecular mechanics force field can be written

$$U = \sum_{\text{stretch}} U_{\text{AB}} + \sum_{\text{bend}} U_{\text{ABC}} + \sum_{\text{dihedral}} U_{\text{ABCD}} + \sum_{\text{out-of-plane}} U_{\text{ABCD}}$$

$$+ \sum_{\text{non-bonded}} U_{\text{AB}} + \sum_{\text{Coulomb}} U_{\text{AB}} \qquad (5.18)$$

or written explicitly in terms of the contributions discussed above

$$
U = \sum_{\text{bonds}} \tfrac{1}{2} k_{\text{AB}} (R_{\text{AB}} - R_{\text{e,AB}})^2 + \sum_{\text{bends}} \tfrac{1}{2} k_{\text{ABC}} (\theta_{\text{ABC}} - \theta_{\text{e,ABC}})^2
$$

$$
+ \sum_{\text{dihedrals}} \frac{U_0}{2} (1 - \cos(n(\chi - \chi_0))) + \sum_{\text{out-of-plane}} \frac{k}{2 \sin^2 \psi_{\text{e}}} (\cos \psi - \cos \psi_{\text{e}})^2
$$

$$
+ \sum_{\text{non-bonded}} \left(\frac{C_{\text{AB}}^{12}}{R_{\text{AB}}^{12}} - \frac{C_{\text{AB}}^{6}}{R_{\text{AB}}^{6}} \right) + \frac{1}{4\pi\epsilon_0} \sum_{\text{charges}} \frac{Q_{\text{A}} Q_{\text{B}}}{R_{\text{AB}}} \tag{5.19}
$$

5.7.1 Variations on a theme

There are a number of variants of this expression in the literature. Some force fields contain mixed terms such as

$$
\frac{k}{2} (R - R_{\text{e}})(\theta - \theta_{\text{e}})
$$

which couple together the bond-stretching modes with angle bending. Others use more complicated expressions for the individual bending and stretching terms. Some force fields allow interactions between lone pairs, which are often referred to as non-atomic interaction centres. In addition, there are specialist force fields that are appropriate for restricted ranges of compounds such as ions, liquid metals and salts.

Force fields are determined by one of two routes. First, in an ideal world, one might calibrate their parameters against accurate quantum mechanical calculations on clusters of small molecules. The alternative is to calibrate against experimental data such as crystal structure, infrared absorption, X-ray measurements and liquid properties such as density, enthalpy of vaporization, Gibbs energies of solvation and the like. To date, almost all modern force fields have been obtained by the latter approach.

The choice of a particular force field for a given application should depend on the type of system for which the force field was designed. For example, some force fields have been calibrated against the solution properties of amino acids. These are obviously the ones to choose when it comes to modelling proteins in solution.

Finally, I must emphasize the importance of the *atom type* (i.e. the chemical environment). The chemical environment of an atom can be distinguished by

1. its hybridization

2. its formal atomic charge

3. its nearest neighbours.

For example, one well-known force field distinguishes five types of oxygen atom:

1. a carbonyl oxygen

2. a hydroxyl oxygen

3. a carboxylic or phosphate oxygen

4. an ester or ether oxygen

5. an oxygen in water.

The interactions are calculated according to atom type, not the 'raw' elements.

5.8 Some Commercial Force Fields

With these principles in mind, it is time to examine some of the common force fields found in professional molecular modelling programs.

5.8.1 DREIDING [6]

This force field is parameterized for all atom types that any chemist would expect for the elements H, C, N, O, P, S, F, Cl, Br and I. In terms of the 'standard' expression we write

$$U = \sum_{\text{bonds}} \tfrac{1}{2} k_{AB} (R_{AB} - R_{e,AB})^2 + \sum_{\text{bends}} \tfrac{1}{2} k_{ABC} (\cos \theta_{ABC} - \cos \theta_{e,ABC})^2$$

$$+ \sum_{\text{dihedrals}} \frac{U_0}{2} (1 - \cos(n(\chi - \chi_0))) + \sum_{\text{out-of-plane}} \frac{k}{2} (\psi - \psi_e)^2$$

$$+ \sum_{\text{non-bonded}} \left(\frac{C_{AB}^{12}}{R_{AB}^{12}} - \frac{C_{AB}^6}{R_{AB}^6} \right) \qquad (5.20)$$

5.8.2 MM1 [7]

In his 1976 Review, Norman Allinger essentially defined what we now call the MM1 force field. He treated hydrocarbons only, ignored the Coulomb terms and used an exp-6 Lennard-Jones potential.

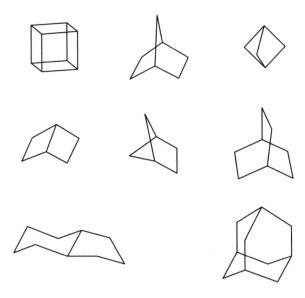

Figure 5.6   Bicyclic and related molecules

Once all the standard cases have been successfully treated, one naturally looks at the difficult ones. In the case of hydrocarbons, these difficult cases comprise strained rings such as cyclobutane. The problem with cyclobutane is this; whilst having all carbon atoms planar can minimize the angular deformation from tetrahedral, the molecule is actually puckered by a substantial angle from planarity. In addition, the C—C bond lengths are unusually large. The obvious solution is to say that a four-membered ring is different from any other hydrocarbon and that the bond angle does not have a natural tetrahedral value, but one then goes down the undesirable path where all difficult cases have their own set of parameters.

Allinger and others introduced a variety of 'mixed' terms into the standard molecular mechanics potential; for example, a bond length–bond angle term and a torsion–bend interaction. Figure 5.6 shows typical bicyclic and related hydrocarbons described in the 1976 review [7].

5.8.3 MM2 (improved hydrocarbon force field)

Allinger introduced MM2 in 1977 [8]. At the time there was a deal of discussion in the literature about how different force fields should represent hydrogen atoms, i.e. as 'hard' or 'soft' atoms. A hard atom was said to be one whose plot of force vs. distance (in the diatom) showed a steep slope. A soft atom was one where the slope was gentler. It all boiled down to the repulsive part of the non-bonded interactions but eventually Allinger decided to retain his exp-6 intermolecular potential.

MM2 differs from MM1 in three main respects:

1. The dihedral term was extended to

$$U = \frac{U_1}{2}(1 + \cos \omega) + \frac{U_2}{2}(1 + \cos 2\omega) + \frac{U_3}{2}(1 + \cos 3\omega) \qquad (5.21)$$

where each of the Us was found by calibration against experiment.

2. The bending term was extended to

$$U_{ABC} = \tfrac{1}{2}k_{ABC}(\theta_{ABC} - \theta_{e,ABC})^2 + \tfrac{1}{2}k'_{ABC}(\theta_{ABC} - \theta_{e,ABC})^6 \qquad (5.22)$$

3. All mention of cross terms between bond stretches and bends were finally dropped.

A great deal of importance was attached to the calculation of enthalpies of formation, and 42 hydrocarbons were treated. The author claimed that his enthalpy results were comparable with experiment in terms of experimental error.

5.8.4 AMBER [9]

AMBER (an acronym for Assisted Model Building and Energy Refinement) is a force field for the simulation of nucleic acids and proteins. It was calibrated against experimental bond lengths and angles obtained from microwave, neutron diffraction and accurate quantum chemical studies. The parameters were then refined with molecular mechanics studies on model compounds such as tetrahydrofuran, deoxyadenosine, dimethyl phosphate, 9-methyladenine-1-methylthymine hydrogen bonded complex and others. The model differs from our standard expression, Equation (5.19), in four ways.

1. Hydrogen bonds were included explicitly with a 12–10 potential

$$U_{\text{H-bonds}} = \sum_{\text{H-bonds}} \left(\frac{C_{12}}{R^{12}} - \frac{C_{10}}{R^{10}} \right)$$

2. An attempt was made to include solvent effects by inclusion of the Coulomb term with a distance-dependent relative permittivity.

3. The AMBER force field is a 'united atom' one, and hydrogen atoms bonded to carbons are not explicitly included. They are absorbed into the atom type parameters for neighbouring atoms.

4. Lone pairs were explicitly included for sulfur hydrogen bonding.

There are a number of different versions of AMBER; the original united atom version was later extended to include all atoms. Just to give you a flavour, one modern software package has the following choices

1. AMBER 2

2. AMBER 3

3. AMBER for saccharides

4. AMBER 94

5. AMBER 96

5.8.5 OPLS (Optimized Potentials for Liquid Simulations) [10]

Like AMBER, OPLS is designed for calculations on amino acids and proteins. The easiest thing is for me to quote part of the Abstract to the keynote paper:

A complete set of inter molecular potential functions has been developed for use in computer simulations of proteins in their native environment. Parameters have been reported for 25 peptide residues as well as the common neutral and charged terminal groups. The potential functions have the simple Coulomb plus Lennard-Jones form and are compatible with the widely used models for water, TIP4P, TIP3P and SPC. The parameters were obtained and tested primarily in conjunction with Monte Carlo statistical mechanics simulations of 36 pure organic liquids and numerous aqueous solutions of organic ions representative of subunits in the side chains and backbones of proteins . . . . Improvement is apparent over the AMBER united-atom force field which has previously been demonstrated to be superior to many alternatives.

I will explain about TIP and Monte Carlo in later chapters. Each atomic nucleus is an interaction site, except that CH_n groups are treated as united atoms centred on the carbon. Hydrogen bonds are not given any special treatment, and no special account is taken of lone pairs.

5.8.6 R. A. Johnson [11]

I mentioned earlier the existence of a number of specialist force fields. The Johnson force field is specific to the pure elements Fe, W and V. The pair potential terms are written

$$
\begin{aligned}
U &= a_1(R - b_1)^3 + c_1R + d_1 && \text{if } \varepsilon_1 < R < \varepsilon_2 \\
&= a_2(R - b_2)^3 + c_2R + d_2 && \text{if } \varepsilon_2 < R < \varepsilon_3 \\
&= a_3(R - b_3)^3 + c_3R + d_3 && \text{if } \varepsilon_3 < R < \varepsilon_4 \\
&= 0 && \text{if } \varepsilon_4 < R
\end{aligned}
\tag{5.23}
$$

where R is the distance between a pair of atoms, the εs are characteristic distances and the as, bs, cs and ds are parameters.

6 The Molecular Potential Energy Surface

For one-dimensional problems, we speak about a molecular potential energy *curve*. The simple potential energy curves we have met so far have all shown a single minimum. From now on, life gets more complicated (or interesting, depending on your viewpoint).

6.1 Multiple Minima

The plot in Figure 6.1 shows how the ethane molecular potential varies with dihedral angle. The figure shows a full rotation of 360°; all the remaining geometrical variables were kept constant. Note that there are three identical minima (and of course three identical maxima), and the differences between maxima and minima are all the same. The '1D' in the figure means that it is a one-dimensional plot. The chemical interpretation is that these three minima correspond to conformers where the hydrogens are as far apart as possible (i.e. in the *trans* position). The maxima correspond to conformers where the C—H bonds eclipse each other.

Multiple minima are common in potential energy surface studies, as we will see. Consider now the substituted ethane CH_2Cl—CH_2Cl. A plot of the potential energy vs. the ClC—CCl dihedral angle gives Figure 6.2. There are three minima, one lower than the other two. The three minima are referred to as *local minima* and the minimum at 180° is called the *global minimum*. The global minimum corresponds to a conformation with the two chlorines as far apart as possible. The two other minima correspond to conformers where each chlorine is *trans* to hydrogen.

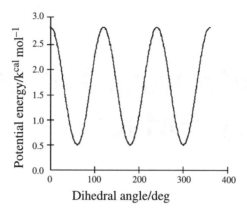

Figure 6.1 Ethane dihedral motion

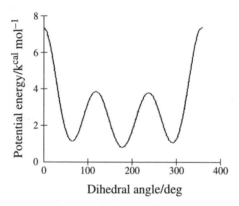

Figure 6.2 CH_2Cl—CH_2Cl

6.2 Saddle Points

One-dimensional potential energy curves show minima, maxima and points of inflexion. Two and higher-dimensional surfaces often show a new feature called a *saddle point*. Saddle points are a maximum in some variables and a minimum in the remainder. The one type of saddle point that interests chemists is where we have a minimum in all variables but one, and a maximum in this remaining one. This corresponds to a transition state in theories of reaction mechanisms. I just mention it here, despite the fact that molecular mechanics cannot properly treat bond breaking or bond making. I am not going to deal in detail with saddle points until I have given you a firm grounding in quantum chemistry, but you should be aware of their existence. There are also points where the molecular potential energy is a maximum. We are rarely interested in these.

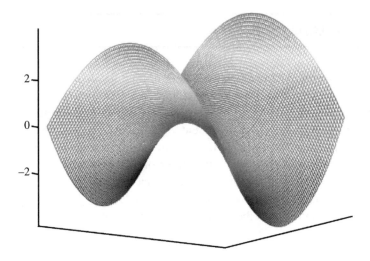

Figure 6.3 Saddle point

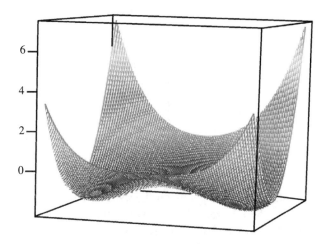

Figure 6.4 Two minima and a saddle point

Figure 6.3 shows a saddle point (for the function $f(x, y) = x^2 - y^2$). The vertical axis shown the function values, and the two horizontal axes are x and y. The apocryphal reason why such points are called saddle points is that the figure above resembles a horse's saddle. Molecular potential energy surfaces generally contain all possible features and you might like to consider the example shown in Figure 6.4, which shows two (equal) minima and a saddle point.

In this Chapter I want to show you how we locate stationary points on the molecular potential energy surface, and how we characterize them. I will start with the second problem, because it is straightforward.

6.3 Characterization

For a linear molecule with N atoms, there are in total $3N$ degrees of freedom. Three of these are translational degrees of freedom, three are rotational (two for a linear molecule) and the remaining $p = 3N - 6$ ($3N - 5$ if linear) are vibrational degrees of freedom. We therefore need p independent variables to describe them, and these are known as the *internal coordinates*. It is traditional to denote such variables by the letter q; if I write them $q_1, q_2, \ldots, q_p$, then the molecular potential energy will depend on these qs and the normal coordinates will be linear combinations of them. For the minute, don't worry about the nature of these variables. In Chapter 5, I showed how to collect such variables into the column vector $\mathbf{q}$ (a matrix of dimension $p \times 1$)

$$
\mathbf{q} = \begin{pmatrix} q_1 \\ q_2 \\ \ldots \\ q_p \end{pmatrix}
$$

and I also defined the gradient of U, $\mathbf{g} = \text{grad } U$, as a column vector, and its hessian $\mathbf{H}$ as a symmetric square matrix

$$
\mathbf{g} = \begin{pmatrix} \dfrac{\partial U}{\partial q_1} \\ \dfrac{\partial U}{\partial q_2} \\ \ldots \\ \dfrac{\partial U}{\partial q_p} \end{pmatrix} ; \quad
\mathbf{H} = \begin{pmatrix} \dfrac{\partial^2 U}{\partial q_1^2} & \cdots & \dfrac{\partial^2 U}{\partial q_1 \partial q_p} \\ \cdots & \cdots & \cdots \\ \dfrac{\partial^2 U}{\partial q_p \partial q_1} & \cdots & \dfrac{\partial^2 U}{\partial q_p^2} \end{pmatrix}
$$

The molecular potential energy depends on the qs, and I will write it as $U(\mathbf{q})$. At a stationary point, the gradient is zero; it's as easy as that. In order to characterize the stationary point, we have to find the eigenvalues of the hessian calculated at that point. If the eigenvalues are all positive, then the point is a minimum. If the eigenvalues are all negative, then the point is a maximum. Otherwise the point is a saddle point. Saddle points of interest to chemists are those that are a minimum in all degrees of freedom except one, and it transpires that the hessian has just one negative eigenvalue at such a point. They are sometimes called *first-order saddle points*.

6.4 Finding Minima

Molecular mechanics tends to be concerned with searching the molecular potential energy surfaces of very large molecules for minima (equilibrium structures). Many mathematicians, scientists and engineers are concerned with related problems. The field of study is now an established part of applied mathematics, and is referred to as *optimization theory*. An optimization problem involves minimizing a function of several variables, possibly subject to restrictions on the values of the variables. There is no single universal algorithm for optimization, and it is necessary for us to spend a little time dealing with the different methods of interest.

I should tell you that one speaks of *optimization* and less commonly of minimization or maximization; the latter two are equivalent, for maximization of the function f is the same as minimization of the function $-f$.

The first molecular mechanics optimization seems to have been carried out by F. H. Westheimer [12] in 1956, who did the calculations 'by hand'. The first computer calculations seem to have been done by J. B. Hendrickson in 1961 [13]. Neither of their methods was generally applicable to molecules.

Many algorithms have been developed over a number of years for the location of stationary points, some of which are suitable for molecular mechanics calculations, some of which are suitable for quantum mechanical calculations and many of which are not particularly suited to either. In molecular mechanics we tend to deal with large molecules and consequently the molecular potential energy function will depend on hundreds if not thousands of variables. On the other hand, evaluation of the potential energy at each point on the hypersurface is relatively simple.

Different considerations apply to quantum chemical calculations, to be dealt with later in the book. Calculation of the energy at points on the surface is far from easy but the number of variables tends to be smaller.

As stated above, transition states are special cases of saddle points; they are stationary points on the surface where the hessian has just one negative eigenvalue. Transition state searching is a hot topic in chemistry, and a number of specialist algorithms have been proposed. I will deal with transition states later in the book.

6.5 Multivariate Grid Search

This is the oldest method for finding minima, and it has a long history. What we do is the following.

1. Choose a suitable grid for the variables.

2. Choose a starting point **A** on the grid.

3. For each variable q_1, q_2, . . . , q_p evaluate the molecular potential energy U at the two points surrounding **A** (as determined by the grid size).

4. Select the new point for which U is a minimum, and repeat steps 3 and 4 until the local minimum is identified.

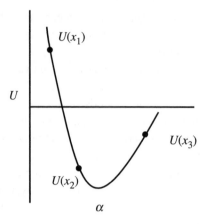

Figure 6.5   Univariate minimization

The method is intuitive, and it is apparent that a local minimum will eventually be found.

6.5.1 Univariate search

This is sometimes called the cyclic search, for we perform successive one-dimensional searches for each of the variables in turn.

1. Choose a starting point **A**.

2. Minimize $U(\mathbf{q})$ for each variable $q_1, q_2, \ldots, q_p$ in turn.

3. Repeat the cycle as necessary.

The second step is the crucial one, and a favourite strategy is quadratic interpolation; suppose that x_1, x_2 and x_3 are three successive values of one of the variables q_i (say), chosen such that they bracket a minimum in that particular variable, as shown in Figure 6.5. We fit the three points to a quadratic, and a little algebra shows that the minimum α is given by

$$\alpha = \frac{1}{2} \frac{U(x_1)(x_3^2 - x_2^2) + U(x_2)(x_1^2 - x_3^2) + U(x_3)(x_2^2 - x_1^2)}{U(x_1)(x_3 - x_2) + U(x_2)(x_1 - x_3) + U(x_3)(x_2 - x_1)} \qquad (6.1)$$

The method is also known as a *line search*.

6.6 Derivative Methods

It is conventional to divide optimization algorithms into those methods that make use of derivatives (gradients and hessians), and those that do not. Neither of the two methods discussed so far makes use of derivatives. Some authors use a further subdivision into *first-order* derivative methods (where we make use of the gradient), and *second-order* derivative methods (where we use both the gradient and the hessian). There are many algorithms in the literature; I can only give you a flavour.

6.7 First-Order Methods

6.7.1 Steepest descent

This first-order derivative scheme for locating minima on molecular potential energy surfaces was put forward by K. Wiberg in 1965 [14]. His basic algorithm can be summarized as follows.

1. Calculate U for the initial structure.

2. Calculate U for structures where each atom is moved along the x, y and z-axes by a small increment. Movement of some atoms will lead to a small change in U, whilst movements of other atoms will lead to a large change in U. (The important quantity is clearly the gradient.)

3. Move the atoms to new positions such that the energy U decreases by the maximum possible amount.

4. Repeat the relevant steps above until a local minimum is found.

Wiberg's method would today be classified as 'steepest descent'; the idea is that we start from point **A** on a molecular potential energy surface and identify the fastest way down the surface to the local minimum. Figure 6.6 below shows a

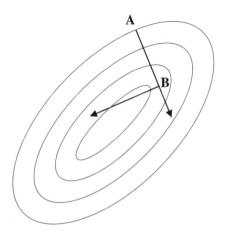

Figure 6.6 Steepest descents

two-dimensional potential energy surface. The negative of grad U at point **A** gives the maximum rate of decrease of U, and also the direction in which to proceed (the gradient vector is perpendicular to the contour at **A**). The only problem is to choose a step length; this can be done by a line search along the vector indicated, and it is clear that U will pass through a linear minimum around point **B**.

We repeat the process from point **B** until the local minimum is (eventually) reached. It can be proved that every step is necessarily at right angles to the one before it, and this leads to one of the infuriating characteristics of the steepest descent method: it takes very many tiny steps when proceeding down a long narrow valley because it is forced to make a right-angled turn at every point, even though that might not be the best route to the minimum.

At this point, I should introduce some notation that is universally understood by workers in the field of optimisation. Like it or not, you will come across it when you read the technical literature.

All the algorithms that I will describe are *iterative*; that is, we start from some initial point **A** on a surface, and then move on in cycles, according to an algorithm, hopefully toward a stationary point. Each cycle of the calculation is called an *iteration*, and people often use the symbol k to keep track of the iteration count. I am going to follow common practice and add a superscript $^{(k)}$ to variables in order to indicate this iteration count. Sometimes I will just write it as ^{k}, when there is no possibility that you will regard the iteration count as meaning 'to the power of'.

6.7.2 Conjugate gradients

Starting from point $\mathbf{q}^{(k)}$ (where k is the iteration count), we move in a direction given by the vector

$$\mathbf{V}^{(k)} = -\mathbf{g}^{(k)} + \gamma^{(k)}\mathbf{V}^{(k-1)} \tag{6.2}$$

where $\mathbf{g}^{(k)}$ is the gradient vector at point $\mathbf{q}^{(k)}$ and $\gamma^{(k)}$ is a scalar given by

$$\gamma^{(k)} = \frac{(\mathbf{g}^{(k)})^{\mathrm{T}}\mathbf{g}^{(k)}}{(\mathbf{g}^{(k-1)})^{\mathrm{T}}\mathbf{g}^{(k-1)}} \tag{6.3}$$

I have used the superscript T to denote the transpose of a matrix. The line search method is then used to determine the distance to be moved along this vector. This algorithm was first proposed by R. Fletcher and C. M. Reeves [15].

Conjugate gradients methods produce a set of directions that overcome the oscillatory behaviour of steepest descents in narrow valleys. Successive directions are not at right angles to each other. Such methods are also referred to as *conjugate direction* methods.

E. Polak and G. Ribière [16] proposed an alternative form for the scalar $\gamma^{(k)}$

$$\gamma^{(k)} = \frac{\left(\mathbf{g}^{(k)} - \mathbf{g}^{(k-1)}\right)^{\mathrm{T}}\mathbf{g}^{(k)}}{\left(\mathbf{g}^{(k-1)}\right)^{\mathrm{T}}\mathbf{g}^{(k-1)}} \tag{6.4}$$

For a purely quadratic function, their method is identical to the original Fletcher–Reeves algorithm. The authors pointed out that most Us encountered in chemistry are only approximately quadratic, and that their method was therefore superior.

6.8 Second-Order Methods

6.8.1 Newton–Raphson

Second-order methods use not only the gradient but also the hessian to locate a minimum. Before launching into any detail about second-order methods, it is worth spending a little time discussing the well-known Newton–Raphson method for finding the roots of an equation of a single variable. This is a simple illustration of a second-order method. For illustration, consider the function

$$f(x) = x\exp\left(-x^2\right)$$

shown as the full curve in Figure 6.7. It is obvious by inspection that our chosen function has roots at $x = 0$ and at plus and minus infinity, but suppose for the sake of

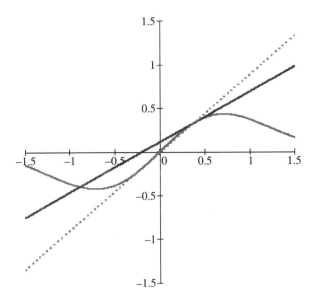

Figure 6.7 Newton–Raphson

argument that we don't know that the function has a root at $x = 0$. We might make a guess that the function has a root at (say) $x = 0.4$. This is our first guess, so I set the iteration count $= 1$ and write the guess $x^{(1)}$ (also, the function has a value of $f^{(1)} = 0.3409$). Thus we have

$$x^{(1)} = 0.4; \qquad f^{(1)} = 0.3409$$

The function can be differentiated to give

$$\frac{df}{dx} = (1 - 2x^2) \exp(-x^2)$$

and the gradient $g^{(1)}$ at $x^{(1)} = 0.4$ has a value 0.5795. The tangent line at $x^{(1)} = 0.4$ is therefore

$$y = 0.5795x + 0.1091$$

which I have shown as a full straight line on Figure 6.7. A little algebra shows that this line crosses the x-axis at

$$x^{(2)} = x^{(1)} - \frac{f^{(1)}}{g^{(1)}}$$

This crosses the x-axis at -0.1883 and this value is taken as the next best estimate $x^{(2)}$ of the root. I then draw the tangent line at $x^{(2)}$, which is shown dotted in Figure 6.7 and so on. Table 6.1 summarizes the iterations.

There are a few points worth noting. First, the convergence is rapid. In fact, for a quadratic function Newton's method will locate the nearest root in just one step. Second, the choice of starting point is crucial. If I start with $x = \pm 0.5$, then the successive estimates simply oscillate between $+0.5$ and -0.5. If I start with $|x| < 0.5$, then the method converges to the root $x = 0$. If I start with $|x| > 0.5$, then we find the infinite roots.

The method can be easily modified to search for stationary points, which are characterized by a zero gradient. The iteration formula

$$x^{(k+1)} = x^{(k)} - f^{(k)}/g^{(k)} \tag{6.5}$$

Table 6.1 Iterations from Figure 6.7

k	$x^{(k)}$	$f^{(k)}$	$g^{(k)}$
1	0.4	0.3409	0.5795
2	-0.1882	-0.1817	0.8968
3	1.4357×10^{-2}	1.4354×10^{-2}	0.9994
4	-5.9204×10^{-6}	-5.9204×10^{-6}	1.0000
5	0.0000	0.0000	1.0000

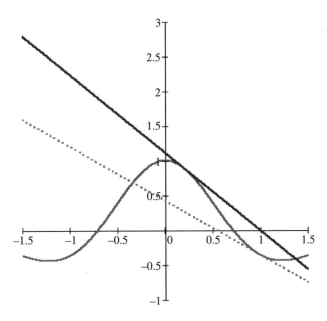

Figure 6.8   Location of the stationary point

becomes

$$x^{(k+1)} = x^{(k)} - g^{(k)}/H^{(k)} \tag{6.6}$$

where $H^{(k)}$ is the value of the second derivative at the current point $x^{(k)}$. For this particular function, we have

$$H(x) = (4x^3 - 6x)\exp(-x^2)$$

and I can now repeat the argument given above, but this time searching for stationary points rather than roots of the equation.

Figure 6.8 shows the gradient (the full curve), and the first two iterations of the Newton–Raphson algorithm for the stationary point, starting at $x^{(1)} = 0.2$. The gradient is zero at infinity and at $x = \pm 0.7071$

In the case of a function of many variables $x_1, x_2, \ldots, x_n$ the Newton–Raphson optimization algorithm can be written

$$\mathbf{X}^{(k+1)} = \mathbf{X}^{(k)} - (\mathbf{H}^{(k)})^{-1}\mathbf{g}^{(k)} \tag{6.7}$$

where I have collected the xs into a column vector $\mathbf{X}$ and so on. The formula can be easily derived from the Taylor expansion.

We therefore have to calculate the gradient and the hessian at each successive point in the iterations. Normally, the algorithm needs an infinite number of steps but it will find the minimum of any quadratic function in just one step, starting at an arbitrary point on the surface. For example, if we consider the function $f(x, y) = x^2 + y^2$, which has a minimum at $x = 0$, $y = 0$, we have

$$\mathbf{g} = \begin{pmatrix} 2x \\ 2y \end{pmatrix}; \qquad \mathbf{H} = \begin{pmatrix} 2 & 0 \\ 0 & 2 \end{pmatrix}$$

$$\mathbf{H}^{-1} = \begin{pmatrix} \frac{1}{2} & 0 \\ 0 & \frac{1}{2} \end{pmatrix}$$

If we start from the point (2, 2), then we calculate

$$\mathbf{X}^{(2)} = \begin{pmatrix} 2 \\ 2 \end{pmatrix} - \begin{pmatrix} \frac{1}{2} & 0 \\ 0 & \frac{1}{2} \end{pmatrix} \begin{pmatrix} 4 \\ 4 \end{pmatrix}$$

which is a null vector.

In practice molecular potential energy surfaces are rarely quadratic and so an infinite number of steps will be required. We simply stop the calculation once a given number of decimal places have been achieved.

6.8.2 Block diagonal Newton–Raphson

There are a number of variations on the Newton–Raphson method, many of which aim to eliminate the need to calculate the full hessian. A widely used algorithm is the *block diagonal Newton–Raphson* method, where just one atom is moved at each iteration. This means that all the elements of the hessian are zero except for a block along the diagonal describing the atom in question.

6.8.3 Quasi-Newton–Raphson

In addition, there are a number of so-called *quasi-Newton–Raphson* methods that gradually build up the inverse hessian in the successive iterations. At each iteration, the vector $\mathbf{X}^{(k)}$ is updated to $\mathbf{X}^{(k+1)}$

$$\mathbf{X}^{(k+1)} = \mathbf{X}^{(k)} - (\mathbf{H}^{(k)})^{-1} \mathbf{g}^{(k)}$$

using the gradient and the current estimate of the inverse hessian. Having made the move to the new position, $\mathbf{H}^{-1}$ is updated from its value at the previous step by an approximate procedure depending on the algorithm employed. The methods of Davidon–Fletcher–Powell, Broyden–Fletcher–Goldfarb–Shanno and Murtaugh–Sargent are often encountered, but there are many others.

6.8.4 The Fletcher–Powell algorithm [17]

The Fletcher–Powell algorithm is a derivative method where elements of the gradient and the hessian are estimated numerically each cycle. Essentially, it assumes that the surface is quadratic around each chosen point, and finds the minimum for the approximate quadratic surface. Steps in the algorithm are as follows.

1. Calculate the energy $U^{(1)}$ for an initial geometry $\mathbf{X}^{(1)}$, and at positive and negative displacements for each of the coordinates.

2. Fit a quadratic for each of the coordinates according to the formula

$$U(\mathbf{X}) = U^k + \sum_{i=1}^{p} \left(g_i^k (X_i - X_i^k) + \tfrac{1}{2} H_{ii}^k (X_i - X_i^k)^2 \right) \tag{6.8}$$

(this essentially gives numerical estimates of the gradient and the hessian; I have dropped the brackets round the iteration count for clarity).

3. Find a minimum of this expression; we have

$$\frac{\partial U}{\partial X_i} = 0$$

$$g_i^k + H_{ii}^k (X_i - X_i^k) = 0$$

$$c_i^k = X_i - X_i^k = -\frac{g_i^k}{H_{ii}^k} \tag{6.9}$$

The last term gives the correction to coordinate x_i; if these are small enough then stop.

4. Calculate the energy at points $\mathbf{X}^k$, $\mathbf{X}^k + \mathbf{c}^k$ and $\mathbf{X}^k + 2\,\mathbf{c}^k$.

5. Fit a quadratic to these three points as above.

6. Find the energy minimum, as above. This gives point $\mathbf{X}^{k+1}$ on the surface.

7. Calculate the gradient $\mathbf{g}^{k+1}$ at this point, increase the iteration count and go back to step 3.

6.9 Choice of Method

The choice of algorithm is dictated by a number of factors, including the storage and computing requirements, the relative speeds at which the various parts of the calculation can be performed and the availability of an analytical gradient and hessian. Analytic first and second derivatives are easily evaluated for molecular mechanics force fields; the only problem might be the physical size of the hessian. For this reason, molecular mechanics calculations on large systems are often performed using steepest descent and conjugate gradients. The Newton–Raphson method is popular for smaller systems, although the method can have problems with structures that are far from a minimum. For this reason, it is usual to perform a few iterations using, for example, steepest descent before switching to Newton–Raphson. The terms 'large' and 'small' when applied to the size of a molecule are of course completely relative and are dependent on the computer power available when you do the calculation.

In cases where the differentiation cannot be done analytically, it is always possible to *estimate* a gradient numerically; for example in the case of a function of one variable

$$\left(\frac{df}{dx}\right)_{x=x_1} \approx \frac{f(x_1 + \Delta) - f(x_1)}{\Delta}$$

where Δ is small. Algorithms that rely on numerical estimates of the derivatives need more function evaluations than would otherwise be the case, so there is a delicate trade-off in computer time. It is generally supposed that gradient methods are superior to non-gradient methods, and it is also generally thought to be advantageous to have an analytical expression for the gradient and hessian.

6.10 The Z Matrix

For a non-linear molecule of N atoms, there are $p = 3N - 6$ vibrational degrees of freedom that should be described by p independent coordinates $q_1, q_2, \ldots, q_p$ and at first sight all should be varied in a geometry optimization. These coordinates are often defined by modelling packages using the so-called Z matrix, which gives a way of building up a molecular geometry in terms of valence descriptors such as bond lengths, bond angles and dihedral angles. A few examples will help to make things clear. First of all water, for which I will take a bond length of 95.6 pm (0.956 Å) and an H—O—H bond angle of 104° (see Figure 6.9).

I can start my Z matrix at any atom, and that atom occupies a line by itself. I chose oxygen, which I have called Oxy. H1 is joined to oxygen with a bond length of ROH (0.956 Å). I have assumed that the bond lengths are equal. H2 is also joined to oxygen with a bond length of 0.956 Å, and the H2—oxygen—H1 bond angle is Ang (104°).

Note that I have subconsciously added a *symmetry constraint* by requiring that the two bond lengths be equal. This means that I am effectively reducing the number of degrees of freedom, for the purposes of geometry optimization. We will see later that this is not necessary.

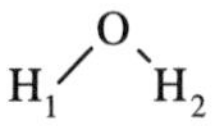

Z matrix

Oxy
H1 Oxy ROH
H2 Oxy ROH H1 Ang

ROH   0.956
Ang   104.

Figure 6.9   Z matrix for water

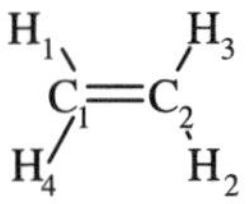

Z matrix

C1
C2   C1   RCC
H1   C1   RCH   C2   AngleHCC
H2   C2   RCH   C1   AngleHCC   H1   180.
H3   C2   RCH   C1   AngleHCC   H1   0.
H4   C1   RCH   C2   AngleHCC   H3   180.

RCC   1.34
RCH   1.08
AngleHCC   120.

Figure 6.10   Ethene Z matrix

I hadn't any particular software in mind when I wrote this example (although it would be fine for the GAUSSIAN suite, as we will see in later chapters). There are many pits for the unwary to fall into, some simple and some subtle. First, two simple ones. I didn't call the first atom 'oxygen' because many packages limit you to a small number of characters for atomic descriptors. Also, I was careful to include a decimal point for the starting values of ROH and Ang. Many packages still expect floating-point numbers rather than integers.

For molecules having more than three atoms, we also have to define the dihedral angles in addition to bond lengths and bond angles. Figure 6.10 shows a Z matrix for ethene, where I have made a lot of assumptions about the molecular geometry. All the C—H bond lengths are equal, and all the HCC bond angles are also taken to be equal. I have also assumed that the molecule is planar (as indeed it is, in its electronic ground state).

6.11 Tricks of the Trade

There are two important tricks of the trade, should you want to input geometries using the Z matrix. The first trick relates to linear structures, the second to symmetrical ring compounds.

6.11.1 Linear structures

Many optimization algorithms run into trouble when trying to optimize geometries having bond angles of 180°. The way around the problem was (and still is) to introduce dummy atoms which are often given a symbol X. Dummy atoms play no part in the calculation, apart from defining the molecular geometry. For example, I am sure you know that ethyne is linear. One correct way to write its Z matrix is as shown in Figure 6.11.

You should by now have picked up a subtle point: I am making use of chemical knowledge to make assumptions about the equality or otherwise of bonds, bond angles and dihedral angles. That's why I took the two C—H bonds to be equal in the ethyne Z matrix above. I am really putting an answer into the question, and I should perhaps have written the ethyne Z matrix as in Figure 6.12.

I freely admit that Z matrices are not easy things to write for large molecules, and in any case downloads of molecular geometries (for example, from the Protein Data Bank) are usually Cartesian coordinates. There is also a problem to be addressed, namely that there are at most $3N - 6$ vibrational degrees of freedom for a non-linear molecule that is described by $3N$ Cartesian (or other) coordinates. The Z matrix has to

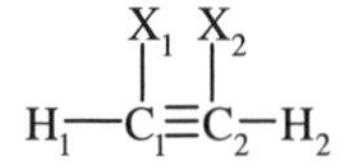

$$H_1\!\!-\!\!C_1\!\!\equiv\!\!C_2\!\!-\!\!H_2$$

with X_1, X_2 above C_1, C_2

Z matrix

```
H1
C1  H1  RCH
X1  C1  10.   H1  90.
C2  C1  RCC  X1  90.   H1  180.
X2  C2  10.   C1  90.   X1  0.
H2  C2  RCH  X2  90.   X1  180.

RCH  1.08
RCC  1.20
```

Figure 6.11   Ethyne Z matrix

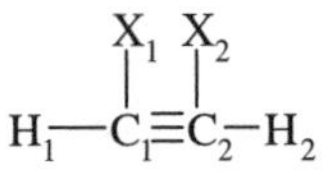

$$\underset{H_1-\underset{|}{C_1}\equiv\underset{|}{C_2}-H_2}{\overset{\overset{X_1\ X_2}{|\ \ |}}{}}$$

Improved Z matrix

```
H1
C1  H1  RC1H1
X1  C1  10.    H1  90.
C2  C1  RCC  X1  90.    H1  180.
X2  C2  10.    C1  90.    X1  0.
H2  C2  RC2H2  X2  90.    X1  180.

RC1H1  1.08
RC2H2  1.09
RCC  1.20
```

Figure 6.12   Improved Z matrix

be correctly written in order to describe these $3N-6$ internal coordinates, and a frequent cause of problems in optimizations specified by a Z matrix was (and still is) the specification of too many or too few internal coordinates. Too few coordinates corresponds to a constraint on the geometry and means that the full surface is not searched. Too many coordinates results in a redundancy leading to zero eigenvalues in the hessian (which cannot then be inverted). If there are no dummy atoms and no symmetry constraints, then the Z matrix must describe $3N-6$ unique internal coordinates, all of which should be varied. When dummy atoms are present, some of the Z matrix parameters are redundant and must be held fixed. Their coordinates are not varied.

6.11.2 Cyclic structures

The second trick is to do with cyclic structures. Imagine furan (Figure 6.13), and you might think to construct the Z matrix by starting (for example) with the oxygen (called O this time, for the sake of variety) and working around the ring.

This Z-matrix will certainly work, but it will lose the symmetry of the molecule (C_{2v}) because of tiny numerical rounding errors. By the time you get back to oxygen, small errors in the bond length–bond angle calculations (much less than 1%) will have destroyed the symmetry of the molecule. This may not matter to you, since most organic molecules have no symmetry elements apart from the identity operation and there is a strong argument in favour of letting the geometry optimization procedure sort out the symmetry, rather than making any assumption about the point group. If the molecule genuinely has C_{2v} symmetry, then this conclusion ought to follow from your calculations. Inclusion of the symmetry constraint, as in Figure 6.14, leads to fewer variables in the optimization.

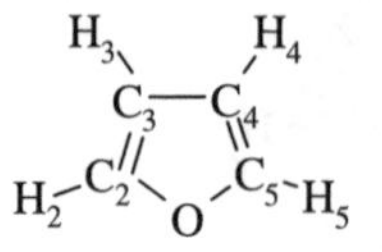

Naive Z matrix

```
O
C2  O   RCO
C3  C2  R23  O   Angle1
C4  C3  R34  C2  Angle2  O 0.0
C5  C4  R23  C3  Angle2  C2 0.0
H2  C2  RCH  C3  Angle4  C4 180.
H3  C3  RCH  C2  Angle5  H2 0.0
H5  C5  RCH  C4  Angle4  C3 180.
H4  C4  RCH  C5  Angle5  H5 0.0

RCO     1.4
R23     1.4
R34     1.4
RCH     1.08
Angle1  120.
Angle2  120.
Angle4  120.
Angle5  120.
```

Figure 6.13   Naive furan

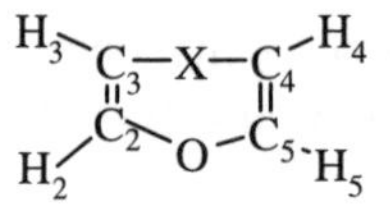

Symmetry-preserving Z matrix

```
X
C3  X  Half
C2  C3  R23  X  90.
O   C2  RCO  C3  Angle1  X  0.0
C4  X   Half  C3  180.  C2  0.0
C5  C4  R23  X  90.  C3  0.0
H2  C2  RCH  O  AA X 180.
H3  C3  RCH  C2  BB H2 0.0
H5  C5  RCH  O  AA X 180.
H4  C4  RCH  C5  BB  O  180.

Half 0.7
R23 1.4
RCO 1.4
R23 1.4
RCH 1.0
Angle1 120.
```

Figure 6.14   Symmetry-preserving furan

In the early days of geometry optimizations it was usual to take these kinds of shortcuts in order to save on computer resource. You probably noticed that I took all CH bonds equal in my furan examples above, yet there are subtle, systematic differences even in CH bond lengths, and these should come out of a respectable calculation. Computer resource is much cheaper than was once the case, and the assumptions of constant bond length, etc. are no longer needed.

The Z matrix method is often claimed to be intuitive to chemists because it uses the everyday concepts of bond lengths, bond angles and so on. On the other hand, many databases give molecular geometries in terms of Cartesian coordinates, not internal ones.

6.12 The End of the Z Matrix

Geometry optimization is of major importance in modern molecular modelling. Most of the early packages used internal coordinates as input by the Z matrix. Virtually all modern (gradient) optimization procedures require calculation of the hessian $\mathbf{H}$ and/or its inverse. In practice, it is usual to make an estimate and update these estimates at every iteration. Sometimes the initial hessian is taken to be the unit matrix, sometimes not. A great strength of the internal coordinate method is that construction of the initial Hessian can be based on chemical ideas; the individual diagonal elements of $\mathbf{H}$ are identified as bond-stretching and bond-bending force constants, etc. Also, the redundant translational and rotational degrees of freedom have already been eliminated.

J. Baker and W. J. Hehre [18] investigated the possibility of performing gradient geometry optimization directly in terms of Cartesian coordinates. Their key finding concerned the identification of a suitable initial hessian, together with a strategy for updates. They reported on a test set of 20 molecules, and argued that optimization in Cartesian coordinates can be just as efficient as optimization in internal coordinates, provided due care is taken of the redundancies.

There are actually two mathematical problems here, both to do with the Newton–Raphson formula. Modern optimizations use gradient techniques, and the formula

$$\mathbf{X}^{(k+1)} = \mathbf{X}^{(k)} - \left(\mathbf{H}^{(k)}\right)^{-1}\mathbf{g}^{(k)}$$

demonstrates that we need to know the gradient vector and the hessian in order to progress iterations. In the discussion above, I was careful to stress the use of p independent variables $q_1, q_2, \ldots, q_p$.

As stressed many times, there are $p = 3N - 6$ independent internal coordinates for a non-linear molecule but $3N$ Cartesian coordinates. The familiar Wilson $\mathbf{B}$ matrix relates these

$$\mathbf{q} = \mathbf{BX} \qquad (6.10)$$

$\mathbf{B}$ has $3N$ rows and p columns, and the rows of $\mathbf{B}$ are linearly dependent. The molecular potential energy depends on the qs, and also on the Xs, but we need to be careful to distinguish between dependent and independent variables. We can certainly write gradient vectors in terms of the two sets of variables

$$\text{grad } U = \begin{pmatrix} \dfrac{\partial U}{\partial q_1} \\ \dfrac{\partial U}{\partial q_2} \\ \ldots \\ \dfrac{\partial U}{\partial q_p} \end{pmatrix} ; \qquad (\text{grad } U)_\text{C} = \begin{pmatrix} \dfrac{\partial U}{\partial X_1} \\ \dfrac{\partial U}{\partial X_2} \\ \ldots \\ \dfrac{\partial U}{\partial X_{3N}} \end{pmatrix}$$

where I have added a subscript 'C' to show that the differentiations are with respect to the $3N$ Cartesian coordinates, but it is grad U that appears in optimization expressions and not $(\text{grad } U)_\text{C}$. We therefore have to relate the two.

Starting from $\mathbf{q} = \mathbf{BX}$ we can show that the gradients are related by

$$\mathbf{B}^\text{T}\text{grad } U = (\text{grad } U)_\text{C}$$

Remember that we want an expression for grad U; if $\mathbf{B}$ were square and invertible, then we would just write

$$\text{grad } U = (\mathbf{B}^\text{T})^{-1}(\text{grad } U)_\text{C}$$

but unfortunately $\mathbf{B}$ is rectangular. We therefore appeal to the mathematical concept of a *generalized inverse*. Consider the $3N \times 3N$ matrix $\mathbf{G} = \mathbf{BuB}^\text{T}$, where $\mathbf{u}$ is an arbitrary non-singular $p \times p$ matrix. The generalized inverse of $\mathbf{G}$, written $\mathbf{G}^{-1}$, is a matrix with the property that $\mathbf{GG}^{-1}$ is a diagonal matrix with a certain number of ones and a certain number of zeros along the diagonal. In our case, $\mathbf{GG}^{-1}$ will have p ones along the diagonal and $3N - p$ zeros. There are many algorithms for calculating such a generalized inverse. The zeros might actually be tiny numbers, depending on the accuracy of the numerical algorithm used.

It can be shown that the gradient vector referred to the independent variables is related to the coordinate gradient vector by

$$\text{grad } U = \mathbf{G}^{-1}\mathbf{Bu} \, (\text{grad } U)_\text{C} \qquad (6.11)$$

Similar considerations apply to the hessian, and formulae are available in the literature.

6.13 Redundant Internal Coordinates

All coordinate systems are equal in principle, but I have stressed above that

1. they should ideally give transferable force constants from molecule to molecule, and

2. they should ideally be capable of being represented as harmonic terms in U; that is, cubic and higher corrections should not be needed.

These requirements can be best satisfied by local internal valence coordinates such as bond lengths, bond angles and dihedral angles. The expression 'local' in this context means that the coordinates should extend to only a few atoms.

P. Pulay and co-workers [19] and H. B. Schlegel *et al.* [20] investigated the use of redundant internal coordinates for gradient optimizations. Pulay defined an internal coordinate system similar to that used by vibrational spectroscopists. It minimizes the number of redundancies by using local symmetry coordinates about each atom and special coordinates for ring deformations, ring fusions, etc. Schlegel used a simpler set of internal coordinates composed of all bond lengths, valence angles and dihedral angles. The mathematical considerations outlined above also apply here.

Packages such as Gaussian98 offer a choice between Z matrix, Cartesian coordinate and redundant internal coordinate optimizations. Perhaps by now you will have an inkling of the theory behind the choices you can make in such powerful packages.

7 A Molecular Mechanics Calculation

In this chapter I will give an example of what can be routinely done at the molecular mechanics (MM) level of theory. As a rough guide, MM is used these days to deduce equilibrium conformations of large molecules. For simplicity, I have used phenylalanine as the example (Figure 7.1), and the HyperChem 6.03 package (the current release is 7.0, see http://www.hyper.com/).

7.1 Geometry Optimization

Geometry optimization requires a starting structure. HyperChem has a database of the amino acids, stored as 'residues', that can be quickly joined together to construct a chain. Of the four force fields available (MM+, AMBER, BIO+ and OPLS) I have chosen MM+ and left all the options set at their default values.

The pK of most amino acids is such that the molecule exists in zwitterion form in aqueous solution, so I edited the residue to give the zwitterion in Figure 7.2. I have also shown the atom numbering for later reference. It can be argued that one should include any 'formal' atom charges, such as the zwitterion charges, explicitly, and so I modified the charge on each oxygen atom to be $-\frac{1}{2}$ electron and added a charge of $+1$ electron on the nitrogen.

There are four options for optimization as shown in the box; the steepest descent algorithm displayed its usual behaviour by making very large initial decreases in the gradient, followed by a seemingly infinite number of small steps near the minimum. The other three options all converged quickly and without a problem. Figure 7.3 shows my answer, and the energy reported is $-0.269\,112\,\text{kcal mol}^{-1}$. Bear in mind that this energy cannot be compared with experiment; it is simply a relative marker for the success or otherwise of our search for the global minimum.

This structure is just one possible local minimum on the molecular potential energy surface. For small molecules, it is possible (but laborious) to search systematically for all the minima but such techniques quickly become impracticable as the molecular size increases.

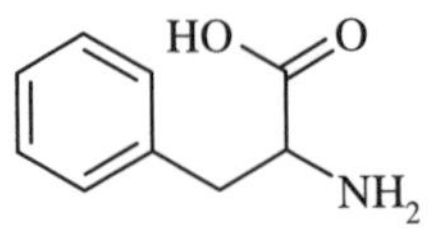

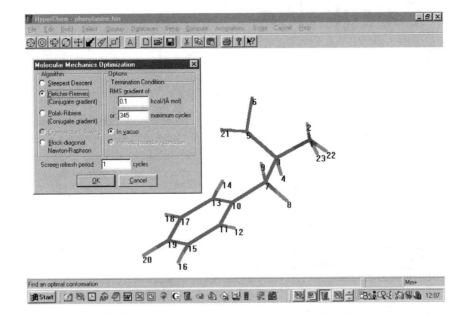

Figure 7.1   Phenylanine

Figure 7.2   Zwitterion before optimization

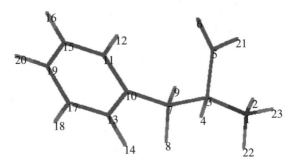

Figure 7.3   Optimized structure 1

7.2 Conformation Searches

Figure 7.4 shows the results we might expect to find from a limited conformational search on a (hypothetical) molecule. Repeating the optimization from different starting points has identified many local minima and the conformation corresponding to energy ε_1 appears to be the global minimum.

Our MM calculations refer to the molecule at 0 K, whilst we would normally concern ourselves with molecules at room temperature. At 0 K, we would find all molecules in the lowest energy conformation but at other temperatures we will find certain fractions of the molecules in higher energy conformations. These fractions are determined by the Boltzmann factor

$$\exp\left(-\frac{\varepsilon}{k_B T}\right)$$

where k_B is the Boltzmann constant. I have therefore drawn a vertical line on the figure of size $k_B T$, and we see that the first four conformations lie close together (compared with the yardstick $k_B T$), and so we expect that all four would make a significant contribution to measured properties of the molecule at temperature T. The remaining groups of conformations are far away from the first four in terms of the yardstick $k_B T$, and so can be discarded.

Several approaches have been applied to the problem of automatically identifying the low-energy conformations of a given molecule. Molecular flexibility is usually associated with rotation of unhindered bond dihedral angles, with little change in bond lengths and bond angles. A common theme is therefore to limit the exploration to those parameters that have the *least* effect on the energy. Another strategy is to automatically scan a possible structure for unfavourably close non-bonded atoms and unlikely ring closure bond lengths before optimization.

A choice then has to be made to decide whether the starting structure should be the same for every optimization, or each optimization should start with the structure found by the previous successful optimization. The latter choice is often favoured, on the grounds that one low-energy conformation is pretty much like another, and starting from one success-ful geometry will tend to keep the search in a low-energy region of the potential surface.

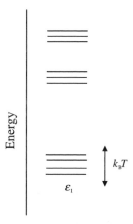

Figure 7.4   Conformation search

Table 7.1   Conformation search on phenylanine

Name of variable	Atoms involved (see Figure 7.5)
Torsion1	10–7–3–5
Torsion2	11–10–7–3

Search Progress:

Conformations:	8	Iterations:	584	Close contacts:	4
Lowest found:	-0.2691555	Optimizations:	500	Torsion tests:	80
Highest kept:	1.905025	Energies:	337459	Wrong chirality:	0
Accept rate:	1	Unconverged:	0	High energy:	0

Conformations Found:

Index	Energy	Gradient	Found	Used	Torsion1	Torsion2
1	-0.2691555	9.487304E-0	3	3	60.41316	69.39847
2	-0.2598099	9.749314E-0	93	110	173.9471	-75.80703
3	-0.1424598	9.900936E-0	8	13	58.22884	76.18497
4	-0.1209122	8.210268E-0	152	173	-61.81094	90.8128
5	0.3740986	9.195589E-0	160	184	173.9406	105.693
6	0.5010667	9.176153E-0	63	71	-67.58146	-80.50834
7	1.849899	9.861183E-0	19	26	-73.16533	18.09213
8	1.905025	9.84622E-03	2	3	-83.90961	42.56096

Figure 7.5   Random Walk Conformation search

Variations in the parameters can be either random or systematic; in the case of systematic variations, it is usual to scan the surface first at 'low resolution', then to refine likely-looking regions at high resolution. In the case of random variations, a common problem is that as the calculation progresses, more and more of the trial structures lead to conformations that have already been examined. Many modern software packages have sophisticated algorithms that can recognize identical conformations, and so reject them from the global search. They also automatically reject conformations that are too high in energy, compared with k_BT.

Table 7.1 shows a run using the *Conformational Search* option of HyperChem 6.03. I had to decide which torsional (dihedral) angles to vary, and I chose just two for this illustrative calculation.

The package did 500 Random Walk iterations (as shown in Figure 7.5), and kept the eight lowest energy conformations. The 'Found' entry shows the number of times that this particular conformation has been found, and 'Used' shows the number of times it has been used as a starting structure. Energies are recorded as kcal mol^{-1}.

7.3 QSARs

Quantitative Structure–Activity Relationships (QSARs) are attempts to correlate molecular structure with chemical or biochemical activity. I will use the remainder of this chapter to give you a flavour for the QSAR properties that are often routinely calculated in MM studies.

7.3.1 Atomic partial charges

Atomic charges can be calculated using the (costly) techniques of molecular quantum mechanics, as we will discover in later chapters. They are often calculated within the spirit of QSAR analyses using the Partial Equalization of Orbital Electronegativity (PEOE) approach of J. Gasteiger and M. Marsili [21].

The key concept of their method is atom *electronegativity* (often given the symbol χ), and an attraction is that their calculation is rapid. L. Pauling and D. M. Yost [22] first introduced this concept in 1932, and they tried to give numerical values to atomic electronegativities based on bond energy values. R. S. Mulliken [23] put the concept on a firmer theoretical footing in 1934 by relating an atomic electronegativity to the ionization energy I and the electron affinity E of the atom concerned

$$\chi = \tfrac{1}{2}(I + E) \tag{7.1}$$

The only problem with Mulliken's definition is that both E and I relate to some mythical state of an atom called the *valence state*. There have been many learned discussions relating to the concept of valence state, and such discussions still occasionally appear in the literature.

Whatever the valence state of an atom is (depending on the atom), it rarely corresponds to a spectroscopic state and so cannot be studied experimentally. Electronegativities and valence states go hand in hand; they are part of our folklore and both are widely quoted in even the most elementary chemistry texts.

R. T. Sanderson [24] proposed that on bond formation, atoms change their electron density until their electronegativities are equal. These ideas were further elaborated

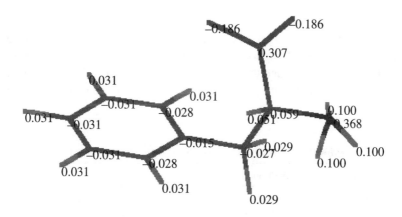

Figure 7.6   Atomic partial charges

by J. Hinze and co-workers [25], who introduced the concept of *orbital electrone-gativity*. This is the electronegativity of a specific orbital in a given valence state. You will understand that I am introducing quantum mechanical concepts before we have finished our discussion of techniques based on classical mechanics; just bear with me.

In the Abstract of their paper, Gasteiger and Marsili state

> *A method is presented for the rapid calculation of atomic charges in σ-bonded and nonconjugated π systems. Atoms are characterised by their orbital electro-negativities. In the calculation, only the connectivities of the atoms are con-sidered. Thus only the topology of a molecule is of importance. Through an iterative procedure partial equalization of orbital electronegativity is obtained. Excellent correlations of the atomic charges with core electron binding energies and with acidity constants are observed.*

These authors decided to relate the electronegativity of orbital ν on atom i ($\chi_{i,\nu}$) to the total charge Q_i on atom i as a quadratic

$$\chi_{i,\nu} = a_{i,\nu} + b_{i,\nu}Q_i + c_{i,\nu}Q_i^2 \tag{7.2}$$

The three unknowns a, b and c were determined from values used at the time for orbital ionization energies and affinities. Typical values are shown in Table 7.2. The calculation is iterative; starting from a reference structure where all the atoms carry zero charge, electric charge is permitted to flow from the less to the more electro-negative atoms. The electronegativity of an atom decreases as it accumulates electric charge, according to the formula, and in the next iteration less charge will flow until eventually the electronegativity of each (charged) atom is equalized. At this point the flow of charge stops.

MM is of course a classical treatment, and orbitals do not appear. The key equation is therefore written in terms of the atoms overall rather than in terms of orbital contributions

$$\chi_i = a_i + b_iQ_i + c_iQ_i^2 \tag{7.3}$$

Atom electronegativities are adjusted at each iteration using the calculated charges, until self-consistency is achieved. Figure 7.6 shows typical results for phenylanine. Note the negative charges on oxygen and the positive charge on nitrogen. Charges calculated in this way depend only on the atomic connectivity and not on the molecular geometry.

Table 7.2 Parameters needed to calculate electronegativity

Atom	Hybridization	a	b	c
H		7.17	6.24	−0.56
C	sp^3	7.98	9.18	1.88
	sp^2	8.79	9.32	1.51
	sp^1	10.39	9.45	0.73

7.3.2 Polarizabilities

I introduced you to polarizabilities in Chapter 3, when we discussed the theory of intermolecular forces. The experimental determination of a molecular polarizability is far from straightforward, especially if the molecule has little or no symmetry. The classical experimental route to the mean polarizability $\langle \alpha \rangle$ is via the refractive index or relative permittivity of a gas. Polarizabilities can be routinely calculated by quantum mechanical techniques, but the calculations are computer-intensive and this route is of no help if your aim in life is the high throughput screening of very many potential pharmaceutical molecules.

An insight into the order of magnitude of polarizabilities can be obtained by considering Figure 7.7, which represents a (mythical) atom. The nucleus of charge Q is at the coordinate origin, and the nucleus is surrounded by an electron cloud of radius a. The electrons are assumed to have a uniform density in space. The total electron charge is $-Q$ and so the atom is electrically neutral.

When an external uniform electric field is applied, the nucleus is displaced a relative distance d in the direction of the field. At this point, the force on the nucleus QE is exactly balanced by the force exerted on the nucleus by the electron cloud.

According to Gauss's electrostatic theorem, this is the same force as would be exerted if all the charge within a sphere of radius d were concentrated at the centre. This charge is $-Q$ times the ratio of the volumes of spheres of radius a and d and so is $-Q d^3/a^3$. This gives a force of magnitude

$$Q^2 \frac{d^3}{a^3} \frac{1}{4\pi\epsilon_0} \frac{1}{d^2} = \frac{Q^2 d}{4\pi\epsilon_0} \frac{1}{a^3}$$

Hence the displacement d satisfies

$$\frac{Q^2 d}{4\pi\epsilon_0 a^3} = QE$$

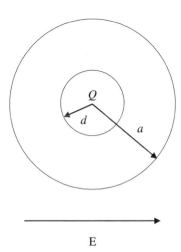

E

Figure 7.7   Construct needed to discuss polarizability

The induced dipole is Qd and the polarizability is Qd/E and so

$$\alpha = 4\pi\epsilon_0\, a^3 \tag{7.4}$$

Apart from the dimensional factor, the volume of the sphere determines the polarizability. Table 7.3 shows the experimental polarizabilities for three inert gas atoms, which illustrates the dependence on volume.

Table 7.4 shows a comparison between typical $\langle\alpha\rangle$ QSAR model calculations for a number of alkanes, together with the result of a serious quantum mechanical calculation. The latter are recorded in the 'Accurate' column. At the chosen level of quantum mechanical theory (BLYP/6-311++G(3d,2p) with optimized geometries, to be discussed in later chapters), the results are at least as reliable as those that can be deduced from spectroscopic experiments. The downside is that the butane calculation took approximately 4 hours on my home PC; the corresponding parameterized calculation took essentially no time at all.

It looks at first sight as if there is an incremental effect. Adding a CH_2 increases the mean polarizability by about $2.13 \times 10^{-40}\,C^2\,m^2\,J^{-1}$ and there has been much speculation over the years as to whether a molecular polarizability can generally be written as a sum of atomic contributions.

Many atomic parameter schemes have been proposed, yet there were always exceptions that needed a complicated sub-rule in order to get agreement with experiment. L. Silberstein [26] wrote in his famous paper of 1917 of '... *the ever growing hierarchy of rules indicating how to treat the exceptions to the law of additivity, although helpful to the chemist, is the clearest confession of non-additivity*'.

The concept of bond polarizability was then introduced in order to try to circumvent the additivity problem. For example, the C—H bond polarizability α_{CH} is taken to be one-quarter of the methane mean polarizability $\langle\alpha\rangle$, and for the alkanes of formula C_nH_{2n+2} we have

$$\langle\alpha(C_nH_{2n+2})\rangle = (n-1)\langle\alpha_{CC}\rangle + (2n+2)\langle\alpha_{CH}\rangle \tag{7.5}$$

Table 7.3 Polarizabilities of inert gases

Inert gas	$\alpha/10^{-40}\,C^2\,m^2\,J^{-1}$
He	0.23
Ne	0.44
Ar	1.83

Table 7.4 Comparison of accurate quantum mechanical calculation with QSAR

Alkane	Accurate $\langle\alpha\rangle/10^{-40}\,C^2\,m^2\,J^{-1}$	QSAR $\langle\alpha\rangle/10^{-40}\,C^2\,m^2\,J^{-1}$
Methane	2.831	2.90
Ethane	4.904	4.94
Propane	7.056	6.99
Butane	9.214	9.02

from which it is possible to deduce a value for α_{CC}. This method actually reproduces experimental data for hydrocarbons to within a few percent. But once again, such additivity schemes fail when applied to molecules containing heteroatoms, and hydrocarbons containing atoms in different states of hybridization.

K. J. Miller and J. A. Savchik [27] proposed a method whereby $\langle \alpha \rangle$ was written as the square of a sum of atomic hybrid components. If they had known about 4-quantity electromagnetic formulae they would have written their formula

$$\langle \alpha \rangle = 4\pi\epsilon_0 \frac{4}{N} \left[\sum_A \tau_A \right]^2 \tag{7.6}$$

but in the event, the $4\pi\epsilon_0$ is missing from their paper. The number of electrons in the molecule is N, and the τs are parameters for each atom in its particular hybrid configuration. The summation is over all atoms in the molecule, and the τs depend on the chemical hybridization but not the chemical environment; they are constant for an sp^2 carbon atom, for example.

The authors based their algorithm on an old quantum mechanical treatment of dipole polarizability discussed in the classic book, *Molecular Theory of Gases and Liquids* [1]. The Miller–Savchik treatment is widely used in QSAR studies, and to give you a flavour I have recalculated the mean polarizabilities and included them in Table 7.4 under the QSAR heading. Given that each calculation took just a few seconds, you can see why QSAR practitioners think so highly of the Miller–Savchik technique. To put things in perspective, we should not be surprised to find such excellent agreement between the two sets of results in Table 7.4; it is heterosubstituted molecules that pose the severe test.

7.3.3 Molecular volume and surface area

Molecular volumes are often calculated by a numerical integration grid technique that I can illustrate by considering the trivial problem of finding the volume of an atom whose van der Waals radius is R (the volume is of course $\frac{4}{3}\pi R^3$).

Figure 7.8 shows a two-dimensional representation of the atom whose van der Waals radius is R, surrounded by a three-dimensional grid of equally spaced points.

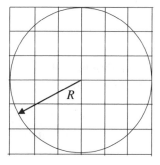

Figure 7.8   Grid around atom

The grid has its centre at the atom centre, and the edges of the grid correspond to the van der Waals radius.

For each grid point in turn we calculate its distance from the centre and determine whether the grid point lies inside or outside the atom. If n is the total number of grid points and n_a the number that lie within the atom whose volume is V, then we have

$$\frac{V}{8R^3} = \frac{n_a}{n}$$

For a polyatomic, we have to give special consideration to grid points that lie in the overlap region. Figure 7.9 shows two atoms, A and B, with radii R_A and R_B. The overlap region is labelled X.

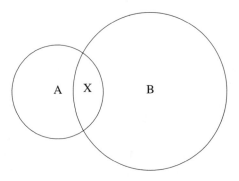

Figure 7.9 Atoms A, B and overlap region X

For atom A, we know that the volume is $\frac{4}{3}\pi R_A^3$. We now surround atom B with a grid, as described above, and test each grid point in turn. If the grid point lies within sphere B, then we test to see if it lies in region X and so has already been counted as part of the volume of atom A. The algorithm proceeds until all atoms have been considered in turn. The molecular volume is found by adding all atomic contributions. There are similar methods for the estimation of molecular surface area.

7.3.4 log(P)

When an organic solute is in equilibrium with water and a polar solvent such as octan-1-ol, it is observed that the ratio of the concentrations

$$P = \frac{[\text{solute in octan-1-ol}]}{[\text{solute in water}]}$$

is roughly constant. P is called the *partition coefficient* and can be used in predicting transmembrane transport properties, protein binding, receptor affinity and pharmacological activity. It is easy to determine P experimentally, but in the process of molecular design we have to deal with a high throughput of possible molecular structures and so a number of attempts have been made to give simple models for predicting P.

In studying the effect of structural variations on P, it was suggested that it had an additive–constitutive character. In the case of a π-substituent, people made use of Hammett's ideas and wrote

$$\pi_X = \log_{10}(P_X) - \log_{10}(P_H) \tag{7.7}$$

where P_H is the partition coefficient for the parent compound and P_X the partition coefficient for a molecule where X has been substituted for H. Workers in the field refer to '$\log(P)$' rather than 'P'.

It was originally envisaged that specific substituents would have the same contribution in different molecules. It has been demonstrated however that this hoped-for additivity does not even hold for many disubstituted benzenes. There are two classical methods for estimating $\log(P)$, both based on the assumed additivity: R. E. Rekker's f-constant method [28] and A. Leo $et\ al.$'s fragment approach [29].

Rekker defined an arbitrary set of terminal fragments using a database of some 1000 compounds with known $\log(P)$. Linear regression was performed, and the regression coefficients designated $group\ contributions$. Deviations from the straight lines were corrected by the introduction of multiples of a so-called 'magic factor' that described special structural effects such as polar groups, etc. $\mathrm{Log}(P)$ is calculated from the fragmental contributions and the correction factors. Leo and Hansch derived their own set of terminal fragments, together with a great number of correction factors.

G. Klopman and L. D. Iroff [30] seem to be the first authors to make use of quantum mechanical molecular structure calculations. They performed calculations at the quantum mechanical MINDO/3 level of theory (to be discussed in Chapter 13) in order to calculate the atomic charge densities of a set of 61 simple organic molecules. They then developed a linear regression model that included the number of C, N, H and O atoms in the given molecule, the atomic charges on C, N and O, and certain 'indicator' variables n_A, n_T and n_M designed to allow for the presence of acid/ester, nitrile and amide functionalities. They found

$$\log_{10}(P) = 0.344 + 0.2078 n_H + 0.093 n_C - 2.119 n_N - 1.937 n_O$$
$$- 1.389 q_C^2 - 17.28 q_N^2 + 0.7316 q_O^2 + 2.844 n_A + 0.910 n_T + 1.709 n_M \tag{7.8}$$

The terms involving q^2 represent the interaction of the solute and solvent.

Klopman and Iroff's method was a great step forward; there are many fewer parameters, it does not produce ambiguous results depending on an arbitrary choice of fragment scheme, and it does not have a complicated correction scheme, apart from the indicator variables. As we will see in Chapter 13, MINDO/3 calculations can be done very quickly even for large molecules.

N. Bodor *et al.* [31] decided to extend and enhance Klopman and Iroff's work by including the molecular volume, weight and surface area. In addition, they included the molecular electric dipole moment in the correlation, all possible products of pairs of charges and a further 57 compounds in the analysis. They used a more up-to-date quantum mechanical model, AM1 (again to be discussed in Chapter 13). The molecular volume and surface area were calculated by a grid method. In their equation

$$
\begin{aligned}
\log_{10}(P) = {}& -1.167 \times 10^{-4}S^2 - 6.106 \times 10^{-2}S + 14.87O^2 - 43.67O + 0.9986I_{\text{alkane}} \\
& + 9.57 \times 10^{-3}M_{\text{w}} - 0.1300D - 4.929Q_{\text{ON}} - 12.17Q_{\text{N}}^4 + 26.81Q_{\text{N}}^2 \\
& - 7.416Q_{\text{N}} - 4.551Q_{\text{O}}^4 + 17.92Q_{\text{O}}^2 - 4.03Q_{\text{O}} + 27.273 \qquad\qquad (7.9)
\end{aligned}
$$

where S (cm^2) is the molecular surface area, O the molecular ovality, I_{alkane} is the indicator variable for alkanes (it is 1 if the molecule is an alkane, 0 otherwise), M_{w} the relative molar mass, D the calculated electric dipole moment, and Q_{ON} the sum of absolute values of the atomic charges on nitrogen and oxygen. The remaining Qs differ from those used by Klopman and Iroff in that they are the square root of the sum of charges on the nitrogen or oxygen atoms. The most significant parameters are the volume and the surface area; the authors claim that this demonstrates that the most important contribution to log(P) is the creation of a hole in the structure of water.

8 Quick Guide to Statistical Thermodynamics

Molecular structure theory tends to deal with the details of individual atoms and molecules, and the way in which a small number of them interact and react. Chemical thermodynamics on the other hand deals with the bulk properties of matter, typically 10^{23} particles. There clearly ought to be a link between the two sets of theories, even though chemical thermodynamics came to maturity long before there was even a satisfactory atomic theory and does not at first sight draw on the concept of a molecule.

Suppose then that we have a macroscopic pure liquid sample, which might consist of 10^{23} particles, and we want to try to model some simple thermodynamic properties such as the pressure, the internal energy or the Gibbs energy. At room temperature, the individual particles making up the sample will be in motion, so at first sight we ought to try to solve the equations of motion for these particles. In view of the large number of particles present, such an approach would be foolhardy. Just to try to specify the initial positions and momenta of so many particles would not be possible, and in any case such a calculation would give too much information.

Even if we could do this impossible task, the next step would be to find a way in which we could relate the individual molecular information to the bulk properties.

For the sake of argument, suppose that the container is a cube. I have shown a two-dimensional slice through the cube as the left-hand side of Figure 8.1, and I have exaggerated the size of the particles by a factor of approximately 10^{10}.

The pressure exerted by a gas on a container wall depends on the rate at which particles collide with the wall. It is not necessary, or even helpful, to know which particle underwent a particular collision. What we need to know are the root mean square speed of the particles, their standard deviation about the mean, the temperature and so on. In chemical thermodynamics, we don't enquire about the behaviour of the individual particles that make up a macroscopic sample; we just enquire about their average properties.

Ludwig Boltzmann and Josiah Willard Gibbs understood all these problems, and invented the subject of *statistical thermodynamics* to get around them.

If we were to measure the pressure exerted on the walls at time intervals $t_1, t_2, \ldots, t_n$ then we might record results $p(t_1), p(t_2), \ldots, p(t_n)$. We could calculate a

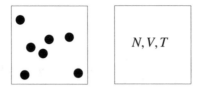

Figure 8.1 Box of particles

sample mean $\langle p \rangle$ and a sample standard deviation using these results

$$\langle p \rangle = \frac{1}{n} \sum_{i=1}^{n} p(t_i)$$

$$\sigma_p = \sqrt{\frac{1}{n} \sum_{i=1}^{n} (p(t_i) - \langle p \rangle)^2}$$

(8.1)

We might expect that the greater the number of measurements, the closer the sample mean would be to the true mean, and the smaller the sample deviation would become.

8.1 The Ensemble

When we considered Figure 8.1, I was careful to draw your attention to the difference between particle properties and bulk properties. I also mentioned that classical thermodynamics is essentially particle-free; all that really matters to such a thermodynamicist are bulk properties such as the number of particles N, the temperature T and the volume of the container V. I have represented this information in the right hand box in Figure 8.1.

Rather than worry about the time development of the particles in the left-hand box in Figure 8.1, what we do is to make a very large number of copies of the system on the right-hand side. We then calculate average values over this large number of replications and according to the *ergodic theorem*, the average value we calculate is exactly the same as the time average we would calculate by studying the time evolution of the original system. The two are the same.

I am not suggesting that all the *cells* in the ensemble are exact replicas at the molecular level; all we do is to ensure that each cell has a certain number of thermodynamic properties that are the same. There is no mention of molecular properties at this stage of the game.

So Figure 8.2 is an ensemble of cells all with the same values of N, V and T. This array of cells is said to form a *canonical ensemble*. There are three other important ensembles in the theory of statistical thermodynamics, and they are named according to what is kept constant in each cell. Apart from the canonical ensemble, where N, V and T are kept constant, statistical thermodynamicists concern themselves also with three others.

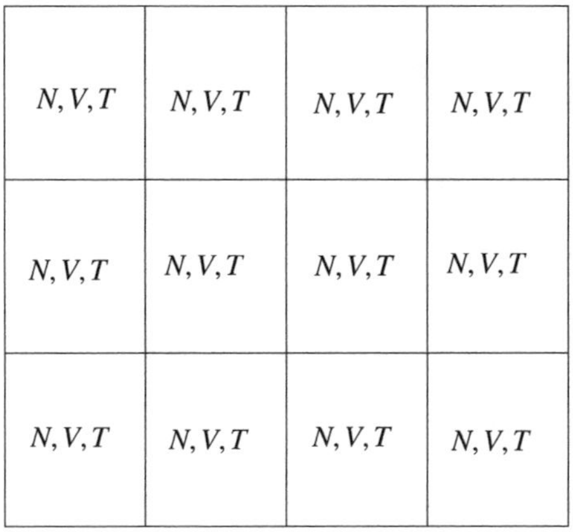

N, V, T	N, V, T	N, V, T	N, V, T
N, V, T	N, V, T	N, V, T	N, V, T
N, V, T	N, V, T	N, V, T	N, V, T

Figure 8.2 Canonical ensemble

In a *microcanonical ensemble*, N, the total energy E and V are kept constant in each cell. In fact, this is a very simple ensemble because energy cannot flow from one cell to another. In an *isothermal–isobaric ensemble*, N, T and the pressure p are kept constant. Finally, we have the *grand canonical ensemble*, where V, T and the chemical potential are kept constant. The grand canonical ensemble is a fascinating one because the number of particles is allowed to fluctuate.

Suppose then that we consider a canonical ensemble of N^* cells, comprising the original cell together with $N^* - 1$ replications. Energy may flow between the cells, but the total energy of the ensemble is constant. Suppose that the possible total energies of the N particles contained in each cell are E_1^*, E_2^*, and so on. We take an energy snapshot, and find a distribution of energies amongst the cells as follows:

N_1^* cells have energy E_1^*,

N_2^* cells have energy E_2^*, etc.

According to Boltzmann, the E^* and the N^* are related by

$$\frac{N_i^*}{N^*} = \frac{\exp\left(-\frac{E_i^*}{k_B T}\right)}{\sum_i \exp\left(-\frac{E_i^*}{k_B T}\right)} \tag{8.2}$$

Be sure to understand that the energies E_i^* are not molecular energies; they are the total energies of the collection of the N particles contained in each cell. Also note that N^* is the number of cells in the ensemble, and that the energies are taken as relative to a common arbitrary zero.

The denominator in the expression above plays an important role in our theory, and so it is given a special name and symbol

$$Q = \sum_i \exp\left(-\frac{E_i^*}{k_B T}\right) \tag{8.3}$$

Q is (in this case) the *canonical partition function*, and it can be used to calculate the usual chemical thermodynamic functions as follows.

8.2 The Internal Energy U_{th}

The IUPAC recommended symbol for thermodynamic internal energy is U, but I have added a subscript 'th' for 'thermodynamic' so that there is no confusion with the total potential energy of a system U (often written Φ). Internal energy is obviously related to the ensemble energy average, but we have to exercise caution. Chemical measurements only give changes in the internal energy, not absolute values. I will therefore write the internal energy as $U_{th} - U_0$, where U_0 is an arbitrary constant. For most purposes we can take U_0 to be zero.

We have, for the ensemble of N^* members

$$U_{th} - U_0 = \frac{\sum_i N_i^* E_i^*}{N^*} \tag{8.4}$$

and according to the ergodic theorem, this is equal to the time average of $U_{th} - U_0$ for any one cell. Using the Boltzmann expression we have

$$U_{th} - U_0 = \frac{\sum_i E_i^* \exp\left(-\frac{E_i^*}{k_B T}\right)}{Q} \tag{8.5}$$

I can tidy up Equation (8.5) by noting

$$\left(\frac{\partial Q}{\partial T}\right)_{V,N} = \frac{1}{k_B T^2} \sum_i E_i^* \exp\left(-\frac{E_i^*}{k_B T}\right) \tag{8.6}$$

and so on substitution

$$U_{th} - U_0 = \frac{k_B T^2}{Q} \left(\frac{\partial Q}{\partial T}\right)_V \tag{8.7}$$

8.3 The Helmholtz Energy A

From the definition of A we have the following

$$A = U - TS$$

$$S = -\left(\frac{\partial A}{\partial T}\right)_{V,N}$$

$$\tag{8.8}$$

$$A = U + T\left(\frac{\partial A}{\partial T}\right)_{V,N}$$

A little manipulation gives

$$\left(\frac{\partial}{\partial T}\left(\frac{A}{T}\right)\right)_{V,N} = -k_B\left(\frac{\partial \ln Q}{\partial T}\right)_{V,N}$$

$$\tag{8.9}$$

$$A - A_0 = -k_B T \ln Q$$

Again, the arbitrary constant A_0 can be taken as zero, since only changes in A are ever measured.

8.4 The Entropy S

Finally, since

$$S = \frac{U - A}{T}$$

we have

$$S = k_B T\left(\frac{\partial \ln Q}{\partial T}\right)_{V,N} + k_B \ln Q$$

$$\tag{8.10}$$

8.5 Equation of State and Pressure

The pressure is related to the Helmholtz energy by

$$p = -\left(\frac{\partial A}{\partial V}\right)_{T,N}$$

and so we find

$$p = k_B T \left(\frac{\partial \ln Q}{\partial V} \right)_{T,N} \tag{8.11}$$

This equation is sometimes called the *equation of state*. The enthalpy and the Gibbs energy can be derived using similar arguments. They turn out to be

$$H = k_B T^2 \left(\frac{\partial \ln Q}{\partial T} \right)_{V,N} + k_B TV \left(\frac{\partial \ln Q}{\partial V} \right)_{T,N}$$

$$G = -k_B T \ln Q + k_B TV \left(\frac{\partial \ln Q}{\partial V} \right)_{T,N} \tag{8.12}$$

8.6 Phase Space

Sophisticated methods such as those due to Hamilton and to Lagrange exist for the systematic treatment of problems in particle dynamics. Such techniques make use of *generalized coordinates* (written $q_1, q_2, \ldots, q_n$) and the *generalized momenta* (written $p_1, p_2, \ldots, p_n$); in Hamilton's method we write the total energy as the *Hamiltonian H*. H is the sum of the kinetic energy and the potential energy, and it is a constant provided that the potentials are time independent. H has to be written in terms of the ps and the qs in a certain way, and systematic application of Hamilton's equations gives a set of differential equations for the system.

To fix our ideas, consider the particle of mass m undergoing simple harmonic motion as discussed in Chapter 4. In this one-dimensional problem I wrote the potential as

$$U = \tfrac{1}{2} k_s (R - R_e)^2$$

so that the total energy is

$$\varepsilon = \tfrac{1}{2} m \left(\frac{dR}{dt} \right)^2 + \tfrac{1}{2} k_s (R - R_e)^2$$

If I put $q = R - R_e$, then the momentum p is $m \, dR/dt$ and I can write the Hamiltonian

$$H = \frac{p^2}{2m} + \frac{k_s q^2}{2} \tag{8.13}$$

We say that the particle moves through *phase space* and in this example the trajectory through phase space is an ellipse (see Figure 8.3), which can be easily seen by

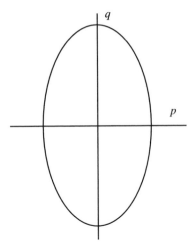

Figure 8.3 Phase space

rewriting Equation (8.13) as

$$\frac{p^2}{a^2} + \frac{q^2}{b^2} = 1$$

Hamilton's equations of motion are

$$\frac{dq_i}{dt} = \frac{\partial H}{\partial p_i}; \qquad \frac{dp_i}{dt} = -\frac{\partial H}{\partial q_i} \qquad (8.14)$$

so for a general problem with N atoms, we have to solve $6N$ first-order differential equations rather than the $3N$ second-order differential equations we would get from straightforward application of Newton's Law.

In the case of a one-particle, three-dimensional system, the Hamiltonian will be a function of the three coordinates $\mathbf{q}$ and the three momenta $\mathbf{p}$, and for a more general problem involving N particles the Hamiltonian will be a function of the $3N$ $\mathbf{q}$s and the $3N$ $\mathbf{p}$s. We say that the $\mathbf{p}$s and the $\mathbf{q}$s together determine a point in $6N$-dimensional phase space, and this point is often denoted Γ.

8.7 The Configurational Integral

Returning now to the canonical partition function, Equation (8.3)

$$Q = \sum_i \exp\left(-\frac{E_i^*}{k_B T}\right)$$

the first thing to note is that all points in phase space contribute to the sum, and the summation has to be replaced by an integral. For an ideal monatomic gas the expression becomes

$$Q = \frac{1}{N!} \frac{1}{h^{3N}} \iint \exp\left(-\frac{E}{k_B T}\right) d\mathbf{p} \, d\mathbf{q} \qquad (8.15)$$

The equation is often written with the Hamiltonian H replacing E, for the reasons discussed above.

The $N!$ term is needed in situations where the particles are completely indistinguishable from one another; for particles that can be distinguished there is no $N!$ term. The integrals have to be done over the spatial variables of all the N particles, and also the momentum variables of the N particles. The integral is therefore a $6N$-dimensional one.

The energy (the Hamiltonian) is always expressible as a sum of kinetic and potential energies, and I have written the mass of each particle m

$$E = \sum_{i=1}^{N} \frac{p_i^2}{2m} + \Phi(\mathbf{q}_1, \mathbf{q}_2, \ldots, \mathbf{q}_N) \qquad (8.16)$$

Kinetic energies depend on the momentum coordinates $\mathbf{p}$. All the potential energies we will meet depend on the spatial coordinates $\mathbf{q}$ but not on the momenta and so the partition function can be factorized into a product of a kinetic part and a potential part

$$Q = \frac{1}{N!} \frac{1}{h^{3N}} \int \exp\left(-\frac{1}{k_B T} \sum_{i=1}^{N} \frac{p_i^2}{2m}\right) d\mathbf{p} \int \exp\left(-\frac{\Phi}{k_B T}\right) d\mathbf{q} \qquad (8.17)$$

The kinetic integral has to be done over the momentum coordinates of all N particles, and it can be seen to be a product of N identical three-dimensional integrals of the type

$$\int \exp\left(-\frac{1}{k_B T} \frac{p_1^2}{2m}\right) d\mathbf{p}_1$$

Each of these is a product of three identical standard integrals of the type

$$\int \exp\left(-\frac{1}{k_B T} \frac{p_x^2}{2m}\right) dp_x$$

and the final result is

$$Q = \frac{1}{N!} \left(\frac{2\pi m k_B T}{h^2}\right)^{3N/2} \int \exp\left(-\frac{\Phi}{k_B T}\right) d\mathbf{q} \qquad (8.18)$$

The $3N$-dimensional integral over the position variables is often referred to as the *configurational integral*. For an ideal gas $\Phi = 0$ and so the configurational integral is V^N, where V is the volume of the container. Some authors include the $N!$ in the definition of the configurational integral.

The canonical partition function for an ideal gas is therefore

$$Q = \frac{V^N}{N!} \left(\frac{2\pi m k_B T}{h^2} \right)^{3N/2} \tag{8.19}$$

The partition function for a real system is often written as the product of an ideal part and an excess part due to non-ideal behaviour

$$Q = Q^{\text{ideal}} Q^{\text{excess}}$$

where

$$Q^{\text{excess}} = \frac{1}{V^N} \int \exp \left(-\frac{\Phi}{k_B T} \right) d\mathbf{q} \tag{8.20}$$

The point of doing this is that thermodynamic properties such as A are often measured experimentally as an ideal and an excess part

$$A = A^{\text{ideal}} + A^{\text{excess}}$$

The ideal part can be related to Q^{ideal} and the excess part to Q^{excess}.

8.8 The Virial of Clausius

Let me focus attention on one particular particle i moving in the box, Figure 8.1. As this particle moves it will be subject to some varying force $\mathbf{F}_i$ and

$$\mathbf{F}_i = m \frac{d\mathbf{v}_i}{dt} \tag{8.21}$$

Taking the scalar product of both sides of this equation with $\mathbf{r}_i$ I get

$$\mathbf{r}_i \cdot \mathbf{F}_i = m\mathbf{r}_i \cdot \left(\frac{d\mathbf{v}_i}{dt} \right) \tag{8.22}$$

Consider now the vector identity

$$\frac{d}{dt}(\mathbf{r}_i \cdot \mathbf{v}_i) = \mathbf{r}_i \cdot \frac{d\mathbf{v}_i}{dt} + \frac{d\mathbf{r}_i}{dt} \cdot \mathbf{v}_i \tag{8.23}$$

which can also be written

$$\frac{d}{dt}(\mathbf{r}_i \cdot \mathbf{v}_i) = \mathbf{r}_i \cdot \frac{d\mathbf{v}_i}{dt} + v_i^2 \tag{8.24}$$

On comparison of Equations (8.22) and (8.24), I have

$$\mathbf{r}_i \cdot \mathbf{F}_i = m\left(\frac{d}{dt}(\mathbf{r}_i \cdot \mathbf{v}_i) - v_i^2\right) \tag{8.25}$$

or

$$-\tfrac{1}{2}\mathbf{r}_i \cdot \mathbf{F}_i = -\tfrac{1}{2}m\frac{d}{dt}\mathbf{r}_i \cdot \mathbf{v}_i + \tfrac{1}{2}mv_i^2 \tag{8.26}$$

The next step is to sum corresponding terms on both sides of the equation for each particle in the box. For N particles each of mass m, this gives

$$-\frac{1}{2}\sum_{i=1}^{N}\mathbf{r}_i \cdot \mathbf{F}_i = -\tfrac{1}{2}m\frac{d}{dt}\sum_{i=1}^{N}\mathbf{r}_i \cdot \mathbf{v}_i + \tfrac{1}{2}m\sum_{i=1}^{N}v_i^2 \tag{8.27}$$

Finally, we take a time average over all the particles in the box, which is assumed to be in an equilibrium state

$$-\frac{1}{2}\left\langle\sum_{i=1}^{N}\mathbf{r}_i \cdot \mathbf{F}_i\right\rangle = -\frac{m}{2}\frac{d}{dt}\left\langle\sum_{i=1}^{N}\mathbf{r}_i \cdot \mathbf{v}_i\right\rangle + \tfrac{1}{2}m\left\langle\sum_{i=1}^{N}v_i^2\right\rangle \tag{8.28}$$

The second term on the right-hand side is obviously the mean kinetic energy of all the particles in the box. This must be $\tfrac{3}{2}Nk_BT$, according to the equipartition of energy principle.

Whatever the value of the first average quantity in brackets on the right-hand side it cannot vary with time because we are dealing with an equilibrium state and so the first time derivative must vanish

$$-\frac{m}{2}\frac{d}{dt}\left\langle\sum_{i=1}^{N}\mathbf{r}_i \cdot \mathbf{v}_i\right\rangle = 0$$

Thus, we have

$$-\frac{1}{2}\left\langle\sum_{i=1}^{N}\mathbf{r}_i \cdot \mathbf{F}_i\right\rangle = \tfrac{1}{2}m\left\langle\sum_{i=1}^{N}v_i^2\right\rangle \tag{8.29}$$

The summation term on the left hand side $-\tfrac{1}{2}\langle\sum_{i=1}^{N}\mathbf{r}_i \cdot \mathbf{F}_i\rangle$ involving the forces and coordinates is often referred to as the *virial of Clausius*.

9 Molecular Dynamics

Molecular mechanics (MM) these days tends to be concerned only with prediction of local minima on molecular potential energy surfaces. QSAR properties are often calculated in order to assist high-volume screening studies in pharmaceuticals applications. Should we want to study the motions of the molecule, all that would be needed would be to investigate the normal modes of vibration (which can be obtained from the hessian). MM does not take account of zero-point vibrations and the calculations refer to a molecule at $0\,\mathrm{K}$, when it is completely at rest. Workers in the modelling field often refer to MM as *energy minimization*.

We now turn our attention to the time development of collections of atoms and molecules, for which the techniques of Molecular Dynamics and Monte Carlo are widely used.

I have stressed in previous chapters the intermolecular potential energy U_{mol} (often written Φ). Assuming pairwise additivity, Φ can be found by summing over all distinct pairs of particles

$$\Phi = \sum_{i=1}^{N-1} \sum_{j=i+1}^{N} U_{ij} \tag{9.1}$$

If the assumption of pairwise additivity is not valid, then we have to include all possible triples, and so on

$$\Phi = \sum_{i=1}^{N-1} \sum_{j=i+1}^{N} U_{ij} + \sum_{i=1}^{N-2} \sum_{j=i+1}^{N-1} \sum_{k=j+1}^{N} U_{ijk} + \cdots \tag{9.2}$$

In this book we will generally be concerned with situations where the potentials (and the forces) are pairwise additive. If we focus on particle A, then the mutual potential energy of A with all the other particles U_{A} is found from

$$U_{\mathrm{A}} = \sum_{j \neq \mathrm{A}} U_{\mathrm{A}j}$$

and we can find the force on particle A, $\mathbf{F}_A$, by differentiating with respect to the coordinates of particle A

$$\mathbf{F}_A = -\text{grad } U_A$$

For example, if we consider a pair of Lennard-Jones particles A and B where

$$U_{AB} = 4\varepsilon \left(\left(\frac{\sigma}{R_{AB}} \right)^{12} - \left(\frac{\sigma}{R_{AB}} \right)^6 \right)$$

we note that the potential only depends on the distance between the particles. The expression for grad U is particularly simple so that

$$\mathbf{F}_A = -\frac{\partial U}{\partial R_A} \frac{\mathbf{R}_{BA}}{R_{BA}}$$

which gives

$$\mathbf{F}_A = 24\varepsilon \left(2 \left(\frac{\sigma}{R_{AB}} \right)^{12} - \left(\frac{\sigma}{R_{AB}} \right)^6 \right) \frac{\mathbf{R}_{AB}}{R_{AB}^2}$$

Newton's second law connects force and acceleration by

$$\mathbf{F}_A = m_A \frac{d^2 \mathbf{R}_A}{dt^2}$$

and in principle we could study the time development of a system by solving this second-order differential equation, one such equation for each of the particles in our system. Calculating the trajectories of N particles therefore appears to involve the solution of a set of $3N$ second-order differential equations. Alternatively, we could use an advanced method such as Hamilton's to solve $6N$ first-order differential equations. For any set of N particles it is always possible to find three coordinates that correspond to translation of the centre of mass of the system, and, if the particles have 'shape', three coordinates that correspond to rotations about three axes that pass through the centre of mass.

Most of the early molecular dynamics studies were directed at the problem of liquid structure, so that is where we will begin our discussion.

9.1 The Radial Distribution Function

Of the three states of matter, gases are the easiest to model because the constituent particles are so far apart on average that we can ignore intermolecular interactions,

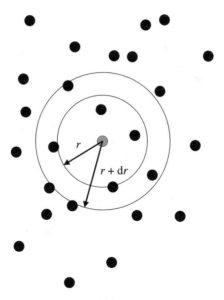

Figure 9.1   Radial distribution function

apart from during their brief collisions. This is why researchers were able to bring the kinetic theory of gases to such an advanced stage by the end of the nineteenth century (and before the existence of a satisfactory theory of molecular structure).

The atoms or molecules in a crystalline solid are arranged in a regular order, and for this reason we usually start a discussion of the solid state from the properties of regular solids. Once such patterns were truly understood at the beginning of the twentieth century, the theory of the solid state made rapid progress.

Liquids are much harder to model and to study experimentally than solids and gases; elementary textbooks usually state that liquids show neither complete order nor complete disorder. The basis of this remark concerns a property called the *radial distribution function* $g(r)$. Consider Figure 9.1, which is a snapshot of the particles in a simple atomic liquid.

We take a typical atom (the grey one, designated i) and draw two spheres of radii r and $r + dr$. We then count the number of atoms whose centres lie between these two spheres, and repeat the process for a large number N of atoms. If the result for atom i is $g_i(r)\, dr$, then the radial distribution function is defined as

$$g(r)\, dr = \frac{1}{N} \sum_{i=1}^{N} g_i(r)\, dr \tag{9.3}$$

This process then has to be repeated for many complete shells over the range of values of r thought to be significant.

In the case of an ideal gas, we would expect to find the number of particles to be proportional to the volume enclosed by the two spheres, which is $4\pi r^2\, dr$. This gives $g(r) = 4\pi r^2$, a simple quadratic curve.

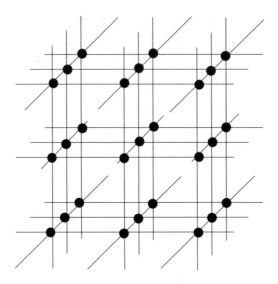

Figure 9.2   Simple cubic lattice

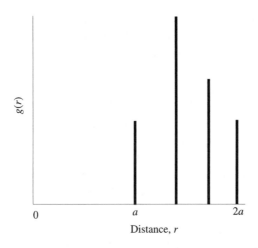

Figure 9.3   First part of the radial distribution function for a simple solid

Consider now the simple cubic solid shown in Figure 9.2 whose nearest neighbour distance is a. Each atom is surrounded by 6 nearest neighbours at a distance a, 12 at a distance $\sqrt{2}\ a$, 8 next-next nearest neighbours at a distance $\sqrt{3}\ a$, 6 at a further distance $2a$ and so on. We would therefore expect to find a radial distribution function similar to the one shown in Figure 9.3. The height of each peak is proportional to the number of atoms a distance r from any given atom.

Radial distribution functions can be deduced experimentally from diffraction studies. In the case of a liquid, Figure 9.4, the curve resembles that expected for a

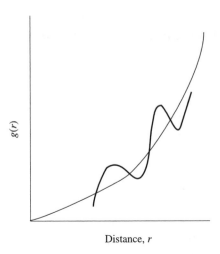

Figure 9.4 Radial distribution function for liquid superimposed on an ideal gas

solid at low temperatures, and at high temperatures it resembles the quadratic expected for an ideal gas. At intermediate temperatures, the two features can be clearly seen; essentially a solid pattern is superimposed on the gas pattern. This gives the experimental basis for the well-known remark about liquid structure quoted above.

9.2 Pair Correlation Functions

The radial distribution function for a gas varies as $4\pi r^2$ and so tends to infinity as r tends to infinity. It is usual to remove the $4\pi r^2$ dependence by defining a related quantity called the *pair correlation function* $g_{AB}(r)$, which gives information about the probability of finding two particles A and B separated by a distance r. If the volume of a system is V and it contains N_A species of type A and N_B species of type B, then the number densities are N_A/V and N_B/V. The fraction of time that the differential volume elements $d\tau_1$ and $d\tau_2$, which are separated by a distance r, simultaneously contain species of type A and B is given by

$$\frac{N_A}{V}\frac{N_B}{V}g_{AB}(r)\,d\tau_1 d\tau_2$$

In a mixture of A and B we would be interested in the three distinct pair correlation functions $g_{AA}(r)$, $g_{BB}(r)$ and $g_{AB}(r)$. These pair correlation functions have a limiting value of 1 for a fluid.

9.3 Molecular Dynamics Methodology

In an ideal gas, the particles do not interact with each other and so the potential Φ is zero. Deviations from ideality are due to the interparticle potential, and most of the early studies were made on just three types of particle: the hard sphere model, the finite square well and the Lennard-Jones model.

9.3.1 The hard sphere potential

The hard sphere potential of Figure 9.5 is the simplest one imaginable; the system consists of spheres of radii σ and $U(r)$ is zero everywhere except when two spheres touch, when it becomes infinite.

The hard sphere potential is of great theoretical interest not because it represents the intermolecular potential of any known substance, rather because any calculations based on the potential are simple. B. J. Alder and T. E. Wainwright introduced the modelling technique now known as Molecular Dynamics to the world in a short *Journal of Chemical Physics* 'Letters to the Editor' article in 1957 [32]. They reported a study of hard disks, the two-dimensional equivalent of hard spheres.

9.3.2 The finite square well

B. J. Alder and T. E. Wainwright's 1959 paper [33] is usually regarded the keynote paper in the field, and you might like to read the Abstract.

A method is outlined by which it is possible to calculate exactly the behaviour of several hundred interacting classical particles. The study of this many-body problem is carried out by an electronic computer that solves numerically the

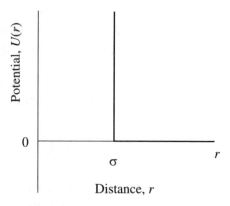

Figure 9.5   Hard sphere potential

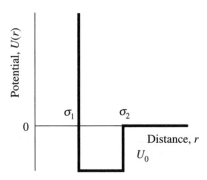

Figure 9.6 Finite square well

simultaneous equations of motion. The limitations of this numerical scheme are enumerated and the important steps in making the program efficient on computers are indicated. The applicability of this method to the solution of many problems in both equilibrium and nonequilibrium statistical thermodynamics is discussed.

In this second paper they chose a three-dimensional system of particles and the finite square well potential shown in Figure 9.6. This potential is especially simple because a given particle does not experience any change in velocity except when it is separated from another particle by σ_1 (when it undergoes an attractive collision) or σ_2 (when it undergoes a repulsive collision). On collision, the velocities are adjusted and the calculation restarts. Statistical data are collected every collision.

In their dynamic calculation all the particles were given initial velocities and positions. In one example, the particles were given equal kinetic energies with the three direction cosines of the velocity vector chosen at random, and initial positions corresponding to a face-centred cubic lattice. Once the initial configuration was set up, they calculated exactly the time at which the first collision occurs. The collision time can be found by evaluating, for every pair in the system, the time taken for the projected paths to reach a separation of σ_1 or σ_2.

If two particles A and B have initial positions $\mathbf{r}_{A,0}$ and $\mathbf{r}_{B,0}$ and velocities $\mathbf{v}_A$ and $\mathbf{v}_B$, then the instantaneous positions at time t will be

$$\mathbf{r}_A = \mathbf{r}_{A,0} + \mathbf{u}_A t$$
$$\mathbf{r}_B = \mathbf{r}_{B,0} + \mathbf{u}_B t$$

giving

$$\mathbf{r}_A - \mathbf{r}_B = \mathbf{r}_{A,0} - \mathbf{r}_{B,0} + (\mathbf{u}_A - \mathbf{u}_B)t$$

and so

$$(\mathbf{r}_A - \mathbf{r}_B)^2 = (\mathbf{r}_{A,0} - \mathbf{r}_{B,0})^2 + 2t(\mathbf{r}_{A,0} - \mathbf{r}_{B,0}) \cdot (\mathbf{u}_A - \mathbf{u}_B) + t^2(\mathbf{u}_A - \mathbf{u}_B)^2$$

If we rewrite the last equation as a quadratic in t as

$$u_{AB}^2 t^2 + 2b_{AB}t + r_{AB,0}^2 = \sigma_\alpha^2$$

where α takes values 1 or 2, then we see that the time required for a repulsive or attractive collision is

$$t_{AB}^{(\alpha)} = \frac{-b_{AB} \pm (b_{AB}^2 - u_{AB}^2(r_{AB}^2 - \sigma_\alpha^2))^{1/2}}{u_{AB}^2} \qquad (9.4)$$

In order to find the first collision time, all pairs have to be analysed. All the particles are then allowed to move for such time, and the velocities of the colliding pair are adjusted according to the equations of motion.

The finite square well occupies an important place in the history of molecular modelling. Real atomic and molecular systems have much more complicated mutual potential energy functions, but the finite square well does at least show a minimum. On the other hand, because of the finite square well potential, the equations of motion are particularly simple and no complicated numerical techniques are needed. There are no accelerations until two particles collide.

9.3.3 Lennardjonesium

The first simulation of a 'real' chemical system was A. Rahman's 1964 study of liquid argon [34]. He studied a system comprising 864 Lennard-Jones particles under conditions appropriate to liquid argon at 84.4 K and a density of 1.374 g cm^{-3}. Once again, there is much to be gained by studying the Abstract, so here is the first part of it.

A system of 864 particles interacting with a Lennard-Jones potential and obeying classical equations of motion has been studied on a digital computer (CDC 3600) to simulate molecular dynamics in liquid argon at 94.4 K and a density of 1.374 g cm^{-3}. The pair correlation function and the constant of self-diffusion are found to agree well with experiment; the latter is 15% lower than the experimental value. The spectrum of the velocity autocorrelation function shows a broad maximum in the frequency range $\omega = 0.25$ $(2\pi k_B T/h)$. The shape of the Van Hove function $G_s(r, t)$ attains a maximum departure from a Gaussian at about $t = 0.3 \times 10^{-2}$ s and becomes a Gaussian again at about 10^{-11} s.

There are several interrelated problems. A sample size has to be chosen; this is usually determined by the available computer resource and the complexity of the potential function, because the potential function has to be calculated very many times during the simulation. The number of particles and the density determine the size of the container. At the same time we need to decide on a potential function; the

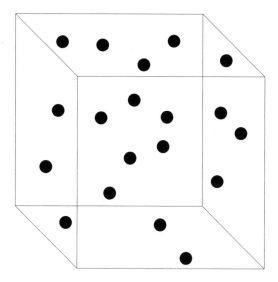

Figure 9.7   Box of argon atoms

natural choice for the inert gases is the Lennard-Jones potential, and we should note that the L-J potential is essentially short range.

So many early papers used the L-J potential that the noun *Lennardjonesium* was coined to describe a non-existent element whose atoms interacted via the L–J potential.

9.4 The Periodic Box

Figure 9.7 shows a suitable virtual box of argon atoms. Examination of the figure reveals two problems. Atoms near the edges of the box will experience quite different resultant forces from the atoms near the centre of the box. Secondly, the atoms will be in motion if the temperature is non-zero. As the system evolves in time, it is quite likely that one of the atoms will pass through the container walls and so disappear from the calculation. This has the undesirable effect of reducing the density.

There is a third subtle point: if the atoms are sufficiently light (He rather than Ar), we would need to take the quantum mechanical zero point energy into effect; even at 0 K, quantum mechanical particles have a residual motion.

The periodic box concept, illustrated in Figure 9.8, gives a solution to the first two problems. We appeal to the ensemble concept of statistical thermodynamics, and surround our system with a large number of *identical* copies. In this case the boxes are truly identical at the atomic level rather than in the usual thermodynamic sense of having N, V and T in common.

Figure 9.8 shows a two-dimensional slice through a small portion of the system (the central box where the atoms are shown grey) and the copies (where the atoms are

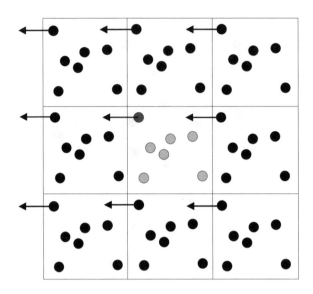

Figure 9.8   Small part of the molecular ensemble

shown black). Each copy is identical at the atomic level, and each atom undergoes the
same time development as its image in every other copy. As the dark grey atom (top
left, central cell) leaves the central cell, its image enters from an adjoining copy,
shown by the vector displacements in the figure. This keeps the density constant.
There are no effects due to the walls because each atom in the central cell is under the
influence of every other atom in the central cell and in all other cells.

   Consider now the dark grey atom (top left in the central cell). We need to calculate
the force on this atom in order to understand its time development. To do this we
should in principle sum the pair potential of the atom with every other atom. Differentiation of the potential with respect to the coordinates of the dark grey atom gives
the force on the particle. This would give an infinite sum.

   In this particular case, there is no great problem because the L-J potential is short
range. We decide on a cut-off distance beyond which the pair potential will be
negligible; this defines a sphere. In order to treat the dark grey atom, we have to
include contributions from all other atoms in the sphere. This is illustrated as a two-
dimensional slice in Figure 9.9.

   Truncation of the intermolecular potential at a cut-off distance introduces two
technical difficulties. First, the pair potential has a discontinuity at the cut-off distance r_c, and secondly, whenever a pair of particles A and B have separation greater
than r_c the total energy is not conserved. The first problem is solved by shifting the
potential function by an amount $U(r_c)$, that is we take

$$U^s(r_{AB}) = \begin{cases} U(r_{AB}) - U(r_c) & \text{if } r_{AB} \leq r_c \\ 0 & \text{if } r_{AB} > r_c \end{cases} \qquad (9.5)$$

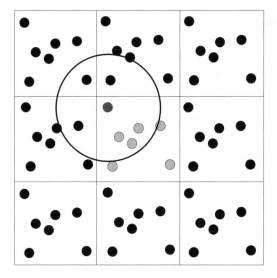

Figure 9.9 The cut-off distance

The second problem can be solved by adding a small linear term to the potential, chosen so that its derivative is zero at the cut-off distance

$$
U^{\mathrm{lin}}(r_{\mathrm{AB}}) =
\begin{cases}
U(r_{\mathrm{AB}}) - U(r_{\mathrm{c}}) - \left(\dfrac{\mathrm{d}U}{\mathrm{d}r_{\mathrm{AB}}} \right)_{R_{\mathrm{c}}} (r_{\mathrm{AB}} - r_{\mathrm{c}}) & \text{if } r_{\mathrm{AB}} \leq r_{\mathrm{c}} \\
0 & \text{if } r_{\mathrm{AB}} > r_{\mathrm{c}}
\end{cases}
\tag{9.6}
$$

9.5 Algorithms for Time Dependence

Once we have calculated the potential and hence the force by differentiation, we have to solve Newton's equation of motion. If $\mathbf{F}_{\mathrm{A}}$ is the force on particle A, whose position vector is $\mathbf{r}_{\mathrm{A}}$ and whose mass is m_{A}, then

$$
\begin{aligned}
\mathbf{F}_{\mathrm{A}} &= m_{\mathrm{A}} \frac{\mathrm{d}^2 \mathbf{r}_{\mathrm{A}}}{\mathrm{d}t^2} \\
&= m_{\mathrm{A}} \mathbf{a}_{\mathrm{A}}
\end{aligned}
$$

This is a second-order differential equation that I can write equivalently as two first-order differential equations for the particle position $\mathbf{r}_{\mathrm{A}}$ and the velocity $\mathbf{v}_{\mathrm{A}}$

$$
\begin{aligned}
\mathbf{F}_{\mathrm{A}} &= m_{\mathrm{A}} \frac{\mathrm{d}\mathbf{v}_{\mathrm{A}}}{\mathrm{d}t} \\
\mathbf{v}_{\mathrm{A}} &= \frac{\mathrm{d}\mathbf{r}_{\mathrm{A}}}{\mathrm{d}t}
\end{aligned}
$$

9.5.1 The leapfrog algorithm

A simple algorithm for integration of these two equations numerically in small time steps Δt can be found by considering the Taylor expansion for $\mathbf{v}(t)$

$$\mathbf{v}_A\left(t+\frac{\Delta t}{2}\right) = \mathbf{v}_A(t) + \left(\frac{d\mathbf{v}_A}{dt}\right)_t \frac{\Delta t}{2} + \frac{1}{2}\left(\frac{d^2\mathbf{v}_A}{dt^2}\right)_t \left(\frac{\Delta t}{2}\right)^2 + \cdots$$

$$\mathbf{v}_A\left(t-\frac{\Delta t}{2}\right) = \mathbf{v}_A(t) - \left(\frac{d\mathbf{v}_A}{dt}\right)_t \frac{\Delta t}{2} + \frac{1}{2}\left(\frac{d^2\mathbf{v}_A}{dt^2}\right)_t \left(\frac{\Delta t}{2}\right)^2 + \cdots \qquad (9.7)$$

Subtracting and rearranging we get

$$\mathbf{v}_A\left(t+\frac{\Delta t}{2}\right) = \mathbf{v}_A\left(t-\frac{\Delta t}{2}\right) + \mathbf{a}_A(t)\Delta t + \cdots \qquad (9.8)$$

I will switch back and forth between, for example, $\mathbf{v}$ and $d\mathbf{r}/dt$ in order to try to improve the readability of the equations. Also, I could have written $\mathbf{v}_A(t)$ or $(\mathbf{v}_A)_t$ to mean the instantaneous velocity of particle A at time t. The acceleration $\mathbf{a}$ is calculated from the force.

Using the same procedure for the Taylor expansion of $\mathbf{r}_A$ at the time point $t + 1/2$ Δt we get

$$\mathbf{r}_A(t+\Delta t) = \mathbf{r}_A(t) + \mathbf{v}_A\left(t+\frac{\Delta t}{2}\right)\Delta t + \cdots \qquad (9.9)$$

Equations (9.8) and (9.9) form the so-called *leapfrog algorithm*, which is reputed to be one of the most accurate and stable techniques for use in molecular dynamics. A suitable time increment Δt for molecular dynamics is a femtosecond (10^{-15} s). In the leapfrog scheme the velocities are first calculated at time $t + 1/2\,\Delta t$. These are used to calculate the positions of the particles at time $t + \Delta t$ and so on. In this way the velocities leap over the positions and then the positions leap over the velocities.

9.5.2 The Verlet algorithm

If instead we start from the Taylor expansion of $\mathbf{r}_A(t)$ we have

$$\mathbf{r}_A(t+\Delta t) = \mathbf{r}_A(t) + \left(\frac{d\mathbf{r}_A}{dt}\right)_t \Delta t + \frac{1}{2}\left(\frac{d^2\mathbf{r}_A}{dt^2}\right)_t (\Delta t)^2 + \cdots$$

$$\mathbf{r}_A(t-\Delta t) = \mathbf{r}_A(t) - \left(\frac{d\mathbf{r}_A}{dt}\right)_t \Delta t + \frac{1}{2}\left(\frac{d^2\mathbf{r}_A}{dt^2}\right)_t (\Delta t)^2 + \cdots \qquad (9.10)$$

which gives (assuming that third-order and higher terms are negligible)

$$\mathbf{r}_A(t + \Delta t) = 2\mathbf{r}_A(t) - \mathbf{r}_A(t - \Delta t) + \left(\frac{d^2\mathbf{r}_A}{dt^2}\right)_t (\Delta t)^2 \qquad (9.11)$$

This is known as the *Verlet algorithm*. The acceleration is obtained from the force experienced by atom A at time t. The velocity does not appear in the expression, but it can be obtained from the finite difference formula

$$\mathbf{v}_A(t) = \frac{\mathbf{r}_A(t + \Delta t) - \mathbf{r}_A(t - \Delta t)}{2\Delta t} \qquad (9.12)$$

The Verlet algorithm uses positions and accelerations at time t and the position at time $t - \Delta t$ to calculate a new position at time $t + \Delta t$. All these have to be stored at every iteration.

A variant is the *velocity Verlet algorithm*, which requires only the storage of positions, velocities and accelerations that all correspond to the same time step. It takes the form

$$\mathbf{r}_A(t + \Delta t) = \mathbf{r}_A(t) + \left(\frac{d\mathbf{r}_A}{dt}\right)_t \Delta t + \frac{1}{2}\left(\frac{d^2\mathbf{r}_A}{dt^2}\right)_t (\Delta t)^2$$

$$\mathbf{v}_A(t + \Delta t) = \left(\frac{d\mathbf{r}_A}{dt}\right)_t + \frac{1}{2}\left(\left(\frac{d^2\mathbf{r}_A}{dt^2}\right)_t + \left(\frac{d^2\mathbf{r}_A}{dt^2}\right)_{t+\Delta t}\right)\Delta t \qquad (9.13)$$

There are many other algorithms in the literature, each with their own strengths and weaknesses.

9.6 Molten Salts

Molten salts (such as the alkali halides) are of great technological interest in the field of metal extraction. The first simulations were done by L. V. Woodcock in 1971 [35]. Molten salts introduce a new problem because the potential energy terms are long range. Consider a (hypothetical) one-dimensional infinite crystal, part of which is shown in Figure 9.10. The unshaded ions have charge $-Q$, the shaded ions have charge $+Q$ and the spacing between the ions is a. Suppose we start with the central (grey) ion at infinity, and all the other ions at their lattice positions as shown. The work done, W, in bringing the grey ion from infinity to its place in the lattice is

$$W = -\frac{2Q^2}{4\pi\epsilon_0 a}\left(1 - \frac{1}{2} + \frac{1}{3} - \frac{1}{4} + \cdots\right)$$

and the term in brackets converges very slowly to its limiting value of ln (2). Such series have to be summed when calculating the force on a given ion in a periodic box

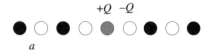

Figure 9.10   Part of a one-dimensional crystal, where the separation between ions is a

Table 9.1   Summation of series for W

No of terms	Sum
1	1.0000
2	0.5000
3	0.8333
4	0.5833
100	0.6882

such as Figure 9.8; in the case of neutral species, the sum is very quickly convergent because of the short-range nature of the forces (see Table 9.1). For neutral systems, a cut-off radius usually is taken beyond which the interactions are set to zero.

9.7 Liquid Water

Water plays a prominent place amongst solvents for obvious reasons. Two-thirds of our planet is covered in water, chemical reactions tend to be done in aqueous solution, and so on. The relatively strong hydrogen bonding in liquid water causes many of its physical properties to be 'anomalous', and the structure of ice has long interested both theoreticians and experimentalists.

Neutron diffraction studies on heavy ice D_2O have shown that water molecules retain their identity in condensed phases with very little distortion of their molecular geometry. This means that water molecules may be treated as rigid asymmetric rotors (with six degrees of freedom) rather than explicitly treating the three nuclei separately (nine degrees of freedom). The classical energy for a collection of N rigid rotor molecules consists of the kinetic energy for translation and rotation, together with the intermolecular potential. Each water molecule is described by six coordinates; three specify the centre of mass and three angles that fix the spatial orientation about the centre of mass. In this section I will denote these coordinates by the six-dimensional vector $\mathbf{X}$. In terms of the linear velocities $\mathbf{v}_i$, the angular velocity vectors ω_i, the moments of inertia $\mathbf{I}_i$ and the coordinates $\mathbf{X}_i$ the energy turns out to be

$$\varepsilon = \frac{1}{2} \sum_{i=1}^{N} \left(m v_i^2 + \omega_i^{\mathrm{T}} \mathbf{I}_i \omega_i \right) + U(\mathbf{X}_1, \mathbf{X}_2, \ldots, \mathbf{X}_N)$$

The question then is the extent to which the intermolecular potential is pairwise additive; such functions may always be resolved into contributions from pairs, triples and higher contributions

$$U(\mathbf{X}_1, \mathbf{X}_2, \ldots, \mathbf{X}_N) = \sum_{i<j}^{N} U^{(2)}(\mathbf{X}_i, \mathbf{X}_j)$$

$$+ \sum_{i<j<k}^{N} U^{(3)}(\mathbf{X}_i, \mathbf{X}_j, \mathbf{X}_k) + \cdots + U^{(N)}(\mathbf{X}_1, \mathbf{X}_2, \ldots, \mathbf{X}_N)$$

$$(9.14)$$

In the case of simple fluids such as liquid argon it is widely believed that U is practically pairwise additive. In other words, the contributions $U^{(p)}$ for $p > 2$ are negligible. However, the local structure in liquid water is thought to depend on just these higher order contributions and so it is unrealistic in principle to terminate the expansion with the pair contributions. It is legitimate to write the potential as a sum of pair potentials, provided one understands that they are effective pair potentials that somehow take higher terms into account.

The classic molecular dynamics study of liquid water is that of A. Rahman and F. H. Stillinger [36]. They wrote the effective pair potential as a sum of two contributions: a Lennard-Jones 12-6 potential U_{LJ} for the two oxygen atoms

$$U_{\mathrm{LJ}}(R) = 4\varepsilon\left(\left(\frac{\sigma}{R}\right)^{12} - \left(\frac{\sigma}{R}\right)^{6}\right)$$

and a function U_{el}, modulated by a function $S(R_{ij})$ that depends sensitively on the molecular orientations about the oxygen atoms

$$U^{(2)}(\mathbf{X}_i, \mathbf{X}_j) = U_{\mathrm{LJ}}(R_{ij}) + S(R_{ij})U_{\mathrm{el}}(\mathbf{X}_i, \mathbf{X}_j) \qquad (9.15)$$

The Lennard-Jones parameters were taken to be those appropriate to neon, $\sigma = 282\,\mathrm{pm}$ and $\varepsilon = 5.01 \times 10^{-22}\,\mathrm{J}$, on the grounds that neon is isoelectronic with water.

Four point charges Q, each of magnitude $0.19\,e$ and each $100\,\mathrm{pm}$ from the oxygen nucleus, are embedded in the water molecule in order to give U_{el}. Two charges are positive, to simulate the hydrogen atoms, whilst the other two are negative and simulate the lone pairs. The four charges are arranged tetrahedrally about the oxygen. The set of 16 Coulomb interactions between two water molecules gives U_{el}. Figure 9.11 shows the

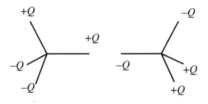

Figure 9.11 Two water molecules

minimum energy conformation, with the two 'molecules' 76 pm apart (i.e. an oxygen–oxygen distance of 276 pm).

The so-called switching function prevents the charges overlapping, and takes values between 0 and 1 depending on the distance between the oxygen atoms. The authors took $N = 216$ water molecules in a cube of side 1862 pm, which corresponds to a density of $1\,\text{g}\,\text{cm}^{-3}$. They took periodic boundary conditions in order to eliminate undesirable wall effects, and a cut-off of distance of 916.5 pm beyond which the intermolecular interactions were set to zero.

The first step in a molecular dynamics calculation is to generate a suitable initial state. In general, random structures created using density values consistent with experiment may give atoms that are very close together. This causes undesirable large repulsive forces in small localized regions of space. One approach is to start from a low-density system with the atoms or molecules placed on simple cell lattices. The density is then increased gradually.

Starting velocities then have to be assigned. One technique is to calculate a value of the speed from equipartition of energy, and then assign random directions to each particle, with the constraint that the system has zero overall momentum. The next step is to let the system progress for a sufficient number of time steps in order to eliminate any undesirable effects due to the choice of initial conditions. After that, the collection of statistics begins. The temperature is inferred from the average values of the translational and rotational kinetic energies, both of which should equal $\frac{3}{2}Nk_{\text{B}}T$.

It is possible in principle to follow a molecular dynamics simulation by displaying the atom coordinates at a series of time points. It proves more convenient to calculate certain statistical quantities. I mentioned the pair correlation function $g^{(2)}(r)$ earlier; in water the three distinct types of nuclear pairs lead to three corresponding functions $g_{\text{OO}}^{(2)}(r)$, $g_{\text{OH}}^{(2)}(r)$ and $g_{\text{HH}}^{(2)}(r)$, which give information about the fraction of time that differential volume elements separated by a distance r simultaneously contain pairs of the nuclei given by the subscripts. These functions all have a limiting value of 1 as $r \to \infty$. The $g_{\text{OO}}^{(2)}(r)$ curve is shown schematically in Figure 9.12. The horizontal axis is actually a reduced distance r/σ.

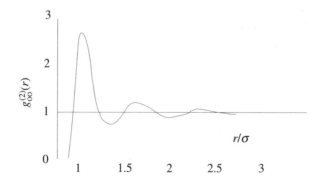

Figure 9.12   O–O pair distribution function

Two important features of the curve are as follows

1. The narrow first peak at 275 pm corresponds to 5.5 neighbouring oxygen atoms.

2. The second peak is low and broad, with a maximum at about 465 pm. The ratio of
   the second peak to the first, 465 pm/275 pm = 1.69, is close to that observed for an
   ideal ice structure (2 $\sqrt{2}/\sqrt{3} = 1.633$).

The *mean square displacement* is often calculated. This indicates the average displacement of an atom during a fixed time period.

9.7.1 Other water potentials

Over the years, a number of authors have tackled the problem of finding a suitable
pair potential for liquid water. All involve a rigid water monomer with a number of
interaction sites. The original TIPS 3-site model proposed by W. L. Jorgensen [37]
has positive charges on the hydrogens and a negative charge ($q_O = -2q_H$) on the
oxygen. The interaction potential is taken as the sum of all intermolecular Coulomb
terms together with a single Lennard-Jones 12-6 term between the oxygens

$$U_{AB} = \frac{1}{4\pi\epsilon_0} \sum_{i \text{ on A}} \sum_{j \text{ on B}} \frac{q_i q_j}{R_{ij}} + \frac{A}{R_{OO}^{12}} - \frac{B}{R_{OO}^6} \tag{9.16}$$

The parameters q_i, A and B were chosen to give reasonable energetic results for gas-
phase complexes of water and alcohols. The author subsequently reoptimized the
parameters to give TIP3P, an improved three-site model.

Four-site models have also been used; the oldest is that due to J. D. Bernal and R.
H. Fowler [38], with a more modern version referred to as TIP4P by W. L. Jorgensen
et al. [39]. In this case the negative charge is moved off the oxygen and towards the
hydrogens at a point along the bisector of the HOH angle. The pair potential is still
calculated according to the equation above, but more distances have to be evaluated.

Stillinger and Rahman's ST2 potential is widely used; this is an improved version
of the one discussed above, but adds a Lennard-Jones 12-6 term between the oxygen
atoms. In all, 17 distances have to be evaluated.

9.8 Different Types of Molecular Dynamics

When Newton's equations of motion are integrated, the energy is conserved. Since
the energy and temperature are related by the equipartition of energy principle, the
temperature should also be constant. Slow temperature drifts do occur as a result of

the numerical integration and also because of the truncation of the forces. Several methods for performing molecular dynamics at constant temperature have been proposed, of which the simplest is known as *rescaling*. If the current temperature at time t is $T(t)$, then for a system of N particles and $3N$ degrees of freedom we have

$$\langle \varepsilon_{\text{kin}} \rangle = \left\langle \sum_{i=1}^{N} \tfrac{1}{2} m_i v_i^2 \right\rangle = \tfrac{3}{2} N k_{\text{B}} T(t)$$

To adjust the temperature to exactly a reference temperature T_{ref} we simply rescale the velocities by a factor

$$\left(\frac{T_{\text{ref}}}{T(t)} \right)^{1/2}$$

Scaling the velocities at appropriate intervals can therefore control the temperature of the system. We speak about *constant-temperature molecular dynamics*.

S. Nose [40] proposed an alternative method that adds an extra degree of freedom, referred to as a heat bath, to the atomic degrees of freedom. Extra kinetic energy and potential energy terms are added to the total energy, and the molecular dynamics is carried out with this one extra degree of freedom. Energy can flow back and forth between the heat bath and the system, and an equation of motion for the extra degree of freedom is solved.

Similar considerations apply to the pressure; in order to control the pressure in a system, any change in volume must be monitored and adjusted. Methods are available, and such calculations are referred to as *constant-pressure molecular dynamics*.

9.9 Uses in Conformational Studies

Molecular dynamics generally is used to study the structure and behaviour of materials, where one is concerned with the intermolecular potential. It is of course also perfectly possible to study single molecules and it is sometimes used as an aid in conformational analysis. As noted previously, there are very many local minima on the potential energy surface and it may be that small variations in torsional angles do not facilitate searching of all regions.

The idea is to start with a molecular mechanics local minimum, which corresponds to a temperature of 0 K. We use molecular dynamics to raise the temperature, let the molecule vibrate for a while at an elevated temperature and cool it back to 0 K. Hopefully the molecule will then have crossed over any potential barriers and a reoptimization will give a new structure.

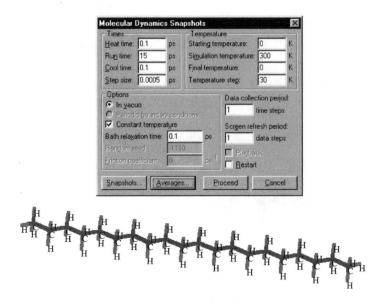

Figure 9.13 Molecular dynamics calculation on hydrocarbon

To give a simple example, consider the straight chain hydrocarbon $C_{20}H_{42}$, shown in Figure 9.13. A MM + optimization starting from the fully stretched structure gave the structure shown. There is restricted rotation about each C—C bond and in a later chapter I will show how theories of polymeric molecules focus attention on the end-to-end distance; the simplest theories predict relationships between this and the number of C—C links in such a molecule. $C_{20}H_{42}$ is barely a polymer, but the principles are the same. There are very many conformers, all similar in energy.

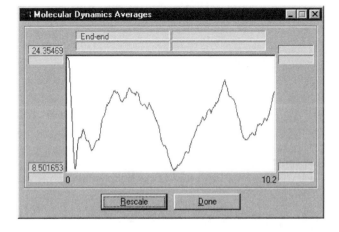

Figure 9.14 End-to-end distance

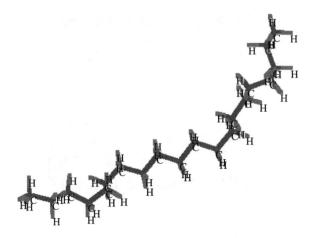

Figure 9.15   C20 hydrocarbon at end of MD experiment

I therefore set up a 15 ps molecular dynamics calculation, as shown in Figure 9.13. The options are shown, and statistics were collected in an MS/EXCEL file. The end-to-end distance varied dramatically over the experiment, from its initial value of 2435 pm through 850 pm. The average end-to-end distance was 1546 pm, as shown in Figures 9.14 and 9.15. The resulting structure was then optimized. One strategy in such calculations is to take snapshots of the structure at regular time intervals and then optimize from each snapshot.

10 Monte Carlo

10.1 Introduction

When discussing QSAR in Chapter 7, I mentioned that the molecular volume was thought to be a useful index and explained how it could be calculated within the framework of a molecular mechanics (MM) study.

Figure 10.1 shows a diatomic molecule, where the atomic radii are R_A and R_B. The problem is to find the volume of the molecule AB. The volume of atom A is $\frac{4}{3}\pi R_A^3$ and the total for AB is given by this plus the part of B that is not shared with atom A. Thus, we have to be careful not to count volume X twice in any calculation.

I suggested that the total could be calculated by surrounding atom B with a regular grid of side $2R_B$, and examining which grid points lay inside B but excluding those lying in region X, and those that lay outside atom B. A simple ratio then gives the contribution to the volume from atom B. The process can be easily extended to any polyatomic system.

Rather than take a regularly spaced cubic array of points around atom B, we can surround atom B by an (imaginary) cube of side $2R_B$ and choose points at random inside this cube. For each point, we examine whether it lies outside atom B, in which case we reject it. If the point lies inside atom B but also inside atom A (i.e. it lies in the X region), we also reject it. Otherwise we increment the number of successful points. A simple proportionality between the number of successful points and the total number tried gives the required volume.

The following BASIC program accomplishes the calculation.

```
REM   MONTE CARLO CALCULATION OF DIATOMIC A−B VOLUME
REM   ATOMIC RADII RA AND RB, SEPARATION RAB.
REM
PI = 3.14159265359#
NPOINTS = 10 000
RA = 2
RB = 1
RAB = 2!
XA = 0!
YA = 0!
```

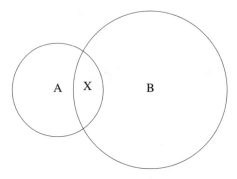

Figure 10.1   A MM diatomic

```
ZA = 0!
XB = 0!
YB = 0!
ZB = RAB
VA = 4/3 * PI * RA^3
VB = 0
FOR I = 1 TO NPOINTS
XI = XB + RB * (2 * RND - 1)
YI = YB + RB * (2 * RND - 1)
ZI = ZB + RB * (2 * RND - 1)
DISTB = SQR ((XI - XB)^2 + (YI - YB)^2 + (ZI - ZB)^2)
IF DISTB > RB THEN 2000
DISTA = SQR ((XI - XA)^2 + (YI - YA)^2 + (ZI - ZA)^2)
IF DISTA < RA THEN 2000
VB = VB + 1
2000 NEXT I
VB = 8 * RB^3 * VB/NPOINTS
VTOTAL = VA + VB
PRINT
PRINT "VOLUME AFTER"; NPOINTS; "POINTS IS"; VTOTAL
END
```

You might have to modify the statements involving RND to make it run with your version of BASIC. I have taken two atoms with radii 2 units and 1 unit, and the atoms are separated by R_{AB} that I have set to 2 units. I have calculated the volume of atom A as $\frac{4}{3}\pi R_A^3$. I then chose 10 000 points at random within a cube of side $2R_B$ surrounding atom B.

If the point lies outside atom B, then we do nothing apart from increment the counter and progress to the next point. If the point lies inside atom B, then we test whether it also lies inside atom A (in which case we have already counted it). I hope the code is transparent; this little program can be easily extended to deal with a polyatomic molecule.

This calculation is an example of the *Monte Carlo* technique (denoted MC). MC is a generic term applied to calculations that involve the use of random numbers for sampling; it became widely used towards the end of the Second World War by physicists trying to study the diffusion of neutrons in fissionable material. In fact, the MC method was first discussed by the French eighteenth-century naturalist Búffon, who discovered that if a needle of length l were dropped at random onto a set of parallel lines with spacing $d > l$, then the probability of the needle crossing a line is $2l/\pi d$.

Straightforward thermodynamic quantities such as the pressure, the internal energy and the Gibbs energy turn out to be impossible to calculate directly for a macroscopic system, simply because of the large number of particles involved. In fact, it is not even sensible to contemplate recording an initial starting point for 10^{23} particles, let alone devising methods for solving the equations of motion. Boltzmann and Gibbs recognized this problem, and invented the subject of *statistical thermodynamics*.

Suppose for the sake of argument that we have a system of N simple atomic particles (such as argon atoms) and we are interested in the total potential energy Φ. If the pair potential is $U_{ij}(R)$ and the potentials are pairwise additive, then we have

$$\Phi = \sum_{i=1}^{N-1} \sum_{j=i+1}^{N} U_{ij}(R_{ij})$$

If I denote the position vectors of the N particles measured at time t by $\mathbf{R}_A(t)$, $\mathbf{R}_B(t), \ldots, \mathbf{R}_N(t)$, then the position vectors will depend on time and the instantaneous value of Φ will depend on the values of the variables at that time. If we make enough measurements $\Phi(t_1), \Phi(t_2), \ldots, \Phi(t_n)$ at times $t_1, t_2, \ldots, t_n$, then the average of these measurements will approach the mean value of Φ

$$\langle \Phi \rangle = \frac{1}{n} \sum_{i=1}^{n} \Phi(t_i)$$

Other properties such as the self-diffusion depend on fluctuations about mean values. In statistical thermodynamics, rather than calculating a time average we consider the average over a large number of replications of the system (an ensemble). The *ergodic theorem* tells us that the two are the same.

In Chapter 8 I considered the case of a canonical ensemble, which consists of a large number of replications (or cells), each of which is identical in the sense that the number of particles, the volume and the temperature are the same. The cells are not identical at the atomic level, all that matters is N, V and T. Energy can flow from one cell to another, but the total energy of the ensemble is constant.

Under these conditions, the chance of finding a cell with energy E is proportional to the Boltzmann factor

$$\exp\left(-\frac{E}{k_B T}\right)$$

The mean value $\langle E \rangle$ can be found by averaging over the replications

$$\langle E \rangle = \frac{\sum E_i \exp\left(-\frac{E_i}{k_B T}\right)}{\sum \exp\left(-\frac{E_i}{k_B T}\right)} \tag{10.1}$$

I explained the importance of the configurational integral, which depends only on exponential terms involving the total mutual potential energy Φ

$$\int \exp\left(-\frac{\Phi(\mathbf{q})}{k_B T}\right) d\mathbf{q} \tag{10.2}$$

and showed how it could be related to the so-called excess thermodynamic functions.

I have written the configurational integral in a simplified way; if there are N particles, then it is actually a $3N$ dimensional integral over the positions of the N particles. We might want to take $N = 10\,000$ particles, but a little thought shows that it is impracticable to carry out such a multidimensional integral by the usual techniques of numerical analysis. Instead we resort to the MC method and generate a representative number of points in conformation space. For each point we calculate a value of Φ. The integral is approximated by a finite sum.

The first serious MC calculation was that of N. Metropolis *et al.* [41]. In fact, this paper marked the birth of computer simulation as a statistical mechanical technique, and it preceded the molecular dynamics studies mentioned in Chapter 9. I can't do better than repeat the words of the authors (i.e. quote the Abstract):

> *A general method, suitable for fast computing machines, for investigating such properties as equations of state for substances consisting of interacting individual molecules is described. The method consists of a modified Monte Carlo integration over configuration space. Results for the two-dimensional rigid-sphere system have been obtained on the Los Alamos MANIAC, and are presented here. These results are compared to the free volume equation of state and to a four-term virial coefficient expansion.*

They took a two-dimensional square array of 224 hard disks with finite radius. Each disk was moved in turn according to the formulae

$$X \rightarrow X + \alpha \xi_1; \qquad Y \rightarrow Y + \alpha \xi_2$$

where α is some maximum allowed distance and the ξ_i are random numbers between 0 and 1. In the case that the move would put one disk on top of another disk, the move is not allowed and we progress to the next disk. This is a primitive form of the technique known as *importance sampling* that is a key feature of modern MC simulations.

In the case of a more complicated problem such as the Lennard-Jones potential, a more sophisticated treatment of importance sampling is needed. The average potential energy can be obtained from the configuration integral

$$\langle \Phi \rangle = \frac{\int \Phi(\mathbf{q}) \exp\left(-\frac{\Phi(\mathbf{q})}{k_B T}\right) d\mathbf{q}}{\int \exp\left(-\frac{\Phi(\mathbf{q})}{k_B T}\right) d\mathbf{q}} \tag{10.3}$$

and each integral is approximated by a finite sum of M terms in MC

$$\langle \Phi \rangle \approx \frac{\sum_{i=1}^{M} \Phi_i(\mathbf{q}) \exp\left(-\frac{\Phi_i(\mathbf{q})}{k_B T}\right)}{\sum_{i=1}^{M} \exp\left(-\frac{\Phi_i(\mathbf{q})}{k_B T}\right)} \tag{10.4}$$

If we calculate the mutual potential energy $\Phi_i(\mathbf{q})$ for a random array of atoms, then the array will have a Boltzmann weight of

$$\exp\left(-\frac{\Phi(\mathbf{q})}{k_B T}\right)$$

and if some atoms are close together, then Φ_i will be large and the Boltzmann factor small.

Most of phase space corresponds to non-physical configurations with very high energies and only for a very small proportion of phase space does the Boltzmann factor have an appreciable value. The authors proposed a modified MC scheme where, instead of choosing configurations randomly and then weighting them with $\exp(-\Phi/k_B T)$, configurations were chosen with a probability distribution of $\exp(-\Phi/k_B T)$ and then weighted equally. This adds a step to the algorithm as follows.

Before making a move, we calculate the energy change of the system caused by the move. If the energy change is negative, then the move is allowed. If the energy change is positive, then we allow the move but with a probability of $\exp(-\Phi/k_B T)$; to do this, we generate a random number ξ_3 between 0 and 1, and if

$$\xi_3 < \exp\left(-\frac{\Phi}{k_B T}\right)$$

then we allow the move to take place and increment the numerator and denominator in Equation (10.1). Otherwise we leave the particle in its old position and move on to the next particle. This summarizes the *Metropolis Monte Carlo* technique.

The box is taken to be periodic, so if a move places the particle outside the box, then an image particle enters the box from the opposite side. The procedure is repeated for very many trials, and the configurational integral is estimated as a sum.

Hard disk systems were the first ones investigated, and they occupy an important place in the history of modelling. The results of the first successful 'realistic' MC simulation of an atomic fluid was published in 1957 by W. W. Wood and F. R. Parker [42]. Again, let the authors tell the story in their own words.

Values obtained by Monte Carlo calculations are reported for the compressibility factor, excess internal energy, excess constant-volume heat capacity, and the radial distribution function of Lennard-Jones (12-6) molecules at the reduced temperature $k_B T/\varepsilon^* = 2.74$, and at thirteen volumes between $v/v^* = 0.75$ and 7.5 (v is the molar volume; $v^* = 2^{-1/2} N_0 r^{*3}$; N_0 is Avogadro's number; ε^* is the depth and r^* the radius of the Lennard-Jones potential well). The results are compared with the experimental observations of Michels ($\approx 150-2000$ atm) and Bridgman ($\approx 2000-15\,000$ atm) on argon at $55\,^\circ C$, using Michels' second virial coefficient values for the potential parameters. Close agreement with Michels is found, but significant disagreement with Bridgman. The Monte Carlo calculations display the fluid–solid transition; the transition pressure and the volume and enthalpy increments are not precisely determined. The Lennard-Jones–Devonshire cell theory gives results which disagree throughout the fluid phase, but agree on the solid branch of the isotherm. Limited comparisons with the Kirkwood–Born–Green results indicate that the superposition approximation yields useful results at least up to $v/v^* = 2.5$.

They therefore studied a three-dimensional Lennard-Jones fluid (i.e. argon) and compared their results with experimental equation of state data. One of the most important technical questions discussed in the paper is the procedure used to truncate a potential in a system with periodic boundary conditions.

10.2 MC Simulation of Rigid Molecules

For rigid non-spherical molecules, it is necessary to vary their orientation as well as the position in space; usually each molecule is translated and rotated by random amounts once per cycle. The simplest approach is to rotate about each Cartesian axis in turn by an angle chosen at random but subject to a suitable maximum allowed variation. For example, if the position vector of an atom is the column vector $\mathbf{R}$, then the position vector after rotation by α about the x-axis is given by

$$\mathbf{R}' = \begin{pmatrix} 1 & 0 & 0 \\ 0 & \cos\alpha & \sin\alpha \\ 0 & -\sin\alpha & \cos\alpha \end{pmatrix} \mathbf{R} \tag{10.5}$$

The three *Euler angles* ϕ, θ and ψ are often used when discussing rotational motion. First of all we rotate by ϕ about the Cartesian z-axis. This changes the x and y-axes to x' and y' as shown in Figure 10.2. Then we rotate by θ about the new x-axis which changes the y and z-axes. Finally, we rotate by ψ about the new z-axis.

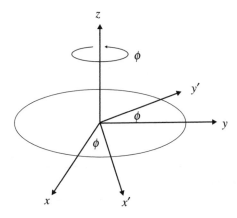

Figure 10.2 First Euler angle

The relationship between a position vector before and after the three rotations is

$$\mathbf{R}' = \begin{pmatrix} \cos\phi\cos\psi - \sin\phi\cos\theta\sin\psi & \sin\phi\cos\psi + \cos\phi\cos\theta\sin\psi & \sin\theta\sin\psi \\ -\cos\phi\sin\psi - \sin\phi\cos\theta\cos\psi & -\sin\phi\sin\psi + \cos\phi\cos\theta\cos\psi & \sin\theta\cos\psi \\ \sin\phi\sin\theta & -\cos\phi\sin\theta & \cos\theta \end{pmatrix} \mathbf{R}$$

$$(10.6)$$

There are two technical problems with the use of Euler angles. First, sampling the angles at random does not give a uniform distribution; it is necessary to sample ψ, $\cos\theta$ and ψ. Second, there are a total of six trigonometric function evaluations per rotation.

A preferred alternative makes use of *quaternions*, which are four-dimensional unit vectors. A quaternion $\mathbf{q}$ is usually written in terms of the scalar quantities q_0, q_1, q_2 and q_3 as

$$\mathbf{q} = \begin{pmatrix} q_0 & q_1 & q_2 & q_3 \end{pmatrix}^{\mathrm{T}}$$

and the components satisfy

$$q_0^2 + q_1^2 + q_2^2 + q_4^2 = 1$$

They can be related to the Euler angles as follows

$$\begin{aligned} q_0 &= \cos\tfrac{1}{2}\theta \cos\tfrac{1}{2}(\phi + \psi) \\ q_1 &= \sin\tfrac{1}{2}\theta \cos\tfrac{1}{2}(\phi - \psi) \\ q_2 &= \sin\tfrac{1}{2}\theta \sin\tfrac{1}{2}(\phi - \psi) \\ q_3 &= \cos\tfrac{1}{2}\theta \sin\tfrac{1}{2}(\phi + \psi) \end{aligned}$$

$$(10.7)$$

The rotation matrix can be written as

$$\mathbf{R'} = \begin{pmatrix} q_0^2 + q_1^2 - q_2^2 - q_3^2 & 2(q_1 q_2 + q_0 q_3) & 2(q_1 q_3 - q_0 q_2) \\ 2(q_1 q_2 - q_0 q_3) & q_0^2 - q_1^2 + q_2^2 - q_3^2 & 2(q_2 q_3 + q_0 q_1) \\ 2(q_1 q_3 + q_0 q_2) & 2(q_2 q_3 - q_0 q_1) & q_0^2 - q_1^2 - q_2^2 + q_3^2 \end{pmatrix} \mathbf{R} \quad (10.8)$$

and all that is necessary is to generate four suitable random numbers.

10.3 Flexible Molecules

Monte Carlo simulations of flexible molecules are difficult to perform unless the system is small or some of the internal degrees of freedom are kept fixed. The simplest way to generate a new configuration is to perform random changes to the Cartesian coordinates of individual atoms in the molecule but it is a common experience that very small changes are needed in order to produce an acceptable Boltzmann factor (in the Metropolis MC sense).

11 Introduction to Quantum Modelling

By the early days of the twentieth century, scientists had successfully developed three of the four great cornerstones of physics; Sir Isaac Newton's mechanics, James Clerk Maxwell's electromagnetic theory and Albert Einstein's theory of special relativity. They had a picture of the physical world where matter was made from point particles and radiation consisted of electromagnetic waves, and this picture seemed to explain all known physical phenomena with just a few untidy exceptions.

These untidy exceptions comprised phenomena such as *the theory of black body radiation, the photoelectric effect, Compton scattering, atomic structure and spectra* and a few other apparently unrelated experimental findings. I don't have space in this book to go into the historical detail; I will simply say that a thorough study of such phenomena led to the fourth cornerstone, quantum theory. Every textbook that has to deal with quantum theory has its own particular treatment and starting point, and this one is no exception; I am going to assume that you have heard of Erwin Schrödinger and his famous equations, and start the discussion at that point. Perhaps I can reassure you by saying that most professional scientists perceive quantum theory as a hard subject (along with electromagnetism). Even Schrödinger didn't fully understand the physical meaning that we now attach to his wavefunctions when he first wrote down his famous equation and solved it for the hydrogen atom.

11.1 The Schrödinger Equation

Consider then the simple case of a particle of mass m constrained to the x-axis by a potential $U(x, t)$ that depends on x and time t. I have allowed the potential to be time dependent in order to cover the possibility of (for example) an electron being influenced by an external electromagnetic wave. Schrödinger's time-dependent equation states that

$$\left(-\frac{h^2}{8\pi^2 m}\frac{\partial^2}{\partial x^2} + U(x,t) \right) \Psi(x,t) = \mathrm{j}\frac{h}{2\pi}\frac{\partial \Psi(x,t)}{\partial t} \tag{11.1}$$

where j is the square root of -1 ($j^2 = -1$). $\Psi(x, t)$ is the *wavefunction* and in general it is a complex valued quantity. The square of the modulus of Ψ can be found from

$$|\Psi(x,t)|^2 = \Psi^*(x,t)\Psi(x,t)$$

where Ψ^* is the complex conjugate of Ψ. The fact that complex wavefunctions are allowed and that the Schrödinger equation mixes up real and imaginary quantities caused Schrödinger and his contemporaries serious difficulties in trying to give a physical meaning to Ψ. According to Max Born, the physical interpretation attaching to the wavefunction $\Psi(x, t)$ is that $\Psi^*(x, t)\Psi(x, t)dx$ gives the probability of finding the particle between x and $x + dx$ at time t. It is therefore the square of the modulus of the wavefunction rather than the wavefunction itself that is related to a physical measurement.

Since $\Psi^*(x, t)\Psi(x, t)dx$ is the probability of finding the particle (at time t) between x and $x + dx$, it follows that the probability of finding the particle between any two values a and b is found by summing the probabilities

$$P = \int_a^b \Psi^*(x,t)\Psi(x,t)dx$$

In three dimensions the equation is more complicated because both the potential U and the wavefunction Ψ might depend on the three dimensions x, y and z (written as the vector **r**) as well as time.

$$\left(-\frac{h^2}{8\pi^2 m}\left(\frac{\partial^2}{\partial x^2} + \frac{\partial^2}{\partial y^2} + \frac{\partial^2}{\partial z^2}\right) + U(\mathbf{r},t)\right)\Psi(\mathbf{r},t) = j\frac{h}{2\pi}\frac{\partial\Psi(\mathbf{r},t)}{\partial t} \qquad (11.2)$$

You are probably aware that the operator in the outer left-hand bracket is the *Hamiltonian operator* and that we normally write the equation

$$\hat{H}\Psi(\mathbf{r},t) = j\frac{h}{2\pi}\frac{\partial\Psi(\mathbf{r},t)}{\partial t} \qquad (11.3)$$

The Born interpretation is then extended as follows

$$\Psi^*(\mathbf{r},t)\Psi(\mathbf{r},t)dxdydz \qquad (11.4)$$

gives the probability that the particle will be found at time t in the volume element $dxdydz$. Such volume elements are often written $d\tau$.

I have not made any mention of time-dependent potentials in this book, but they certainly exist. You will have met examples if you have studied electromagnetism where the *retarded potentials* account for the generation of electromagnetic radiation. For the applications we will meet in this book, the potentials are time independent; in such cases the wavefunction still depends on time, but in a straightforward way as I will now demonstrate.

11.2 The Time-Independent Schrödinger Equation

I am going to use a standard technique from the theory of differential equations called *separation of variables* to simplify the time-dependent equation; I want to show you that when the potential does not depend on time, we can 'factor out' the time dependence from the Schrödinger time-dependent equation and just concentrate on the spatial problem. I will therefore investigate the possibility that we can write $\Psi(\mathbf{r}, t)$ as a product of functions of the spatial ($\mathbf{r}$) and time (t) variables, given that the potential only depends on $\mathbf{r}$. We write

$$\Psi(\mathbf{r}, t) = \psi(\mathbf{r})T(t) \tag{11.5}$$

and substitute into the time-dependent Schrödinger equation to get

$$\left(-\frac{h^2}{8\pi^2 m}\left(\frac{\partial^2}{\partial x^2} + \frac{\partial^2}{\partial y^2} + \frac{\partial^2}{\partial z^2}\right) + U(\mathbf{r}, t)\right)\psi(\mathbf{r})T(t) = j\frac{h}{2\pi}\frac{\partial\psi(\mathbf{r})T(t)}{\partial t}$$

Note that the differential operators are partial ones and they only operate on functions that contain them. So we rearrange the above equation

$$T(t)\left(-\frac{h^2}{8\pi^2 m}\left(\frac{\partial^2}{\partial x^2} + \frac{\partial^2}{\partial y^2} + \frac{\partial^2}{\partial z^2}\right) + U(\mathbf{r}, t)\right)\psi(\mathbf{r}) = j\psi(\mathbf{r})\frac{h}{2\pi}\frac{dT(t)}{dt}$$

divide by $\Psi(\mathbf{r}, t)$ and simplify to get

$$\frac{1}{\psi(\mathbf{r})}\left(-\frac{h^2}{8\pi^2 m}\left(\frac{\partial^2}{\partial x^2} + \frac{\partial^2}{\partial y^2} + \frac{\partial^2}{\partial z^2}\right) + U(\mathbf{r}, t)\right)\psi(\mathbf{r}) = j\frac{1}{T(t)}\frac{h}{2\pi}\frac{dT(t)}{dt}$$

The left-hand side of the equation depends on the spatial variable $\mathbf{r}$ and the time variable t. Suppose now that the potential depends only on $\mathbf{r}$ and not on t. That is to say

$$\frac{1}{\psi(\mathbf{r})}\left(-\frac{h^2}{8\pi^2 m}\left(\frac{\partial^2}{\partial x^2} + \frac{\partial^2}{\partial y^2} + \frac{\partial^2}{\partial z^2}\right) + U(\mathbf{r})\right)\psi(\mathbf{r}) = j\frac{1}{T(t)}\frac{h}{2\pi}\frac{dT(t)}{dt} \tag{11.6}$$

Under these circumstances, the left-hand side of the equation depends only on $\mathbf{r}$ and the right-hand side only on t. The spatial and the time variables are completely independent and so the right-hand side and the left-hand side must both equal a constant. In the theory of differential equations, this constant is called the *separation constant*. In this case it is equal to the total energy so I will write it as ε

$$\frac{1}{\psi(\mathbf{r})}\left(-\frac{h^2}{8\pi^2 m}\left(\frac{\partial^2}{\partial x^2}+\frac{\partial^2}{\partial y^2}+\frac{\partial^2}{\partial z^2}\right)+U(\mathbf{r})\right)\psi(\mathbf{r})=\varepsilon$$

$$j\frac{1}{T(t)}\frac{h}{2\pi}\frac{dT(t)}{dt}=\varepsilon$$

The second equation can be solved to give

$$T(t)=A\exp\left(-\frac{2\pi j\varepsilon t}{h}\right) \tag{11.7}$$

and so the full solution to the time-dependent equation in the special case that the potential is time independent can be written

$$\Psi(\mathbf{r},t)=\psi(\mathbf{r})\exp\left(-\frac{2\pi j\varepsilon t}{h}\right) \tag{11.8}$$

Probabilities then become time independent since

$$\Psi^*(\mathbf{r},t)\Psi(\mathbf{r},t)=\psi^*(\mathbf{r})\exp\left(+\frac{2\pi j\varepsilon t}{h}\right)\psi(\mathbf{r})\exp\left(-\frac{2\pi j\varepsilon t}{h}\right)$$

$$=\psi^*(\mathbf{r})\psi(\mathbf{r})$$

11.3 Particles in Potential Wells

I have mentioned a number of model potentials in earlier chapters when dealing with Molecular Dynamics and Monte Carlo. A model potential of historical interest in quantum mechanical studies is the 'infinite well' (see Figure 11.1).

11.3.1 The one-dimensional infinite well

A particle of mass m is constrained to a region $0 \leq x \leq L$ by a potential that is infinite everywhere apart from this region, where it is equal to a constant U_0. This might be a very crude model for the conjugated electrons in a polyene, or for the conduction electrons in a metallic conductor. (In fact, there is a technical problem with this particular model potential caused by the infinities, as I will explain shortly.)

To solve the problem we examine the Schrödinger equation in each of the three regions and then match up the wavefunction at the boundaries. The wavefunction

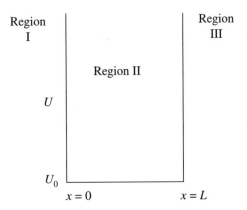

Figure 11.1 One-dimensional infinite well

must be continuous at any boundary. In regions I and III the potential is infinite and so the wavefunction must be zero in order to keep the Schrödinger equation finite. In region II the potential is a finite constant U_0 and we have

$$\left(-\frac{h^2}{8\pi^2 m}\frac{d^2}{dx^2} + U_0\right)\psi(x) = \varepsilon\psi(x)$$

which can be rewritten

$$\frac{d^2\psi}{dx^2} + \frac{8\pi^2 m(\varepsilon - U_0)}{h^2}\psi = 0 \tag{11.9}$$

The standard solution to this second-order differential equation is

$$\psi = A\sin\sqrt{\frac{8\pi^2 m(\varepsilon - U_0)}{h^2}}x + B\cos\sqrt{\frac{8\pi^2 m(\varepsilon - U_0)}{h^2}}x$$

where A and B are constants of integration. We know that $\psi = 0$ in regions I and III so the wavefunction must also be zero at $x = 0$ and at $x = L$ in order to be continuous at these points. Substituting $\psi = 0$ at $x = 0$ shows that $B = 0$. If we substitute $\psi = 0$ at $x = L$ we get

$$0 = A\sin\sqrt{\frac{8\pi^2 m(\varepsilon - U_0)}{h^2}}L$$

This means that either $A = 0$ or the argument of sin must be an integral multiple of π (written n). A cannot be 0, for otherwise the wavefunction would be 0 everywhere and so, on rearranging

$$\varepsilon_n = U_0 + \frac{n^2 h^2}{8mL^2}$$

$$\psi_n = A\sin\left(\frac{n\pi}{L}x\right) \tag{11.10}$$

The energy is quantized, and so I have added a subscript to the ε for emphasis. It generally happens that the imposition of boundary conditions leads to energy quantization. It is often necessary when tackling such problems to match up both the wavefunction and its gradient at the boundaries, but in this particular case the solution appeared just by considering the wavefunction. (In fact, the infinities in this particular potential mean that the first derivative is not continuous at the boundaries; 'real' potentials don't contain such sudden jumps to infinity.) n is the *quantum number* and at first sight the allowed values are $n = \ldots -2, -1, 0, 1, 2, \ldots$. The value $n = 0$ is not allowed for the same reason that A cannot be 0. We can deduce the constant of integration A by recourse to the Born interpretation of quantum mechanics; the probability of finding the particle between x and $x + dx$ is

$$\psi_n^2(x)\, dx$$

Probabilities have to sum to 1 and so we require

$$\int_0^L \psi_n^2(x)\, dx = 1$$

from which we deduce

$$\psi_n = \pm\sqrt{\frac{2}{L}} \sin\left(\frac{n\pi x}{L}\right)$$

We note also that the values of $\pm n$ give wavefunctions that differ only by multiplication by -1, since

$$\sin\left(-\frac{n\pi x}{L}\right) = -\sin\left(\frac{n\pi x}{L}\right)$$

According to the Born interpretation, both give the same probability density and so we only need one set. It is conventional to take the quantum numbers as positive and the value of A as the positive square root so the results can be collected together as

$$\boxed{\begin{aligned} \varepsilon_n &= U_0 + \frac{n^2 h^2}{8mL^2} \\ \psi_n &= \sqrt{\frac{2}{L}} \sin\left(\frac{n\pi x}{L}\right) \\ n &= 1, 2, 3 \ldots \end{aligned}} \qquad (11.11)$$

Plots of the squares of the wavefunctions for $n = 1$ and $n = 3$ are shown in Figure 11.2. For a particle with $n = 1$ (the full curve) there is zero probability of finding the particle at the end points and a maximum probability of finding the particle at the centre point. For $n = 3$ there are three equal maxima of probability.

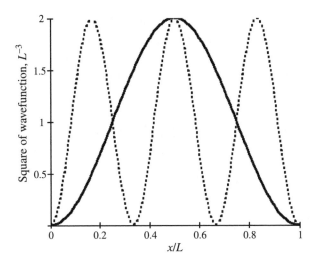

Figure 11.2   Squares of wavefunction for $n = 1$ and $n = 3$

11.4 The Correspondence Principle

In 1923, Neils Bohr formulated a principle that he had already used in his treatment of the hydrogen atom, called the *Correspondence Principle*; this states that quantum and classical predictions have to agree with each other in the limit, as the quantum number(s) get large. In other words, the results of classical physics may be considered to be the limiting cases of quantum mechanical calculations with large quantum

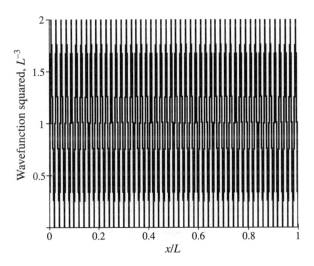

Figure 11.3   Square of wavefunction $n = 30$

numbers. A plot of ψ_{30}^2 shows the operation of the Correspondence Principle; the probabilities appear much more evenly spread out over the region (see Figure 11.3). In the case of a 'classical' particle in such a well, we would expect an equal probability of finding the particle at any point along the line.

11.5 The Two-Dimensional Infinite Well

In the two-dimensional version of this problem, we consider a single particle of mass m constrained to a region of the xy plane by a potential that is infinite everywhere except for a region where $0 \leq x \leq \infty$ and $0 \leq y \leq \infty$, where it is equal to a constant U_0 as shown in Figure 11.4. I have taken the region to be square, but this is not necessary.

The (time-independent) Schrödinger equation is

$$\frac{\partial^2 \psi}{\partial x^2} + \frac{\partial^2 \psi}{\partial y^2} + \frac{8\pi^2 m(\varepsilon - U_0)}{h^2}\psi = 0 \tag{11.12}$$

and the wavefunction now depends on x and y. Straightforward application of the 'separation of variables' technique discussed above gives the following wavefunctions and energies

$$\boxed{\begin{aligned} \varepsilon_{n,k} &= U_0 + \frac{n^2 h^2}{8mL^2} + \frac{k^2 h^2}{8mL^2} \\ \psi_{n,k} &= \frac{2}{L}\sin\left(\frac{n\pi x}{L}\right)\sin\left(\frac{k\pi y}{L}\right) \\ n,k &= 1,2,3\ldots \end{aligned}} \tag{11.13}$$

There are now two quantum numbers that I have decided to call n and k. Each quantum number can equal a positive integer. The wavefunction is a product of sine

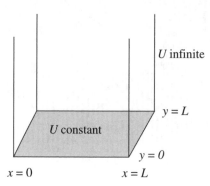

Figure 11.4 Two-dimensional infinite well

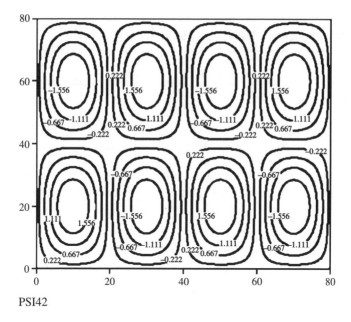

PSI42

Figure 11.5 Wavefunction, contour view

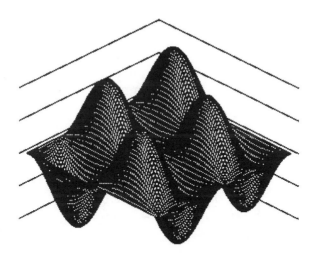

Figure 11.6 Two-dimensional box, surface plot

functions, and we now have to think about visualization. Two popular ways to visualize such objects are as contour diagrams or as surface plots. The two representative figures, Figures 11.5 and 11.6, refer to the solution $n = 4$ and $k = 2$. The horizontal axes are x and y, the vertical axis ψ.

11.6 The Three-Dimensional Infinite Well

It should be obvious by now how we proceed to three dimensions. The relevant Schrödinger equation is

$$\frac{\partial^2 \psi}{\partial x^2} + \frac{\partial^2 \psi}{\partial y^2} + \frac{\partial^2 \psi}{\partial z^2} + \frac{8\pi^2 m(\varepsilon - U_0)}{h^2}\psi = 0 \qquad (11.14)$$

which can be solved by the standard techniques discussed above. Application of the boundary conditions gives three quantum numbers that I will write n, k and l. The solutions are

$$\boxed{\begin{aligned} \varepsilon_{n,k,l} &= U_0 + \frac{n^2 h^2}{8mL^2} + \frac{k^2 h^2}{8mL^2} + \frac{l^2 h^2}{8mL^2} \\ \psi_{n,k,l} &= \left(\frac{2}{L}\right)^{3/2} \sin\left(\frac{n\pi x}{L}\right)\sin\left(\frac{k\pi y}{L}\right)\sin\left(\frac{l\pi y}{L}\right) \\ n, k, l &= 1, 2, 3 \ldots \end{aligned}} \qquad (11.15)$$

Figure 11.7 summarizes the one-dimensional and the cubic three-dimensional infinite well problems. The energies are plotted in units of $h^2/8mL^2$. The one-dimensional

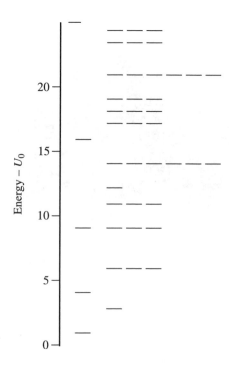

Figure 11.7 One-dimensional and cubic three-dimensional wells

energy levels are shown as the left-hand 'column' of the figure. They simply increase as n^2.

The three-dimensional problem shows a much more interesting pattern. First of all, many of the energy levels are *degenerate* (that is to say, there are many *quantum states* with the same energy). Some of the degeneracies are *natural*, for example $\psi_{1,2,3}$, $\psi_{1,3,2}$, $\psi_{2,1,3}$, $\psi_{2,3,1}$, $\psi_{3,1,2}$ and $\psi_{3,2,1}$ all have energy $14h^2/(8mL^2)$. There are also many *accidental* degeneracies; for example there are nine degenerate quantum states with energy $41h^2/(8mL^2)$ because $1^2 + 2^2 + 6^2$ and $4^2 + 4^2 + 3^2$ are both equal to 41. Secondly, as the energy increases, the energy levels crowd together and the number of quantum states of near equal energy is seen to increase. This crowding together does not happen in the one-dimensional case.

The square two-dimensional case lies somewhere in between, with fewer degeneracies. The degeneracies do not occur if we take arbitrary sides for the two- and three-dimensional regions. Nevertheless, their quantum states still crowd together as the quantum numbers increase.

11.7 Two Non-Interacting Particles

Consider now the case of two non-interacting particles A and B in a one-dimensional well. Writing x_A and x_B for their position coordinates along the x-axis, the time-independent Schrödinger equation is (taking the constant of integration to be zero, for simplicity)

$$-\frac{h^2}{8\pi^2 m}\left(\frac{\partial^2}{\partial x_A^2} + \frac{\partial^2}{\partial x_B^2}\right)\Psi(x_A, x_B) = E\Psi(x_A, x_B) \qquad (11.16)$$

I have written capital Ψ and E to emphasize that the equation refers to a system of two particles, and I will use this convention from now on. Equation (11.16) looks like a candidate for separation of variables, so we investigate the conditions under which the total wavefunction can be written as a product of one-particle wavefunctions

$$\Psi(x_A, x_B) = \psi_A(x_A)\psi_B(x_B)$$

Substitution and rearrangement gives

$$-\frac{h^2}{8\pi^2 m}\frac{1}{\psi_A}\frac{\mathrm{d}^2\psi_A}{\mathrm{d}x_A^2} - \frac{h^2}{8\pi^2 m}\frac{1}{\psi_B}\frac{\mathrm{d}^2\psi_B}{\mathrm{d}x_B^2} = E$$

Applying the now familiar argument, we equate each term on the left-hand side to a (different) separation constant ε_A and ε_B, which have to sum to E. The total wavefunction is therefore a product of wavefunctions for the individual particles, and the total energy is the sum of the one-particle energies. There are two quantum numbers that I will call n_A and n_B and we find

$$\Psi_{n_A,n_B} = \frac{2}{L} \sin\left(\frac{n_A \pi x_A}{L}\right) \sin\left(\frac{n_B \pi x_B}{L}\right)$$

$$E_{n_A,n_B} = (n_A^2 + n_B^2)\frac{h^2}{8mL^2} \tag{11.17}$$

$$n_A, n_B = 1, 2, 3 \ldots$$

We often summarize the results of such calculations, where the solution turns out to be a product of individual one-particle solutions, by a simple extension of Figure 11.7.

Figure 11.8 shows one possible solution, where particle A has quantum number 2 and particle B has quantum number 1. There is more to this simple problem than meets the eye; you might be wondering how it is possible to 'label' one particle A and one B. If A and B had been two motorcars travelling along the M6 motorway, then the labelling could be done by the vehicle number plate. But how can we possibly label electrons, or any other atomic particles for that matter? This isn't an easy question; I will return to it later in the book.

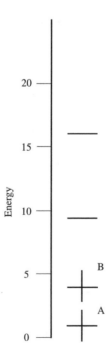

Figure 11.8   Two particles in a one-dimensional box

11.8 The Finite Well

The infinite well occupies an important place in the history of molecular quantum mechanics, and we will meet it again in Chapter 12. We discussed the finite well in Chapters 9 and 10, but from the viewpoint of classical mechanics. The finite well potential gives a rather crude model for a diatomic molecule, and it is shown again in Figure 11.9. It keeps an infinite potential for the close approach of the two nuclei, but allows for a 'valence region' where the potential is constant.

The potential is infinite for $-\infty \leq x \leq 0$, zero (or a constant) for the valence region where $0 \leq x \leq L$ and a constant D for $L \leq x \leq \infty$. As usual, we divide up the x-axis into three regions and then match up ψ at the two boundaries. It is also necessary to match up the first derivative of ψ at the $x = L$ boundary.

In region I, the potential is infinite and so $\psi = 0$. This means that $\psi = 0$ at the boundary when $x = 0$. In region II, the potential is constant (zero) and so we have

$$-\frac{h^2}{8\pi^2 m}\frac{d^2\psi(x)}{dx^2} = \varepsilon\psi(x)$$

This can be solved to give

$$\psi = A \sin\sqrt{\frac{8\pi^2 m\varepsilon}{h^2}}x + B\cos\sqrt{\frac{8\pi^2 m\varepsilon}{h^2}}x$$

where A and B are constants of integration. We can eliminate the B term straight away by the argument of Section 11.3. In region III we have

$$\left(-\frac{h^2}{8\pi^2 m}\frac{d^2}{dx^2} + D\right)\psi(x) = \varepsilon\psi(x)$$

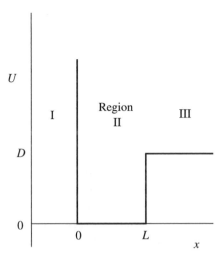

Figure 11.9 The finite well

For now, let us interest ourselves only in those solutions where the energy is less than or equal to D. Such solutions are called *bound states*. We will shortly see that solutions exist where the energy is greater than D and these are called *unbound states*. I can rewrite the Schrödinger equation for region III as

$$\frac{d^2\psi}{dx^2} - \frac{h^2}{8\pi^2 m}(D - \varepsilon)\psi = 0$$

The standard solution to this second-order differential equation is

$$\psi = E\exp\left(\sqrt{\frac{8\pi^2 m(D-\varepsilon)}{h^2}}x\right) + F\exp\left(-\sqrt{\frac{8\pi^2 m(D-\varepsilon)}{h^2}}x\right)$$

where E and F are constants of integration.

We now have to apply boundary conditions for the bound state solutions where ε is less than D. If we want the wavefunction to be zero at infinity, then E must be zero. The wavefunction and its first derivative must also be equal at the boundary $x = L$. Thus we have

$$A\sin\left(\sqrt{\frac{8\pi^2 m\varepsilon}{h^2}}L\right) = F\exp\left(-\sqrt{\frac{8\pi^2 m(D-\varepsilon)}{h^2}}L\right)$$

$$\sqrt{\frac{8\pi^2 m\varepsilon}{h^2}}A\cos\left(\sqrt{\frac{8\pi^2 m\varepsilon}{h^2}}L\right) = -\sqrt{\frac{8\pi^2 m(D-\varepsilon)}{h^2}}F\exp\left(-\sqrt{\frac{8\pi^2 m(D-\varepsilon)}{h^2}}L\right)$$

Dividing one equation by the other and rearranging gives a formula involving the energy ε

$$\tan\left(\sqrt{\frac{8\pi^2 m\varepsilon}{h^2}}L\right) = -\sqrt{\frac{\varepsilon}{D-\varepsilon}} \tag{11.18}$$

The energy is quantized and the precise values of ε depend on D and L.

11.9 Unbound States

I mentioned in the discussion that we would be concerned only with solutions where the energy ε was less than the well depth D. Such states are called *bound states*, which means that the particle is somehow localized to the potential well in the sense that the probability of finding the particle between x and $x + dx$ should decrease the farther we move away from the potential well (you probably noticed that I *assumed* the wavefunction should fall off exponentially, and so I deliberately excluded the possibility that the probability could get larger with distance).

Depending on L and D, a one-dimensional finite well can have one, many or an infinite number of bound state solutions. Finite wells can also have *unbound* solutions that are perfectly acceptable in so far as both ψ and its first derivative are finite and continuous for all values of x, but they do not necessarily have the property that they tend to zero for large values of x.

Taken individually these solutions cannot represent the state of a particle because they do not vanish for large values of x. This causes a real problem with the Born interpretation because they cannot be made to satisfy the normalization condition. The solution to this problem is to combine waves into *wave packets*; this topic is discussed in more advanced textbooks.

11.10 Free Particles

The simplest example of unbound states are afforded by studying free particles; consider again the one-dimensional problem of a particle (of mass m) and a potential that is a constant which I will take to be zero. The time-independent Schrödinger equation is

$$\left(-\frac{h^2}{8\pi^2 m}\frac{\mathrm{d}^2}{\mathrm{d}x^2} \right)\psi(x) = \varepsilon\psi(x)$$

which has solutions that I can write

$$\psi(x) = A\exp(\mathrm{j}kx), \quad \text{where } k = \pm\frac{\sqrt{8\pi^2 m\varepsilon}}{h}$$

Here k can take any value; the energy ε is positive, but now it has a continuum of possible values. The time-dependent solution is therefore

$$\Psi(x,t) = A\exp(\mathrm{j}kx)\exp\left(-\frac{2\pi\mathrm{j}\varepsilon t}{h} \right)$$

which can be written

$$\Psi(x,t) = A\exp(\mathrm{j}(kx - \omega t)),$$

where $\omega = 2\pi\varepsilon/h$. This is a classic one-dimensional wave equation, and it corresponds to a de Broglie wave. In three dimensions we write

$$\Psi(\mathbf{r},t) = A\exp(\mathrm{j}(\mathbf{k}\cdot\mathbf{r} - \omega t)) \tag{11.19}$$

where $\mathbf{k}$ is the wavevector. You will have encountered the wavevector if you have studied electromagnetism beyond elementary level.

We note then that for each value of the energy, there are two possible wave-functions

$$
\psi(x) = A \exp\left(+\mathrm{j}\frac{\sqrt{8\pi^2 m\varepsilon}}{h}x \right)
$$
$$
\psi(x) = A \exp\left(-\mathrm{j}\frac{\sqrt{8\pi^2 m\varepsilon}}{h}x \right)
$$

(11.20)

that is, there is a double degeneracy. A feature of free particle wavefunctions is that they cannot be made to satisfy the Born normalization condition because there is no value of A that will satisfy the 'equation'

$$
|A|^2 \int_{-\infty}^{+\infty} |\psi(x)|^2 \mathrm{d}x = 1
$$

This is not a trivial problem, and workers in the field often interpret solutions of the form of Equation (11.20) as an *infinite beam* of non-interacting particles moving to the right or left (according to the $+$ or $-$ sign) with a constant number of particles per length.

11.11 Vibrational Motion

I gave you the classical treatment of a diatomic molecule in Chapters 4 and 5, where I assumed that the relevant potential was Hooke's Law. I demonstrated that the vibrations of a diatomic molecule (atomic masses m_1 and m_2) were exactly equivalent to the vibrations of a single atom of reduced mass μ where

$$
\mu = \frac{m_1 m_2}{m_1 + m_2}
$$

If I write the 'spring' extension $R - R_e$ as x, then the vibrational Schrödinger equation can be written

$$
\frac{\mathrm{d}^2\psi}{\mathrm{d}x^2} + \frac{8\pi^2\mu}{h^2}\left(\varepsilon - \tfrac{1}{2}k_s x^2 \right) = 0
$$

(11.21)

where k_s is the force constant. The energy ε and wavefunction ψ now describe the nuclear vibrations. This is a more difficult differential equation than the ones discussed earlier in this chapter, but fortunately it is well known in many areas of engineering, physics and applied mathematics. A detailed solution of the vibrational Schrödinger equation is given in all advanced texts, and so to cut a long story short we interest ourselves in bound states and thus impose a condition $\psi \to 0$ as $x \to \pm\infty$.

This gives rise to energy quantization described by a single quantum number that is written v. The general formula for the vibrational energies is

$$\varepsilon_v = \frac{h}{2\pi}\sqrt{\frac{k_s}{\mu}}(v + \tfrac{1}{2}) \quad v = 0, 1, 2 \ldots \tag{11.22}$$

The harmonic oscillator is not allowed to have zero vibrational energy. The smallest allowed energy is when $v = 0$, and this is referred to as the *zero point energy*. This finding is in agreement with Heisenberg's Uncertainty Principle, for if the oscillating particle had zero energy it would also have zero momentum and would be located exactly at the position of the minimum in the potential energy curve.

The normalized vibrational wavefunctions are given by the general expression

$$\psi_v(\xi) = \left(\frac{1}{2^v v!}\sqrt{\frac{\beta}{\pi}}\right)^{1/2} H_v(\xi)\exp\left(-\frac{\xi^2}{2}\right) \tag{11.23}$$

where

$$\beta = \frac{2\pi\sqrt{k_s\mu}}{h} \quad \text{and} \quad \xi = \sqrt{\beta}x$$

The polynomials H_v are called the *Hermite polynomials*; they are solutions of the second-order differential equation

$$\frac{d^2 H}{d\xi^2} - 2\xi\frac{dH}{d\xi} + 2vH = 0 \tag{11.24}$$

and are most easily found from the recursion formula

$$H_{v+1}(\xi) = 2\xi H_v(\xi) - 2vH_{v-1}(\xi)$$

The first few are shown in Table 11.1. Squares of the vibrational wavefunctions for $v = 0$ and $v = 5$ (with numerical values appropriate to $^{14}N_2$) are shown as Figures 11.10 and 11.11. The classical model predicts that the maximum of probability is when the

Table 11.1 The first few Hermite polynomials, $H_v(\xi)$

v	$H_v(\xi)$
0	1
1	2ξ
2	$4\xi^2 - 2$
3	$8\xi^3 - 12\xi$
4	$16\xi^4 - 48\xi^2 + 12$
5	$32\xi^5 - 160\xi^3 + 120\xi$

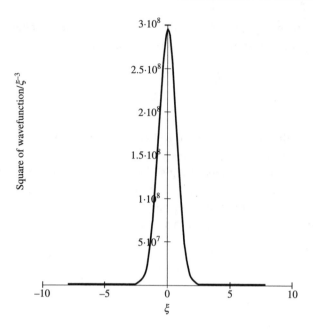

Figure 11.10   Square of vibrational wavefunction $v = 0$ for dinitrogen

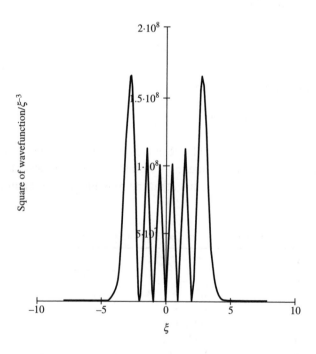

Figure 11.11   Square of vibrational wavefunction $v = 5$ for dinitrogen

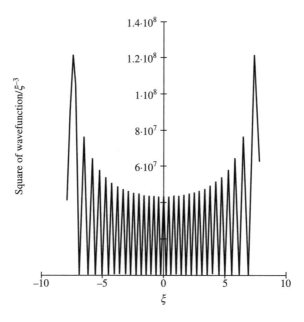

Figure 11.12 Square of vibrational wavefunction $v = 30$ for dinitrogen

particle is at the ends of its vibrations (when it is stationary). The quantum mechanical model seems quite different, since the $v = 0$ solution has a probability maximum half way along the bond. The operation of the correspondence principle should be clear once we consider higher quantum numbers. Even by $v = 5$ it is becoming clear that the maxima are moving toward the ends of the vibration. Examination of $v = 30$, as shown in Figure 11.12, confirms the trend, which in any case can be deduced analytically.

12 Quantum Gases

In Chapter 11 I discussed the quantum mechanical description of a particle in a selection of potential wells, including the three-dimensional infinite cubic box of side L. For a single particle of mass m, I showed that the energy is quantized and that each quantum state is described by three quantum numbers, n, k and l, as follows

$$\varepsilon_{n,k,l} = U_0 + \frac{n^2 h^2}{8mL^2} + \frac{k^2 h^2}{8mL^2} + \frac{l^2 h^2}{8mL^2}$$

$$\psi_{n,k,l} = \left(\frac{2}{L}\right)^{3/2} \sin\left(\frac{n\pi x}{L}\right) \sin\left(\frac{k\pi y}{L}\right) \sin\left(\frac{l\pi y}{L}\right) \tag{12.1}$$

$$n,k,l = 1,2,3\ldots$$

We often set the constant of integration U_0 to zero, without any loss of generality. We will make this simplification for the rest of the chapter.

Consider now an ideal gas comprising N such particles, constrained to the cubic three-dimensional infinite potential well. Ideal gas particles don't interact with each and the total wavefunction is therefore a product of one-particle wavefunctions (orbitals), as discussed in Chapter 11. Also, the total energy is the sum of the one-particle energies.

Suppose we arrange that the system is thermally isolated and then make a measurement on the system to find how the energy is distributed amongst the particles at (say) 300 K. We will find some particles with low quantum numbers and some with high. The kinetic energy of particle i is given by

$$\varepsilon_i = (n_i^2 + k_i^2 + l_i^2) \frac{h^2}{8mL^2}$$

and the total kinetic energy is therefore

$$E_{\text{kin}} = \frac{h^2}{8mL^2} \sum_{i=1}^{N} (n_i^2 + k_i^2 + l_i^2) \tag{12.2}$$

This can be identified with the thermodynamic internal energy U_{th}

$$U_{\text{th}} = \frac{h^2}{8mL^2} \sum_{i=1}^{N} (n_i^2 + k_i^2 + l_i^2) \tag{12.3}$$

We do not yet know how the kinetic energy will be divided out amongst the particles, all we know is that the summation will be a complicated function of the quantum numbers. For the sake of argument, call the summation A. Since the box is a cube of side L, the volume V of the box is L^3 and so we can write

$$U_{th} = \frac{h^2}{8m} A V^{-2/3} \tag{12.4}$$

We know from elementary chemical thermodynamics that for a thermally isolated system

$$p = -\left(\frac{\partial U_{th}}{\partial V}\right)_N$$

and so the particle in a box model gives the following relationship between pressure and volume

$$p = \frac{2}{3}\frac{h^2}{8m} A V^{-5/3} \tag{12.5}$$

which can be rearranged together with the expression for U_{th} to give

$$pV = \tfrac{2}{3} U_{th} \tag{12.6}$$

That is to say, the pressure times the volume is a constant at a given temperature. Not only that, but if the internal energy of the sample (of amount n) is given by the equipartition of energy expression $\tfrac{3}{2} nRT$, then we recover the ideal gas law.

12.1 Sharing Out the Energy

Statistical thermodynamics teaches that the properties of a system in equilibrium can be determined in principle by counting the number of states accessible to the system under different conditions of temperature and pressure. There is as ever some interesting small print to this procedure, which I will illustrate by considering three collections of different particles in the same cubic infinite potential well. First of all, I look at distinguishable particles such as the ones discussed above that have mass and quantized energy. I raised the question in Chapter 11 as to whether or not we could truly distinguish one particle from another, and there is a further problem in that some particles have an intrinsic property called *spin* just as they have a mass and a charge.

We will then extend the ideas to photons (which have momentum but no mass), and finally consider the case where the particles are electrons in a metal.

We noted in Chapter 11 two differences between a one-dimensional and a three-dimensional infinite well; first, the three-dimensional well has degeneracies, both natural and accidental, and secondly the three-dimensional quantum states tend to crowd together as the quantum numbers increase. If we consider a box of $1\,dm^3$ containing, for example, ^{20}Ne atoms at 300 K, then each atom will have $\frac{3}{2}k_BT$ kinetic energy, according to the equipartition of energy principle. A short calculation suggests that we will find typically quantum numbers as high as 10^9; energy is certainly quantized, but the separation between the energy levels is minute in comparison with k_BT. Under these circumstances, it is possible to treat the energy levels and the quantum numbers as if they were continuous rather than discrete. This is called the *continuum approximation*. Direct calculation shows how the number of quantum states lying in a given small interval increases with energy (Table 12.1).

What matters in many experiments is not the precise details of an energy level diagram, but rather a quantity called the *density of states*, $D(\varepsilon)$, defined such that the number of states between ε and $\varepsilon + d\varepsilon$ is $D(\varepsilon)\,d\varepsilon$. Figure 12.1 shows a plot of the number of states vs. the square root of the energy$/(h^2/8\,mL^2)$, and it seems that there is a very good linear relationship.

Table 12.1 Properties of quantum states. Here $E = h^2/8\,mL^2$

Number of quantum states with energy lying between	
(100 and 110) E	85
(1000 and 1010) E	246
(10 000 and 10 010) E	1029
(100 000 and 100 010) E	2925
(1 000 000 and 1 000 010) E	8820

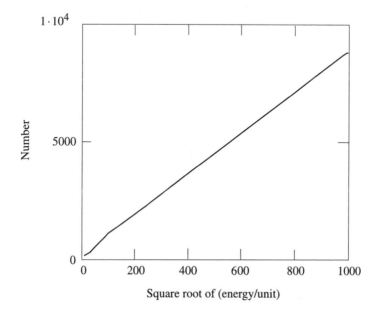

Figure 12.1 Square root relationship

I want now to establish that this is indeed the general case, and I can do this using a process called *Rayleigh counting*.

12.2 Rayleigh Counting

First of all I plot out the quantum numbers n, k and l along the x, y and z Cartesian axes, as shown in Figure 12.2. For large quantum numbers the points will be very close together and there will be very many possible combinations of the three particular quantum numbers n, k and l that correspond to the same energy

$$\varepsilon_{n,k,l} = (n^2 + k^2 + l^2)\frac{h^2}{8mL^2}$$

According to Pythagoras's theorem, each combination will lie on the surface of a sphere of radius r, where

$$r^2 = n^2 + k^2 + l^2$$

We therefore draw the positive octant of a sphere of radius r, as shown. The volume of the octant also contains all other combinations of the three quantum numbers that have energy less than or equal to the value in question. We only need consider this octant, since all quantum numbers have to be positive. The number of states with energy less than or equal to ε is

$$\frac{1}{8}\frac{4\pi}{3}(n^2 + k^2 + l^2)^{3/2}$$

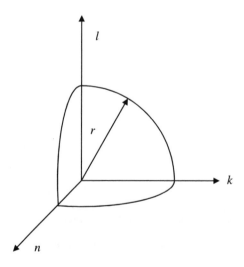

Figure 12.2   Rayleigh counting

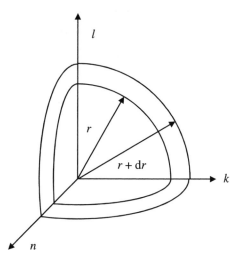

Figure 12.3   Rayleigh counting continued

or in terms of the energy

$$\frac{1}{8}\frac{4\pi}{3}\frac{8mL^2}{h^2}\varepsilon^{3/2}$$

I now draw a second octant of radius $r+dr$, shown in Figure 12.3, which contains all quantum states having energy less than or equal to $\varepsilon+d\varepsilon$. The volume of the outer sphere is

$$\frac{1}{8}\frac{4\pi}{3}\frac{8mL^2}{h^2}(\varepsilon+d\varepsilon)^{3/2}$$

The number of states having energy between ε and $\varepsilon+d\varepsilon$ is given by the volume contained between the two spheres

$$\frac{1}{8}\frac{4\pi}{3}\frac{8mL^2}{h^2}(\varepsilon+d\varepsilon)^{3/2}-\frac{1}{8}\frac{4\pi}{3}\frac{8mL^2}{h^2}\varepsilon^{3/2} \tag{12.7}$$

which is by definition $D(\varepsilon)\,d\varepsilon$. Use of the binomial theorem in Equation (12.7) gives

$$\begin{aligned}
D(\varepsilon)\,d\varepsilon &= \frac{1}{8}2\pi\frac{8mL^2}{h^2}\varepsilon^{1/2}\,d\varepsilon \\
&= \frac{2\pi mL^2}{h^2}\varepsilon^{1/2}\,d\varepsilon
\end{aligned} \tag{12.8}$$

I will write the result as

$$D(\varepsilon)=B\varepsilon^{1/2} \tag{12.9}$$

12.3 The Maxwell Boltzmann Distribution of Atomic Kinetic Energies

It is possible to perform experiments that measure the kinetic energies of atoms in gaseous samples. Once again we note that there are so many particles present that the quantities of interest are the mean and standard deviation, rather than the properties of every individual particle. We normally study the variation of a quantity

$$f = \frac{1}{N} \frac{dN}{d\varepsilon} \qquad (12.10)$$

with energy ε at a fixed thermodynamic temperature. Here, N is the number of atoms in the sample. The related quantity

$$g = dN/d\varepsilon$$

is referred to as the *distribution function* of the kinetic energy. When the experiments are carried out, we find a characteristic spread of kinetic energies as shown in Figure 12.4.

The distribution is independent of the type of atom; the solid curve refers to a temperature of 500 K and the dashed curve refers to a temperature of 1500 K. The

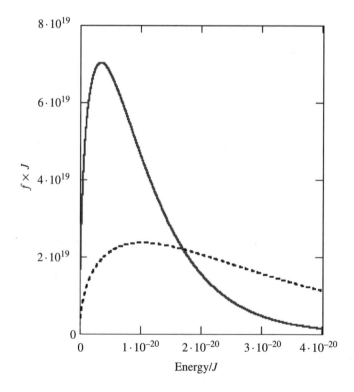

Figure 12.4   Distribution of atomic kinetic energies

peak moves to higher energy as the temperature increases, but the area under the curve remains constant.

The quantity $g(\varepsilon)$ gives the number of atoms whose energies lie between ε and $\varepsilon + \mathrm{d}\varepsilon$. The number of atoms whose energies lie between ε_A and ε_B is therefore

$$\int_{\varepsilon_\mathrm{A}}^{\varepsilon_\mathrm{B}} g(\varepsilon) \, \mathrm{d}\varepsilon$$

and the total number of atoms is given by

$$N = \int_0^\infty g(\varepsilon) \, \mathrm{d}\varepsilon$$

I can give a very simple explanation of Figure 12.4, on the assumption that the sample of particles is constrained to a cubic three-dimensional infinite well. My explanation depends on the continuum approximation and so will only work for ordinary temperatures, for at low temperatures the spacing between quantum states will be comparable with $k_\mathrm{B}T$ and so the continuum approximation will not apply. Also, it will not work for anything but an atom since molecules have rotational and vibrational energies in addition to translational kinetic energy.

Apart from the density of states mentioned above, $D(\varepsilon) = B\varepsilon^{1/2}$, we have to consider the average occupancy of each quantum state. This is given by the Boltzmann factor

$$AN \exp\left(-\frac{\varepsilon}{k_\mathrm{B}T}\right)$$

where A is a constant and N the number of atoms. Combining the two expressions we get

$$\frac{1}{N}\frac{\mathrm{d}N}{\mathrm{d}\varepsilon} = A \exp\left(-\frac{\varepsilon}{k_\mathrm{B}T}\right) B\varepsilon^{1/2}$$

We can eliminate the constants A and B in terms of N by use of the equations above and we find, after rearrangement,

$$\frac{1}{N}\frac{\mathrm{d}N}{\mathrm{d}\varepsilon} = \frac{2}{\sqrt{\pi}(k_\mathrm{B}T)^{3/2}} \exp\left(-\frac{\varepsilon}{k_\mathrm{B}T}\right) \varepsilon^{1/2} \tag{12.11}$$

which gives exact agreement with the experimental curves. It can be established by differentiation of Equation (12.11) that the peak occurs at $\varepsilon = \frac{1}{2}k_\mathrm{B}T$, and it can also be established by integration that the average energy per particle $\langle \varepsilon \rangle$ is $\frac{3}{2}k_\mathrm{B}T$, in accord with the equipartition of energy principle.

12.4 Black Body Radiation

A study of black body radiation was a milestone in the path to our modern theory of quantum mechanics. The term black body radiation refers to the electromagnetic radia-

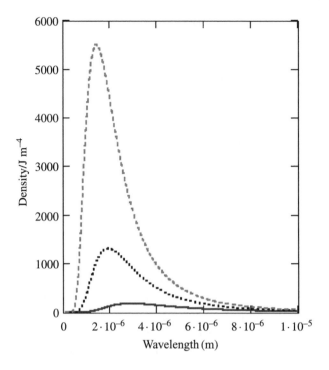

Figure 12.5   Black body radiation

tion emitted by a 'perfect emitter' (usually a heated cavity). The curves in Figure 12.5 relate to the energy U emitted by a black body of volume V per wavelength λ at temperatures of 1000 K, 1500 K and 2000 K. The top curve relates to the highest temperature and the bottom curve relates to the lowest. The quantity plotted on the y-axis is

$$\frac{1}{V}\frac{dU}{d\lambda}$$

and it is observed that

- each of the curves has a peak at a certain maximum wavelength λ_{max},

- λ_{max} moves to a shorter wavelength as the temperature increases.

These curves have a quite different functional form to the atomic kinetic energy curves shown in Figure 12.4. Max Planck studied the problem and was able to deduce the following expression that gives a very close fit to the experimental data

$$\frac{1}{V}\frac{dU}{d\lambda} = \left(\frac{8\pi hc_0}{\lambda^5}\right)\frac{1}{\exp\left(\frac{hc_0}{\lambda k_B T}\right) - 1} \tag{12.12}$$

For present purposes, it proves profitable to think of the problem in the following way: we know that the heated cavity contains photons, and that the energy of a photon of wavelength λ is hc_0/λ. We also know that $dU = \varepsilon\, dN = \varepsilon\, g(\varepsilon)\, d\varepsilon$. The experiment actually gives a distribution function $g(\varepsilon)$ such that the number of photons with energies between ε and $\varepsilon + d\varepsilon$ is $g(\varepsilon)\, d\varepsilon$. Equation (12.12) can be developed to give

$$g(\varepsilon) = \frac{8\pi V}{h^3 c_0^3} \varepsilon^2 \frac{1}{\exp\left(\frac{\varepsilon}{k_B T}\right) - 1} \tag{12.13}$$

If we interpret this expression as

$$g(\varepsilon) = (\text{density of states}) \times (\text{average occupancy of a state})$$

we see that the density of states for photons is proportional to ε^2 whilst the average occupancy of a state is proportional to

$$\frac{1}{\exp\left(\frac{\varepsilon}{k_B T}\right) - 1}$$

rather than the Boltzmann factor. This factor is referred to as the *Bose factor* after S. N. Bose.

I can explain the density of states quite easily; photons have zero mass and so the energy formula (12.1) for a particle in a three-dimensional infinite well

$$\varepsilon_{n,k,l} = \frac{h^2}{8mL^2}(n^2 + k^2 + l^2)$$

cannot apply. The correct formula turns out to be

$$\varepsilon_{n,k,l} = \frac{hc_0}{2L}(n^2 + k^2 + l^2)^{1/2} \tag{12.14}$$

There are still three quantum numbers, but the formula does not include a mass. As the energy increases, the quantum states become numerous and cram together. We can evaluate the density of states just as before by taking a narrow range of energies ε and $\varepsilon + d\varepsilon$, and counting the number of quantum states. We find

$$D(\varepsilon) \propto \varepsilon^2$$

that is to say, the photon states crowd together more quickly than for a classical particle in such a well. A detailed calculation gives the proportionality constant

$$D(\varepsilon) = \frac{4\pi V}{h^3 c_o^3} \varepsilon^2 \tag{12.15}$$

where V is the volume of the container. This explains the density of states, but not why the Bose factor is different from the Boltzmann factor.

12.5 Modelling Metals

We usually divide solids into *conductors* (materials such as copper that conduct electricity), *insulators* (materials such as glass that don't conduct electricity), and *semiconductors* that have properties in between. We visualize metallic conductors as comprising a rigid array of cations surrounded by a sea of mobile electrons, as shown in Figure 12.6.

12.5.1 The Drude model

The first attempt to model metals was the *free electron* model developed independently by P. Drude and by Lorentz. According to this model, each metallic atom in the lattice donates its valence electrons to form a sea of electrons. So a sodium atom would lose one electron and an aluminium atom three; these conduction electrons move freely about the cation lattice points, rather like the atoms in an ideal gas.

The model had some early successes; electrical conduction can be explained because the application of an electric field to the metal will cause the conduction electrons to move from positions of high to low electrical potential and so an electric

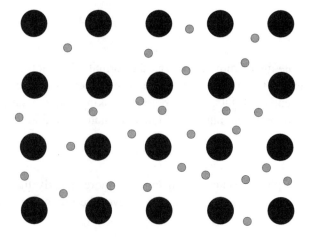

Figure 12.6 Simple model of a metallic conductor

current can be established. In fact, Ohm's Law can be recovered from the Drude model with a little analysis.

Eventually, the Drude model gradually fell into disrepute. An obvious failing was its inability to explain the heat capacity of metals. If the model of solid aluminium is correct and each aluminium atom loses three valence electrons, we might expect that these valence electrons would make a contribution to the internal energy of $3 \times \frac{3}{2}RT$, with a corresponding molar heat capacity of $\frac{9}{2}R$ (i.e. $32\,\mathrm{J\,K^{-1}\,mol^{-1}}$). In fact, Al obeys the Dulong–Petit rule with a molar heat capacity of about $25\,\mathrm{J\,K^{-1}\,mol^{-1}}$, so the equipartition of energy principle seems to overestimate the contribution from the conduction electrons.

The electrons are of course in motion, and we should ask about their energy distribution just as we did in earlier sections. There is no experimental technique that can measure the spread with the same accuracy, as can be done with photons in the black body experiment, or with particles in the kinetic theory experiments. A variation on the theme of the photoelectric effect (called high-energy photoemission) can however be used to probe electron kinetic energies in a metal. If I write the familiar expression for the distribution function $g(\varepsilon)$ as

$$g(\varepsilon) = (\text{density of states}, D(\varepsilon)) \times (\text{average occupancy of a state}, p(\varepsilon))$$

then it turns out that the density of states is proportional to $\varepsilon^{1/2}$, just as it was for the spread of kinetic energies in an ideal gas. The average occupancy factor is quite different to anything we have met so far

$$p(\varepsilon) = \frac{1}{\exp\left(\frac{(\varepsilon - \varepsilon_{\mathrm{F}})}{k_{\mathrm{B}}T}\right) + 1} \tag{12.16}$$

ε_{F} is a parameter called the *Fermi energy*, and I have illustrated the factor for the case of a Fermi energy of $2.5\,\mathrm{eV}$ at $600\,\mathrm{K}$ and $6000\,\mathrm{K}$ in Figure 12.7. The lower temperature shows a factor of 1 for all energies until we reach the Fermi level, when the factor becomes zero. The lower the temperature, the sharper the cut-off. The corresponding Boltzmann factors are shown in Figure 12.8, from which it is seen that the two sets of results are radically different.

The experimental distribution function is therefore

$$g(\varepsilon) = B\varepsilon^{1/2}\,\frac{1}{\exp\left(\frac{(\varepsilon - \varepsilon_{\mathrm{F}})}{k_{\mathrm{B}}T}\right) + 1} \tag{12.17}$$

and the average occupancy of states factor cannot be explained by the Drude model. On the other hand, it could be argued that the ideal gas model is inappropriate because charged particles exert tremendous long-range forces on each other. Not only that, there are frequent collisions between all the charged particles. However,

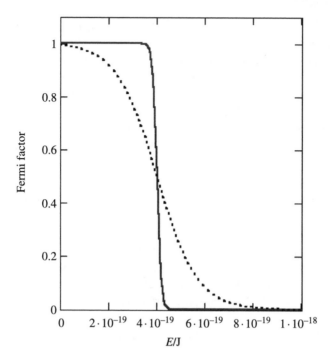

Figure 12.7 The Fermi factor

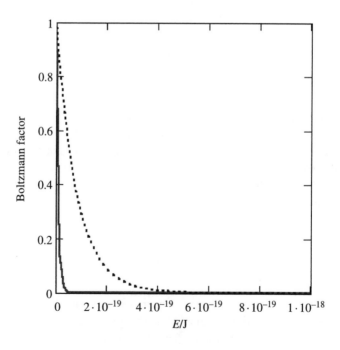

Figure 12.8 Corresponding Boltzmann factors

analysis shows that the problem lies with the *classical* Drude treatment, not with the ideal gas model.

12.5.2 The Pauli treatment

It was W. Pauli who devised a quantum mechanical treatment of the electron gas; in Pauli's model, the conduction electrons are still treated as an ideal gas of non-interacting particles, but the analysis is done according to the rules of quantum mechanics. My starting point is the energy level diagram for a cubic three-dimensional infinite potential well, Figure 12.9.

Pauli assumed that the metal could be modelled as an infinite potential well, and that the electrons did not interact with each other. He also assumed that each quantum state could hold no more than two electrons. At 0 K, the N conduction electrons fill up the $N/2$ lowest energy quantum states. I have shown 8 conduction electrons on the figure, for illustration. The highest occupied quantum state corresponds to the Fermi level, and the Fermi factors are therefore

$$p(\varepsilon) = \begin{cases} 1 & \text{for } \varepsilon \leq \varepsilon_F \\ 0 & \text{for } \varepsilon > \varepsilon_F \end{cases}$$

At higher temperatures, the electrons will be spread amongst the available quantum states, but subject still to the restriction that each quantum state can hold a maximum

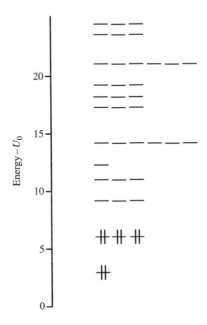

Figure 12.9   Electrons in metal

Table 12.2 Density of states and average occupancies

Particles	Density of states proportional to	Average occupancy proportional to
'Classical' ideal gas	$\varepsilon^{1/2}$	$\exp(-\frac{\varepsilon}{k_B T})$
Photons in a cavity	ε^2	$1/(\exp(\frac{\varepsilon}{k_B T}) - 1)$
Electrons in a metal	$\varepsilon^{1/2}$	$1/(\exp(\frac{(\varepsilon-\varepsilon_F)}{k_B T}) + 1)$

of two electrons. For electron energies much above the Fermi level so that $\varepsilon - \varepsilon_F \gg k_B T$, the Fermi factor can be approximated by ignoring the 1 in the denominator to give

$$p(\varepsilon) \approx \exp\left(-\frac{\varepsilon - \varepsilon_F}{k_B T}\right)$$

which is very similar to the Boltzmann factor. For sufficiently large electron energies, the Fermi factor reduces to the Boltzmann factor. For electron energies below the Fermi level, the factor can be written

$$p(\varepsilon) \approx 1 - \exp\left(\frac{\varepsilon - \varepsilon_F}{k_B T}\right)$$

which simply tells us that nearly all the states are fully occupied.

We have therefore reached a point where three different types of particles in the same three-dimensional infinite potential well show very different behaviour (see Table 12.2).

I have been able to explain the density of states, but not the average occupancies. It is now time to tidy up this loose end.

12.6 The Boltzmann Probability

I have made extensive use of Boltzmann factors throughout the text so far, without giving any justification. Let me now give some insight as to where the formula came from. I am not trying to give a rigorous mathematical proof, just appealing to your knowledge gained by studying the text so far.

Consider once again an ideal gas of N non-interacting particles contained in a thermally insulated container of volume V. This means that N, the volume V and the total energy U are constant. No energy can flow into or out of the container.

For the sake or argument, I am going to assume energy quantization for the particles in the potential well, and that there is a particularly simple quantum state level diagram (Figure 12.10). Each particle can have energy nD, where $n = 1, 2, 3 \ldots$ and D is a constant. There are no interactions between the particles and so the total energy U of the system is given by the sum of the individual particle energies. For the

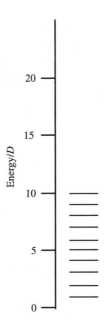

Figure 12.10   Simple quantum state diagram

sake of argument, I will take U as $9D$ and the number of particles $N = 3$. I will label the particles A, B and C, and we have to enumerate the ways in which three particles with these simple quantum state energies can achieve a total energy of $9D$.

The rows of Table 12.3 indicate every possible combination of integers that make up 9; to save space, I have only included one of each possible arrangements after the first; so, for example, particle A could have $7D$ with particles B and C having $1D$, particle B could have $7D$ with particles A and C having $1D$, or particle C could have $7D$ with particles A and B having $1D$. This is the origin of the integer 3 in the final column; it is the number of ways we can achieve the total energy with each individual distinguishable particle having a certain amount of energy. For clarity, I have omitted the 'D' from all entries except the first.

Table 12.3   Arrangements and probabilities

A	B	C	W = no. of ways	Probability
7D	1D	1D		
1	7	1	3	3/28
1	1	7		
6	2	1	6	6/28
5	2	2	3	3/28
5	3	1	6	6/28
4	4	1	3	3/28
4	3	2	6	6/28
3	3	3	1	1/28

The numbers in the third column are obviously related to factorials; if each particle has a different energy, then there are 3! ways in which we can allocate the energies amongst the three particles. If two energies are the same, then we have to divide the 3! by 2! and so on. If we were to make a spot check of the quantum state occupancy, then we would find on average a situation where one particle had $7D$ and the other two $1D$, for 3/28th of our measurements (on the other hand, since this is a quantum system, I had better be a bit more precise and say that if we were to prepare a very large number of such systems and measure the energy distribution in each copy, then we would find $7D$, $1D$, $1D$ in 3/28th of the systems). Each arrangement is known as a *configuration*.

Consider now a macroscopic system comprising N identical but distinguishable particles, each of which has a similar quantum state diagram to that shown in Figure 12.10. The particles are non-interacting. We can fix the total energy U and the number of particles, but we cannot fix how the energy is distributed amongst the available energy levels (and it doesn't make much sense to ask about the energy distribution amongst N individual particles, given that N might be 10^{23}). The most sensible thing is to say that there are N_1 particles with energy ε_1, N_2 with energy ε_2 and so on. Suppose also that the highest possible quantum state p has energy ε_p.

Consideration of our simple example above shows that the number of ways in which we can allocate the particles to the quantum states is

$$W = \frac{N!}{N_1! N_2! \cdots N_p!} \tag{12.18}$$

The fact that a particular quantum state may have no particles (or, in other words, it has a zero probability) is accounted for by the fact that $0! = 1$. It turns out that for large N there is one particular configuration that dominates, in which case the values of the N_i are those that maximize W. At first sight they can be found by setting the gradient of W to zero, but the N_i cannot vary freely; they are constrained by the fact that the number of particles and the total energy U are both constant

$$\begin{aligned} N &= N_1 + N_2 + \cdots + N_p \\ U &= N_1\varepsilon_1 + N_2\varepsilon_2 + \cdots + N_p\varepsilon_p \end{aligned} \tag{12.19}$$

A standard method for tackling *constrained variational* problems of this kind is *Lagrange's method of undetermined multipliers*, which is discussed in detail in all the advanced texts. I will just outline the solution without justification.

The first step in our derivation is to take logs of either side of expression (12.28)

$$\ln W = \ln N! - \sum_{i=1}^{p} \ln N_i!$$

For large values of n it is appropriate to make use of the Stirling formula for factorials

$$\ln n! = \tfrac{1}{2}\ln 2\pi + (n + \tfrac{1}{2})\ln n - n + \frac{1}{12n} + O\!\left(\frac{1}{n^2}\right) + \cdots$$

In practice all terms except $n \ln n$ are usually omitted and we have

$$\ln W = \ln N! - \sum_{i=1}^{p} (N_i \ln N_i! - N_i) \tag{12.20}$$

We therefore look for a maximum of $\ln W$. We know from elementary differential calculus that

$$d \ln W = \sum_{i=1}^{p} \frac{\partial \ln W}{\partial N_i} dN_i = 0 \tag{12.21}$$

at a stationary point. We cater for the constraints by differentiating Equations (12.19)

$$0 = dN_1 + dN_2 + \cdots + dN_p$$
$$0 = dN_1 \varepsilon_1 + dN_2 \varepsilon_2 + \cdots + dN_p \varepsilon_p$$

multiplying by arbitrary undetermined multipliers that I will call α and $-\beta$, and then adding

$$d \ln W = \sum_{i=1}^{p} \frac{\partial \ln W}{\partial N_i} dN_i + \alpha \sum_{i=1}^{p} dN_i - \beta \sum_{i=1}^{p} \varepsilon_i dN_i \tag{12.22}$$

All the dN_i can now be treated as independent variables and at a stationary point

$$\frac{\partial \ln W}{\partial N_i} + \alpha - \beta \varepsilon_i = 0 \tag{12.23}$$

All that remains is to simplify Equation (12.23) using the Stirling formula, and the final result turns out as

$$N_i = \exp(\alpha) \exp(-\beta \varepsilon_i) \tag{12.24}$$

Next we have to evaluate the undetermined multipliers α and β; if we sum over all the allowed quantum states then we have

$$N = \sum_{i=1}^{p} N_i = \exp(\alpha) \sum_{i=1}^{p} \exp(-\beta \varepsilon_i)$$

which gives

$$\exp(\alpha) = \frac{N}{\sum\limits_{i=1}^{p} \exp(-\beta\varepsilon_i)} \tag{12.25}$$

The value of β is not quite as straightforward. What we need to do is to use the expression to calculate a quantity such as the heat capacity of an ideal monatomic gas which can be compared with experiment. When this is done we find

$$N_i = \frac{N \exp\left(-\frac{\varepsilon_i}{k_B T}\right)}{\sum\limits_{i=1}^{p} \exp\left(-\frac{\varepsilon_i}{k_B T}\right)} \tag{12.26}$$

The formula is correct even when the quantum states are degenerate, as it refers to the individual quantum states. It also holds in an equivalent integral form when the energies form a continuum.

12.7 Indistinguishability

Let me now return to the infinite one-dimensional well, with just two non-interacting particles present. I labelled the particles A and B, and showed in Chapter 11 that the solutions were

$$\Psi_{n_A, n_B} = \frac{2}{L} \sin\left(\frac{n_A \pi x_A}{L}\right) \sin\left(\frac{n_B \pi x_B}{L}\right)$$

$$E_{n_A, n_B} = (n_A^2 + n_B^2)\frac{h^2}{8mL^2} \tag{12.27}$$

$$n_A, n_B = 1, 2, 3 \ldots$$

Suppose one particle has a quantum number of 1, whilst the other has a quantum number of 2. There are two possibilities

$$\Psi_{1,2} = \frac{2}{L} \sin\left(\frac{1\pi x_A}{L}\right) \sin\left(\frac{2\pi x_B}{L}\right)$$

$$\Psi_{2,1} = \frac{2}{L} \sin\left(\frac{2\pi x_A}{L}\right) \sin\left(\frac{1\pi x_B}{L}\right) \tag{12.28}$$

and they are degenerate (they have the same energy). In such situations it turns out that any linear combination of degenerate solutions is also an acceptable solution. This is easily seen by rewriting the problem in Hamiltonian operator form

$$\hat{H}\Psi_{1,2} = E\Psi_{1,2}$$

$$\hat{H}\Psi_{2,1} = E\Psi_{2,1}$$

$$\hat{H}(a\Psi_{1,2} + b\Psi_{2,1}) = E(a\Psi_{1,2} + b\Psi_{2,1})$$

In the above equation, a and b are non-zero scalars that could be complex.

We live in a world of mass production where items that come off a production line appear to be identical. A shopping trip to the supermarket might yield four tins of dog food, all with the same label, the same contents and the same price, but on close examination it usually turns out that there are differences. One tin might have a scratch, one might be slightly dented, and the label might be peeling from the third and so on. In the everyday world, objects that seem at first sight to be identical may very well turn out to be distinguishable. Even if the cans were exactly identical, we could label them A, B, C and D with an indelible marker and so make them distinguishable.

Things are very different in the world of atoms and molecules. A visit to MolMart®️ for four hydrogen atoms would reveal that all hydrogen atoms are exactly the same; there is no question of one atom having a scratch or a dent, and it is certainly not possible to label a hydrogen atom with an indelible marker. We say that the four hydrogen atoms are *indistinguishable*. This simple observation has far-reaching consequences.

I mentioned in earlier sections that wavefunctions themselves do not have any physical interpretation, but that the modulus squared of a wavefunction is an experimentally observable probability density. So if $\psi(x)$ describes a single particle, then

$$|\psi(x)|^2 \, dx$$

gives the chance of finding the particle between x and $x + dx$ and this can be measured in experiments such as X-ray and neutron diffraction.

The wavefunctions $\Psi_{1,2}$ and $\Psi_{2,1}$ above involve the coordinates of two particles, and in this case the physical interpretation is that, for example,

$$|\Psi(x_A, x_B)|^2 \, dx_A dx_B$$

represents the probability of simultaneously finding particle A between x_A and $x_A + dx_A$ with particle B between x_B and $x_B + dx_B$. If we now add the condition that the particles are truly indistinguishable, then we should get exactly the same probability on renaming the particles B and A. So if we consider $\Psi_{1,2}^2$

$$(\Psi_{1,2})^2 = \left(\frac{2}{L} \sin\left(\frac{1\pi x_A}{L}\right) \sin\left(\frac{2\pi x_B}{L}\right) \right)^2$$

it is clear that the probability density treats the two particles on a different footing and so implies that they are distinguishable. The acid test is to write down the probability density and then interchange the names of the particles. The probability density should stay the same.

The two wavefunctions $\Psi_{1,2}$ and $\Psi_{2,1}$ do not satisfy this requirement and so they are not consistent with our ideas about indistinguishability. I can illustrate the problem by plotting the squares of the wavefunctions as density maps, Figures 12.11 and 12.12. The plots are of the square of the wavefunction (the 'out-of-plane' axis) versus the x coordinates of the two electrons along the in-plane axes. The diagrams imply quite

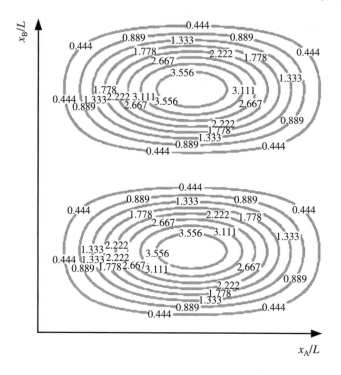

Figure 12.11 $\Psi_{1,2}^2$ vs. x_A and x_B

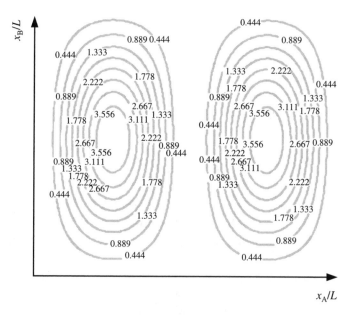

Figure 12.12 $\Psi_{2,1}^2$ vs. x_A and x_B

different probability densities for an arbitrary choice of the pair of coordinates: the probability contours ought to be symmetrical about their diagonals. The problem can be removed if we make use of the degeneracy mentioned above and construct a wave-function that is either the sum or the difference of the product functions with x_A and x_B interchanged. Each of these two combinations gives a probability density that is unaffected by an interchange of the names of the two particles. Allowing for the correct normalization, these combinations are

Symmetric

$$\Psi_s(x_A, x_B) = \sqrt{\tfrac{1}{2}}(\Psi_{1,2}(x_A, x_B) + \Psi_{2,1}(x_A, x_B)) \qquad (12.29)$$

Antisymmetric

$$\Psi_a(x_A, x_B) = \sqrt{\tfrac{1}{2}}(\Psi_{1,2}(x_A, x_B) - \Psi_{2,1}(x_A, x_B)) \qquad (12.30)$$

The labels *symmetric* and *antisymmetric* arise from the fact that interchange of the particle names leaves the first combination unchanged but leads to a reversal of sign in the second combination. The resulting probability maps are shown in Figures 12.13 and 12.14.

In the symmetric state the maximum probability occurs at the points for which $x_A = x_B = L/4$ or $3L/4$ and quite generally the probability is large only if the difference between the particles is small.

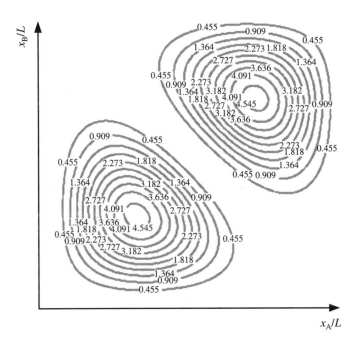

Figure 12.13 Symmetric wavefunction squared

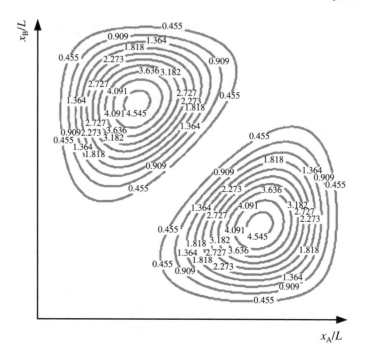

Figure 12.14 Antisymmetric wavefunction squared

The antisymmetric state is quite different; the probability is zero everywhere along the diagonal (which corresponds to $x_A = x_B$) and the probability maxima occur at positions where the two particles are far apart. I must emphasize that this apparent attraction and repulsion between particles has nothing whatever to do with the forces between them; indeed, we assumed at the start that the particles did not interact with each other. It certainly isn't an additional force in nature; it is simply to do with the indistinguishability of particles and the symmetry of the wavefunction describing them. Like the zero-point vibrational energy, this is a purely quantum mechanical effect and has no 'explanation' in classical physics.

12.8 Spin

It turns out to be an experimental fact that many particles have an intrinsic property called *spin*, just as they have a mass, a charge and so on. If the particles concerned are electrons, then we refer to the phenomenon as *electron spin*. Electron spin is expressed in two possible configurations usually referred to as *spin up* and *spin down* with respect to a given axis of quantization. I'll have more to say about electron spin in a later chapter; it doesn't appear from the Schrödinger treatment, just accept it as an experimental fact for the minute.

The two states are written conventionally α and β and authors often use the symbol s for the spin variable. So in the case of electrons, we might write $\alpha(s_A)$ to mean that particle A was in the spin up state and $\beta(s_A)$ to mean that particle A was in the spin down state. In the case of two electrons we have four possibilities and we label the quantum states individually $\alpha(s_A)$, $\beta(s_A)$, $\alpha(s_B)$ and $\beta(s_B)$. The combination $\alpha(s_A)\beta(s_B)$ describes a situation where electron A has spin up and electron B has spin down and there are four possible combination states

$$\alpha(s_A)\alpha(s_B) \qquad \alpha(s_A)\beta(s_B) \qquad \beta(s_A)\alpha(s_B) \qquad \beta(s_A)\beta(s_B)$$

The first and the last are automatically symmetric with respect to interchange of the two electron names, but the middle two are not satisfactory because they imply that the two electrons can somehow be treated differently. Bearing in mind the discussion of Section 12.7, we construct two other spin functions that have definite spin symmetry properties

Symmetric

$$\alpha(s_A)\alpha(s_B)$$

$$\sqrt{\tfrac{1}{2}}(\alpha(s_A)\beta(s_B) + \beta(s_A)\alpha(s_B))$$

$$\beta(s_A)\beta(s_B)$$

Antisymmetric

$$\sqrt{\tfrac{1}{2}}(\alpha(s_A)\beta(s_B) - \beta(s_A)\alpha(s_B))$$

If we had been considering two spinless particles, e.g. two alpha particles, then there would be no spin states and the above complications would not have arisen. We now have to combine the space and the spin wavefunctions. We do this by multiplying together a space part and a spin part, and it should be clear that there are two possible results (see Table 12.4) and to decide between them we have to resort to experiment. For many-electron systems, the only states that are ever observed are those that correspond to antisymmetric total wavefunctions. W. Heisenberg recognized this fact in 1926, and this led to a detailed understanding of the electronic spectrum and structure of the helium atom [43]. The condition of antisymmetry

Table 12.4 Combining space and spin functions

Space part	Spin part	Overall wavefunction
Symmetric	Symmetric	Symmetric
Antisymmetric	Symmetric	Antisymmetric
Symmetric	Antisymmetric	Antisymmetric
Antisymmetric	Antisymmetric	Symmetric

eliminates half of the possible states that would otherwise be possible. (When considering more than two particles, we have to consider all possible permutations amongst the space and spin functions).

12.9 Fermions and Bosons

Not all wavefunctions for identical particles are antisymmetric, but detailed studies show that, for a given kind of particle, states of only one overall symmetry exist; the states are *either* all symmetric *or* all antisymmetric with respect to interchange of particle names. Particles whose overall wavefunction is antisymmetric are called *fermions* (after E. Fermi), and particles whose overall wavefunction is symmetric are called *bosons* (after S. N. Bose). Typical fermions and bosons are shown in Table 12.5.

12.10 The Pauli Exclusion Principle

The famous *Pauli Exclusion Principle* is a statement of our discussion above; in its general form it says that fermion wavefunctions must be antisymmetric to exchange the names of two fermions, whilst boson wavefunctions must be symmetric. The principle is stated in different ways in different books, for example the world's most popular physical chemistry text [44] says: 'No more than two electrons may occupy any orbital, and if two do occupy it their spin directions must be opposite.'

In principle we must invoke the symmetry or antisymmetry of a wavefunction when dealing with systems of two or more identical particles. For example, if we wish to describe two ground-state hydrogen atoms, then the total wavefunction must be antisymmetric to exchange the names of the two electrons, even if an infinite distance separates the two atoms. Common sense suggests that this is not necessary, and the criterion is to ask by how much the two wavefunctions overlap each other. If the overlap is appreciable, then we must worry about the exclusion principle. If the overlap is negligible, then we can disregard the symmetry or otherwise of the total wavefunction.

Table 12.5 Fermions and bosons

Fermion	Boson
Electron	Photon
Proton	Deuteron
Neutron	Alpha particle
^{3}He atom	^{4}He atom

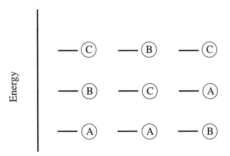

Figure 12.15   Boltzmann's counting method

12.11 Boltzmann's Counting Rule

We must now return to the three probabilities discussed in Section 12.5. Consider a simple case of three particles A, B and C, each of which has a different energy. Figure 12.15 shows three of the six possible arrangements of the three *distinguishable* particles, and according to the Boltzmann counting procedure, we take account of all six and they are all weighted equally when we calculate the number of ways, *W*.

This procedure is not consistent with the principles of quantum mechanics discussed above. We certainly cannot label particles and make them distinguishable, and it is not meaningful to enquire which particle is in one particular state. All we can do is to say how many particles are in each state, and instead of six possibilities there is just one (with one particle in each state).

We then have to consider whether the particles are fermions or bosons; if they are fermions with spin $\frac{1}{2}$ (like electrons), then each quantum state can hold no more than 2 particles, one of α and one of β spin. If they are bosons, then the exclusion principle does not apply and all the particles can crowd together in any one state, should they so wish. This is the basis of the three different probabilities discussed in Section 12.5.

That leaves just one loose end; if all particles are either fermions or bosons, why does the Boltzmann distribution work so well in explaining the properties of 'ordinary' gases? The overwhelming experimental evidence is that the particles in an ordinary gas can be treated as if they were distinguishable particles, subject to the Boltzmann probability law. The answer is quite simply that under many normal circumstances the quantum states are very sparsely occupied and so fermions do not need to be aware of the exclusion principle, whilst bosons cannot display their ability to all crowd together into the lowest state.

13 One-Electron Atoms

The hydrogen atom occupies an important place in the history of molecular modelling and quantum theory. It is the simplest possible atom, with superficially simple properties, yet it was a thorn in the side of late nineteenth- and early twentieth-century science. Rutherford had discovered the nucleus by 1911, and had a rough idea of its size from his scattering studies and so people pictured the hydrogen atom as a negative charge (the electron) making a circular orbit with constant angular velocity round the nucleus (a positive charge). This is shown in Figure 13.1.

I should mention a small technical point: if we want to eliminate the overall translational motion of the atom from our discussion, the atomic system, nucleus plus electron, actually rotates round the centre of gravity. This rotational motion of the electron (mass m_e) and the nucleus (mass M) is equivalent to rotation of a single particle with reduced mass μ about the origin. Here

$$\frac{1}{\mu} = \frac{1}{M} + \frac{1}{m_e}$$

On account of its circular motion, the electron is accelerating. According to Maxwell's classical electromagnetic theory, accelerating electric charges emit energy as electromagnetic radiation and so the atom ought not to be stable since it should lose all its energy as it orbits the nucleus and emits radiation. Also, the frequency of this emitted radiation should equal the frequency of revolution.

13.1 Atomic Spectra

The problems were that atoms are certainly stable, and that a hydrogen atom can emit or absorb electromagnetic radiation having only certain specific wavelengths. For example, Balmer had discovered in 1885 that a series of five spectral lines in the visible region with wavelengths $\lambda = 656, 486, 434, 410$ and 397 nm fitted the formula

$$\lambda_n = 364.6 \frac{n^2}{n^2 - 4} \, \text{nm} \tag{13.1}$$

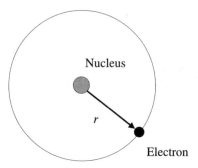

Figure 13.1   A one-electron atom

It was eventually realized that the reciprocal of the wavelength was the more fundamental quantity and that the equation was better written

$$\frac{1}{\lambda_n} = R_H \left(\frac{1}{2^2} - \frac{1}{n^2} \right) \tag{13.2}$$

where R_H is the *Rydberg constant for hydrogen*. Similar series of spectral lines were
observed for He$^+$ where the proportionality constant is (very nearly) $\frac{1}{4} R_H$, for Li^{2+}
(with a proportionality constant of almost $\frac{1}{9} R_H$) and so on. The important point is that
the value of $1/\lambda$ for every spectral line is the difference between two terms that we
now know correspond directly to particular energy levels in the atom.

13.1.1 Bohr's theory

Neils Bohr is reputed to have said that once he saw the Balmer formula, the whole
problem of the hydrogen atom became clear to him. In any event, in 1913 he was able
to devise a spectacularly successful theory that could account for the spectral lines.
He equated the centrifugal force for circular motion with the attractive electrostatic
force between the nucleus (charge Ze) and the electron (charge $-e$)

$$\frac{Ze^2}{4\pi\epsilon_0 r^2} = \frac{\mu v^2}{r} \tag{13.3}$$

This gives two unknowns, r and the speed v. He assumed that the orbital angular
momentum was quantized in units of $h/2\pi$

$$\mu v r = n \frac{h}{2\pi} \tag{13.4}$$

where the quantum number n could take values $1, 2, 3 \dots$. This assumption is exactly
equivalent to requiring that the circumference of the orbit is just the right value to

permit the electron to behave as a standing wave (i.e. a de Broglie wave). In other words, the circumference has to be an integral number n of de Broglie wavelengths, that is to say

$$2\pi r = n\frac{h}{p} \tag{13.5}$$

where $\mathbf{p}$ is the momentum $= \mu \mathbf{v}$. Simultaneous solution of Equations (13.3) and (13.4) gives allowed values for the radius and the speed

$$r_n = \frac{\epsilon_0 h^2 n^2}{\pi \mu Z e^2}$$

$$v_n = \frac{Z e^2}{2\epsilon_0 h n} \tag{13.6}$$

and also quantized energies given by

$$\begin{aligned}
\varepsilon_n &= -\frac{Z e^2}{4\pi\epsilon_0 r_n} + \frac{1}{2}\mu v_n^2 \\
&= -\frac{\mu Z^2 e^4}{8 h^2 \epsilon_0^2} \frac{1}{n^2}
\end{aligned} \tag{13.7}$$

Bohr referred to these orbits as *stationary states* because, according to his theory, the electron did not radiate when it was in any one such state. When the electron passed from one state (the initial state) to another (the final state), radiation was emitted or absorbed according to the Planck–Einstein formula

$$|\varepsilon_f - \varepsilon_i| = \frac{h c_0}{\lambda} \tag{13.8}$$

Bohr's theory predicted the existence of many series of spectral lines for the H atom (for which $Z=1$) whose wavelengths fitted the generic formula

$$\frac{1}{\lambda_{if}} = R_H \left(\frac{1}{n_f^2} - \frac{1}{n_i^2} \right)$$

$$n_f = 1, 2, 3 \ldots$$

$$n_i > n_f$$

The Balmer series corresponds to $n_f = 2$, and the series corresponding to $n_f = 3$ had already been observed in the infrared region of the electromagnetic spectrum by Paschen. Soon after Bohr published his theory, Lyman identified the series with $n_f = 1$ in the ultraviolet. The $n_f = 4$ (Brackett) and $n_f = 5$ (Pfund) series were also identified. Bohr's theory also predicted a first ionization energy (given by $R_H h c_0$, approximately 2.16806×10^{-18} J or 13.5 eV) that agreed well with experiment, and a physically reasonable value (52.9 pm) for the radius of the lowest energy orbit.

Many series with high values of n have since been observed. For example, radio astronomers are very interested in the $n_i = 167$ to $n_f = 166$ emission that occurs at a wavelength of 21.04 cm.

Bohr's theory gives the following expression for R_H, in agreement with experiment

$$R_H = \frac{\mu e^4}{8\epsilon_0^2 h^3 c_0}$$

13.2 The Correspondence Principle

Bohr made much use of the Correspondence Principle discussed in Chapter 11, which says that quantum mechanical results must tend to those obtained from classical physics, in the limit of large quantum numbers. For example, if we consider the transition from level n to level $n - 1$, the emitted frequency is

$$\nu = \frac{\mu e^4 Z^2}{8\epsilon_0^2 h^3} \left(\frac{1}{(n-1)^2} - \frac{1}{n^2} \right)$$

For large n this approximates to

$$\nu = \frac{\mu e^4 Z^2}{8\epsilon_0^2 h^3} \frac{2}{n^3} \tag{13.9}$$

According to Maxwell's electromagnetic theory, an electron moving in a circular orbit should emit radiation with a frequency equal to its frequency of revolution $v/2\pi r$. Using the Bohr expressions for v and r we deduce an expression in exact agreement with the frequency (13.9) above.

Despite its early successes, Bohr's theory had many failings. For example, it could not explain the structure and properties of a helium atom. Many ingenious attempts were made to improve the model, for example by permitting elliptic orbits rather than circular ones, but the theory has gradually faded into history. It is sometimes referred to as the *old quantum theory*.

13.3 The Infinite Nucleus Approximation

I started the chapter by correctly considering the motion of a one-electron atom about the centre of mass, and pointed out that this was equivalent to the motion of a single particle of reduced mass

$$\frac{1}{\mu} = \frac{1}{M} + \frac{1}{m_e}$$

Because the mass of the nucleus is so much greater than that of the electron, the reduced mass of the atom is almost equal to (and slightly less than) the mass of the electron. For that reason, workers in the field often treat the nucleus as the centre of coordinates and the electron as rotating round the nucleus, which is taken to have infinite mass; this is called the *infinite nucleus approximation*. It's just a small correction, the two masses μ and m_e are equal to 1 part in 10^4 which is usually good enough, but as we have seen, atomic spectroscopic data are known to incredible accuracy and it is sometimes necessary to take account of this difference. We write, for an infinite mass nucleus,

$$\varepsilon_n = -\frac{m_e e^4}{8h^2 \epsilon_0^2} \frac{1}{n^2}$$

$$R_\infty = \frac{m_e e^4}{8\epsilon_0^2 h^3 c_0} \quad \text{and} \quad a_0 = \frac{\epsilon_0 h^2}{\pi m_e e^2}$$

(13.10)

where a_0 is called the *first Bohr radius* and R_∞ is 'the' Rydberg constant, with value

$$R_\infty = 10\ 973\ 731.568\ 5458\ \text{m}^{-1}$$

13.4 Hartree's Atomic Units

One problem we encounter in dealing with atomic and molecular properties is that large powers of 10 invariably appear in equations, together with complicated combinations of physical constants such as those in Equations (13.10). For simplicity, we tend to work in a system of units called the Hartree units, or atomic units (we will soon come across the two Hartrees in Chapter 14). The 'atomic unit of length' is the *bohr* and is equal to a_0 above. The 'atomic unit of energy' is the *hartree* and is equal to $2 R_\infty h c_0$. It is written E_h and it is also equal to the mutual potential energy of a pair of electrons separated by distance a_0

$$E_h = \frac{1}{4\pi\epsilon_0} \frac{e^2}{a_0}$$

Other atomic units can be defined, as Table 13.1 shows.

The Born interpretation of quantum mechanics tells us that $\Psi^*(\mathbf{r})\,\Psi(\mathbf{r})\,d\tau$ represents the probability that we will find the particle whose spatial coordinate is $\mathbf{r}$ within the volume element $d\tau$. Probabilities are real numbers and so the dimensions of the wavefunction must be $(\text{length})^{-3/2}$. In the atomic system of units, the unit of wavefunction is therefore $a_0^{-3/2}$ (or $\text{bohr}^{-3/2}$).

Table 13.1 Hartree's atomic units

Physical quantity	Symbol	X	Value of X
Length	l, x, y, z, r	a_0	5.2918×10^{-11} m
Mass	m	m_e	9.1094×10^{-31} kg
Energy	ε	E_h	4.3598×10^{-18} J
Charge	Q	e	1.6022×10^{-19} J
Electricdipole moment	p_e	ea_0	8.4784×10^{-30} Cm
Electric quadrupole moment	θ_e	$ea_0{}^2$	4.4866×10^{-40} C m^2
Electricfield	E	$E_h e^{-1} a_0{}^{-1}$	5.1422×10^{11} V m^{-1}
Electricfield gradient	$-V_{zz}$	$E_h e^{-1} a_0{}^{-2}$	9.7174×10^{21} V m^{-2}
Magnetic induction	B	$(h/2\pi)e^{-1}a_0{}^{-2}$	2.3505×10^{5} T
Electricdipole polarizability	α	$e^2 a_0{}^{-2} E_h{}^{-1}$	1.6488×10^{-41} C^2 m^2 J^{-1}
Magnetizability	ξ	$e^2 a_0{}^{-2} m_e{}^{-1}$	7.8910×10^{-29} J T^{-2}

13.5 Schrödinger Treatment of the H Atom

The wavefunction of a hydrogen atom Ψ_{tot} depends on the coordinates of the electron x, y, z and the nucleus X, Y, Z. The time-independent Schrödinger equation is

$$-\frac{h^2}{8\pi^2 m_e}\left(\frac{\partial^2 \Psi_{tot}}{\partial x^2} + \frac{\partial^2 \Psi_{tot}}{\partial y^2} + \frac{\partial^2 \Psi_{tot}}{\partial z^2}\right) - \frac{h^2}{8\pi^2 M}\left(\frac{\partial^2 \Psi}{\partial X^2} + \frac{\partial^2 \Psi}{\partial Y^2} + \frac{\partial^2 \Psi}{\partial Z^2}\right) + U\Psi_{tot} = \varepsilon_{tot}\Psi_{tot}$$

(13.11)

where m_e is the electron mass, M the nuclear mass and U the mutual electrostatic potential energy of the nucleus and the electron

$$U = \frac{-Ze^2}{4\pi\epsilon_0}\frac{1}{\sqrt{(x-X)^2 + (y-Y)^2 + (z-Z)^2}}$$

(13.12)

I have temporarily added a subscript 'tot' to show that we are dealing with the total atom, nucleus plus electron, at this stage. The equation is not at all simple in this coordinate system and it proves profitable to make a change of variable. We take the coordinate origin as the centre of gravity of the atom (coordinates x_c, y_c, z_c) and use spherical polar coordinates for the nucleus and the electron. If the spherical polar coordinates of the electron are a, θ, ϕ and those of the nucleus b, θ, ϕ where

$$a = \frac{M}{M + m_e} r$$

$$b = -\frac{m_e}{M + m_e} r$$

then, if μ is the reduced mass

$$\mu = \frac{m_e M}{m_e + M}$$

$$X = x_c - \frac{\mu}{M} r \sin\theta \cos\phi; \qquad x = x_c + \frac{\mu}{m_e} r \sin\theta \cos\phi$$

$$Y = y_c - \frac{\mu}{M} r \sin\theta \sin\phi; \qquad y = y_c + \frac{\mu}{m_e} r \sin\theta \sin\phi \qquad (13.13)$$

$$Z = z_c - \frac{\mu}{M} r \cos\theta; \qquad z = z_c + \frac{\mu}{m_e} r \cos\theta$$

With this change of variable we find

$$-\frac{h^2}{8\pi^2(m_e + M)}\left(\frac{\partial^2 \Psi_{tot}}{\partial x_c^2} + \frac{\partial^2 \Psi_{tot}}{\partial y_c^2} + \frac{\partial^2 \Psi_{tot}}{\partial z_c^2}\right) \qquad (13.14)$$

$$-\frac{h^2}{8\pi^2\mu}\left(\frac{1}{r^2}\frac{\partial}{\partial r}\left(r^2\frac{\partial \Psi_{tot}}{\partial r}\right) + \frac{1}{r^2\sin\theta}\frac{\partial}{\partial\theta}\left(\sin\theta\frac{\partial\Psi}{\partial\theta}\right) + \frac{1}{r^2\sin^2\theta}\frac{\partial^2\Psi_{tot}}{\partial\phi^2}\right)$$

$$+ U\Psi_{tot} = \varepsilon_{tot}\Psi_{tot} \qquad (13.15)$$

and also in this coordinate system

$$U = -\frac{1}{4\pi\epsilon_0}\frac{Ze^2}{r}$$

It is apparent that we can separate the wave equation into two equations, one referring to the centre of mass and involving the total mass of the atom, and the other containing r, θ and ϕ. We put $\Psi_{tot} = \psi_e\psi_c$ and follow through the separation of variables argument to obtain

$$-\frac{h^2}{8\pi^2(m_e + M)}\left(\frac{\partial^2\psi_c}{\partial x_c^2} + \frac{\partial^2\psi_c}{\partial y_c^2} + \frac{\partial^2\psi_c}{\partial z_c^2}\right) = (\varepsilon_{tot} - \varepsilon_e)\psi_c \qquad (13.16)$$

$$-\frac{h^2}{8\pi^2\mu}\left(\frac{1}{r^2}\frac{\partial}{\partial r}\left(r^2\frac{\partial\psi_e}{\partial r}\right) + \frac{1}{r^2\sin\theta}\frac{\partial}{\partial\theta}\left(\sin\theta\frac{\partial\psi_e}{\partial\theta}\right) + \frac{1}{r^2\sin^2\theta}\frac{\partial^2\psi_e}{\partial\phi^2}\right)$$

$$+ U\psi_e = \varepsilon_e\psi_e \qquad (13.17)$$

The first equation relates to the translation of the atom as a whole, and I have dealt with such equations in earlier chapters. The second equation is usually called the *electronic equation*.

It should be clear from this discussion that in the treatment of any atomic or molecular system the translational degree(s) of freedom may always be separated from the internal degrees of freedom and so need not be considered in general. Also, from now on, I will drop the subscript 'e' from the electronic equation.

In the special case of a one-electron atom, the electronic wavefunction depends only on the coordinates of (the) one electron and so it is technically described as an *atomic orbital*. In view of the spherical symmetry, we might expect that it would prove profitable to consider a further separation of variables

$$\psi(r, \theta, \phi) = R(r)Y(\theta, \phi) \qquad (13.18)$$

This proves to be the case and $R(r)$ usually is called the *radial function* (although I should warn you that many authors call $P(r) = rR(r)$ the radial function; we will see why this proves a useful choice shortly). The Ys turn out to be eigenfunctions of the orbital angular momentum, and again we will see why in due course.

It turns out that most of the wavefunction solutions of Equation (13.18) are not spherically symmetrical, and you may find this odd given that the mutual potential energy U depends only on the scalar distance r (we call such fields *central fields*). The fact is that $U(r)$ being spherically symmetrical does not imply that the solutions are also spherical; the gravitational field is also a central field, and you probably know that planetary motion is not spherically symmetrical.

13.6 The Radial Solutions

Before launching into a discussion of the full solutions, I want to spend a little time considering just those solutions that depend only on r. That is, those solutions that have a constant angular part; this is not the same thing as finding $R(r)$ for the general case. I will also make the infinite nucleus approximation from this point on.

A little rearrangement and manipulation of Equation (13.17) above gives, for functions that depend only on r,

$$-\frac{h^2}{8\pi^2 m_e}\frac{d^2(rR)}{dr^2} - \frac{Ze^2}{4\pi\epsilon_0 r}(rR) = \varepsilon(rR) \qquad (13.19)$$

The form of this equation suggests that we make the substitution $P = rR$, which explains why some authors focus on $P(r)$ rather than $R(r)$ and refer to $P(r)$ as the radial function. This substitution gives

$$-\frac{h^2}{8\pi^2 m_e}\frac{d^2 P}{dr^2} - \frac{Ze^2}{4\pi\epsilon_0 r}P = \varepsilon P$$

It is usual in many scientific and engineering applications to try to simplify such complicated equations by use of 'reduced variables'. In this case we divide each variable by its atomic unit, so for example

$$r_{red} = r/a_0$$
$$\varepsilon_{red} = \varepsilon/E_h$$
$$\frac{d^2}{dr_{red}^2} = a_0^2 \frac{d^2}{dr^2}$$

and simplify to give

$$-\frac{1}{2}\frac{d^2 P_{red}}{dr_{red}^2} - \frac{Z}{r_{red}}P_{red} = \varepsilon_{red}P_{red}$$

It gets a bit tedious writing the subscript 'red', so most authors choose to state their equations in dimensionless form such as

$$-\frac{1}{2}\frac{d^2 P}{dr^2} - \frac{Z}{r}P = \varepsilon P \tag{13.20}$$

with the tacit understanding that the variables are reduced ones and therefore dimensionless. Sometimes authors say (incorrectly) that the equation is written 'in atomic units', and occasionally we come across statements of the impossible (for example $m_e = 1$ and $h/2\pi = 1$, etc.).

We therefore have to solve the differential equation, Equation (13.20), and we follow standard mathematical practice by first examining the limiting behaviour for large r

$$\frac{d^2 P}{dr^2} = -2\varepsilon P$$

If we limit the discussion to bound states (for which ε is negative), then we note that

$$P(r) \approx \exp(-\beta r) \tag{13.21}$$

where β is a positive real constant. This suggests that we look for a wavefunction of the form

$$P(r) = \Phi(r)\exp(-\beta r) \tag{13.22}$$

where $\Phi(r)$ is a polynomial in r. Substitution of (13.22) into the radial equation, Equation (13.20), then gives

$$\frac{d^2 \Phi}{dr^2} - 2\beta\frac{d\Phi}{dr} + \left(\beta^2 + 2\varepsilon + \frac{2Z}{r}\right)\Phi = 0 \tag{13.23}$$

Suppose for the sake of argument we choose

$$\Phi(r) = \alpha_2 r^2 + \alpha_1 r + \alpha_0 \tag{13.24}$$

where the coefficients α have to be determined. Substitution of (13.24) into (13.23) gives

$$2\alpha_2 - 2\beta(2\alpha_2 r + \alpha_1) + \left(\beta^2 + 2\varepsilon + \frac{2Z}{r}\right) \times (\alpha_2 r^2 + \alpha_1 r + \alpha_0) = 0$$

For this equation to hold for all values of the variable r, the coefficients of all powers of r must separately vanish; thus

$$(\beta^2 + 2\varepsilon)\alpha_2 = 0$$
$$(\beta^2 + 2\varepsilon)\alpha_1 + 2(Z - 2\beta)\alpha_2 = 0$$
$$(\beta^2 + 2\varepsilon)\alpha_0 + 2(Z - \beta)\alpha_1 + 2\alpha_2 = 0$$
$$2Z\alpha_0 = 0$$

It follows that

$$\varepsilon = -\frac{\beta^2}{2}$$
$$\beta = \frac{Z}{2}$$
$$Z\alpha_1 + 2\alpha_2 = 0$$
$$\alpha_0 = 0$$

The first two equations give an energy of $-Z^2/8$ and because ε is actually the energy divided by an atomic unit of energy E_h we have

$$\varepsilon = -\frac{Z^2}{8}E_h$$
$$= -\frac{1}{2}\frac{Z^2}{2^2}E_h$$

The third and fourth equations determine all the coefficients but one, which we are free to choose. If we take $\alpha_1 = 1$, then $\alpha_2 = -Z/2$ and we have

$$\Phi(r) = -\frac{Z}{2}r^2 + r \tag{13.25}$$

13.7 The Atomic Orbitals

We must now examine the general case where the wavefunction depends on three variables. Substitution of

$$\psi(r, \theta, \phi) = R(r)Y(\theta, \phi) \tag{13.26}$$

into the electronic Schrödinger equation, Equation (13.17), gives

$$\frac{1}{R}\frac{\partial}{\partial r}\left(r^2\frac{\partial R}{\partial r}\right) + \frac{8\pi^2 m_e}{h^2}(\varepsilon - U)r^2 = -\frac{1}{Y\sin\theta}\frac{\partial}{\partial\theta}\left(\sin\theta\frac{\partial Y}{\partial\theta}\right) - \frac{1}{Y\sin^2\theta}\frac{\partial^2 Y}{\partial\phi^2} \tag{13.27}$$

By our usual argument, both sides of this equation must be equal to a constant that we will call λ. We then have two equations

$$\frac{1}{r^2}\frac{d}{dr}\left(r^2\frac{dR}{dr}\right) + \left(\frac{8\pi^2 m_e}{h^2}(\varepsilon - U) - \frac{\lambda}{r^2}\right)R = 0$$

$$\frac{1}{\sin\theta}\frac{\partial}{\partial\theta}\left(\sin\theta\frac{\partial Y}{\partial\theta}\right) + \frac{1}{\sin^2\theta}\frac{\partial^2 Y}{\partial\phi^2} + \lambda Y = 0$$

(13.28)

I discuss the second equation in the Appendix when considering angular momentum; the allowed solutions are the spherical harmonics $Y_{l,m_l}(\theta, \phi)$ where $\lambda = l(l+1)$ and l and m_l integers. Introducing the value for λ into the first equation of (13.28) and expanding the first term gives

$$\frac{d^2 R}{dr^2} + \frac{2}{r}\frac{dR}{dr} + \left(\frac{8\pi^2 m_e}{h^2}(\varepsilon - U) - \frac{l(l+1)}{r^2}\right)R = 0 \qquad (13.29)$$

or, in terms of $P(r)$ introduced above (where $P(r) = rR(r)$)

$$\frac{d^2 P}{dr^2} + \left(\frac{8\pi^2 m_e}{h^2}(\varepsilon - U) - \frac{l(l+1)}{r^2}\right)P = 0 \qquad (13.30)$$

The term in $l(l+1)$ is called the *centrifugal potential*; it adds to the Coulomb term to give an *effective potential*.

The radial equation, Equation (13.30), is more complicated than Equation (13.17) because of the $l(l+1)/r^2$ term, but in essence the same method of solution can be used. The details are given in standard traditional quantum chemistry texts such as Eyring, Walter and Kimball. The radial solutions are a set of functions from mathematical physics called the *associated Laguerre polynomials*. (The Laguerre polynomial $L_\alpha(x)$ of degree α in x is defined as

$$L_\alpha(x) = \exp(x)\frac{d^\alpha}{dx^\alpha}(x^\alpha \exp(-x))$$

and the βth derivative of $L_\alpha(x)$ is called an associated Laguerre polynomial.)

13.7.1 $l = 0$ (s orbitals)

The atomic orbitals are given names depending on the three quantum numbers. The first three s orbitals are given in Table 13.2; they are conventionally written in terms of the variable $\rho = Zr/a_0$ (where $Z = 1$ for hydrogen). They are often presented as plots of ψ vs. ρ or of $4\pi r^2\psi^2$ vs. ρ. The latter is known as the radial distribution function and it gives the probability of finding the electron in between two shells or radii r and $r + dr$ surrounding the nucleus. Representative plots for hydrogen are

Table 13.2   First few s-orbitals

n, l, m	Symbol	Normalized wavefunction
1, 0, 0	1s	$\frac{1}{\sqrt{\pi}} \left(\frac{Z}{a_0}\right)^{3/2} \exp(-\rho)$
2, 0, 0	2s	$\frac{1}{4\sqrt{2\pi}} \left(\frac{Z}{a_0}\right)^{3/2} (2 - \rho) \exp\left(-\frac{\rho}{2}\right)$
3, 0, 0	3s	$\frac{2}{81\sqrt{3\pi}} \left(\frac{Z}{a_0}\right)^{3/2} (27 - 18\rho + 2\rho^2) \exp\left(-\frac{\rho}{3}\right)$

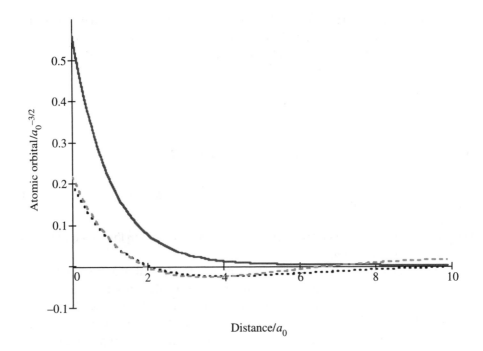

Figure 13.2   Hydrogen 1s, 2s and 3s orbital plots

shown in Figures 13.2 and 13.3. The dimension of ψ is $(\text{length})^{-3/2}$ and so the quantities plotted are, respectively, $a_0^{3/2}\psi$ and $a_0(4\pi r^2 \psi^2)$ vs. r/a_0.

All the curves are asymptotic to the horizontal axis at infinity. The 2 s curve crosses the axis once (it has a *single radial node*), the 3s curve crosses the axis twice and so on. There is nothing else particularly remarkable about such plots of decreasing exponential functions. The radial distribution curves are a little more interesting.

The radial distribution curve shows a maximum at the first Bohr radius for the 1s orbital. Bohr's theory stated that the electron would be in a fixed circular orbit around the centre of mass with exactly this radius! As the principal quantum number increases, so does the average value of the radial distribution function. Elementary chemistry texts attach much importance to diagrams of these kinds.

There are other ways to visualize the atomic orbitals. Figures 13.4 and 13.5 show contour diagrams for the 1s and 2s hydrogen atomic orbitals. The particular software package I used marks the axes with somewhat arbitrary units depending on the range

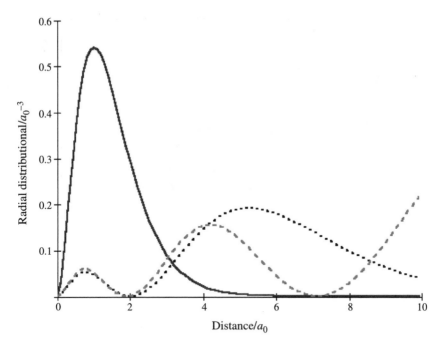

Figure 13.3   Radial distribution curves for the hydrogen 1s, 2s and 3s orbitals

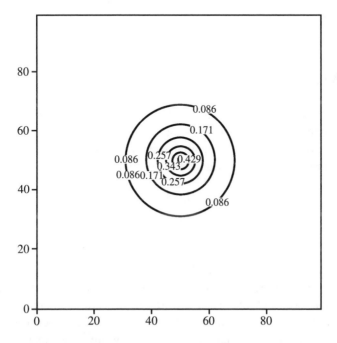

Figure 13.4   Hydrogen 1s contour diagram (nucleus at centre)

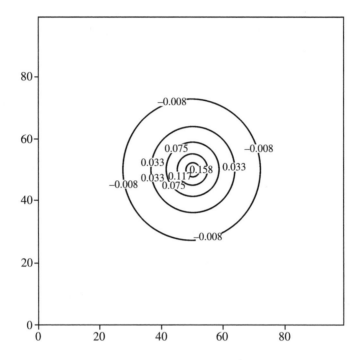

Figure 13.5 Hydrogen 2s contour diagram

of points chosen, although the contour values themselves are correct and the nucleus
is at the centre of the plane.

13.7.2 The p orbitals

For angular momentum quantum number $l = 1$, there are three possible values of m_l,
namely -1, 0 and $+1$. The angular factors for $m_l = \pm 1$ are complex (they involve
the square root of -1 in an exponential factor) and it is usual to make use of the
de Moivre theorem in order to visualize the orbitals

$$\cos \phi = \tfrac{1}{2}(\exp(j\phi) + \exp(-j\phi))$$
$$\sin \phi = \frac{1}{2j}(\exp(j\phi) - \exp(-j\phi)) \tag{13.31}$$

The *real equivalent* p orbitals are therefore taken as linear combinations of the
$m_l = +1$ and -1. Despite the name, they are not completely equivalent to the
complex orbitals because only the latter are eigenfunctions of the z component of
the angular momentum. Luckily, this only matters in the presence of an external
magnetic field. The real equivalent 2p and 3p orbitals are shown in Table 13.3. These
are usually represented as contour diagrams; the orbital shapes are memorable

Table 13.3 The first few p orbitals

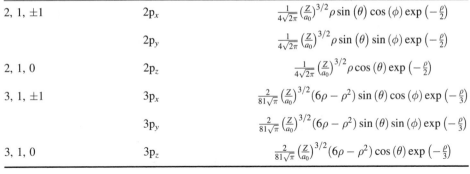

n, l, m	Symbol	Normalized wavefunction
2, 1, ±1	$2p_x$	$\frac{1}{4\sqrt{2\pi}}\left(\frac{Z}{a_0}\right)^{3/2}\rho\sin(\theta)\cos(\phi)\exp\left(-\frac{\rho}{2}\right)$
	$2p_y$	$\frac{1}{4\sqrt{2\pi}}\left(\frac{Z}{a_0}\right)^{3/2}\rho\sin(\theta)\sin(\phi)\exp\left(-\frac{\rho}{2}\right)$
2, 1, 0	$2p_z$	$\frac{1}{4\sqrt{2\pi}}\left(\frac{Z}{a_0}\right)^{3/2}\rho\cos(\theta)\exp\left(-\frac{\rho}{2}\right)$
3, 1, ±1	$3p_x$	$\frac{2}{81\sqrt{\pi}}\left(\frac{Z}{a_0}\right)^{3/2}(6\rho-\rho^2)\sin(\theta)\cos(\phi)\exp\left(-\frac{\rho}{3}\right)$
	$3p_y$	$\frac{2}{81\sqrt{\pi}}\left(\frac{Z}{a_0}\right)^{3/2}(6\rho-\rho^2)\sin(\theta)\sin(\phi)\exp\left(-\frac{\rho}{3}\right)$
3, 1, 0	$3p_z$	$\frac{2}{81\sqrt{\pi}}\left(\frac{Z}{a_0}\right)^{3/2}(6\rho-\rho^2)\cos(\theta)\exp\left(-\frac{\rho}{3}\right)$

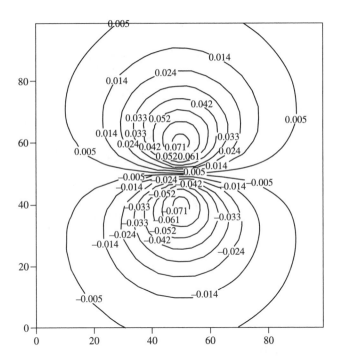

Figure 13.6 H atom 2pz orbital

because they are equivalent and point along the appropriate axes. Figure 13.6 shows a hydrogenic $2p_z$ orbital, which points along the vertical z-axis.

13.7.3 The d orbitals

Similar comments apply to the $l = 2$ orbitals. There are five in all, four of which come in complex pairs. We combine the corresponding values of m_l and $-m_l$ just as for the

Table 13.4 The 3d orbitals

n, l, m_l	Symbol	Normalized wavefunction
3, 2, 0	$3d_{zz}$	$\frac{1}{81\sqrt{6\pi}}\left(\frac{Z}{a_0}\right)^{3/2}\rho^2\left(3\cos^2(\theta)-1\right)\exp\left(-\frac{\rho}{3}\right)$
3, 2, ±1	$3d_{xz}$	$\frac{\sqrt{2}}{81\sqrt{\pi}}\left(\frac{Z}{a_0}\right)^{3/2}\rho^2\sin(\theta)\cos(\theta)\cos(\phi)\exp\left(-\frac{\rho}{3}\right)$
	$3d_{yz}$	$\frac{\sqrt{2}}{81\sqrt{\pi}}\left(\frac{Z}{a_0}\right)^{3/2}\rho^2\sin(\theta)\cos(\theta)\sin(\phi)\exp\left(-\frac{\rho}{3}\right)$
3, 2, ±2	$3d_{x2-y2}$	$\frac{1}{81\sqrt{2\pi}}\left(\frac{Z}{a_0}\right)^{3/2}\rho^2\sin^2(\theta)\cos(2\phi)\exp\left(-\frac{\rho}{3}\right)$
	$3d_{xy}$	$\frac{1}{81\sqrt{2\pi}}\left(\frac{Z}{a_0}\right)^{3/2}\rho^2\sin^2(\theta)\sin(2\phi)\exp\left(-\frac{\rho}{3}\right)$

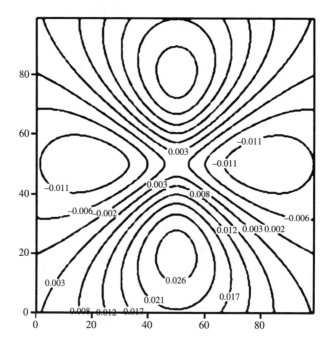

Figure 13.7 Hydrogen $3d_{zz}$ orbital

p orbitals to give real equivalents. Table 13.4 records the 3d orbitals. Once again these are usually represented as contour diagrams, see Figures 13.7 and 13.8.

13.8 The Stern–Gerlach Experiment

If we suspend a compass needle in the earth's magnetic field, then it aligns itself along the magnetic field lines. A compass needle is an example of a magnetic dipole, and the strength of the interaction between the compass needle and this external magnetic field is determined by the *magnetic dipole moment*, $\mathbf{p}_m$. This is a vector

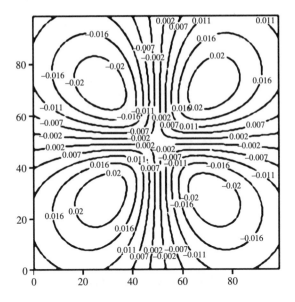

Figure 13.8   Hydrogen $3d_{xz}$ orbital

quantity pointing by convention from the south pole of the needle to the north pole. The interaction between the dipole and the field is determined by the magnetic potential energy

$$U_m = -\mathbf{p}_m \cdot \mathbf{B}(\mathbf{r})$$ (13.32)

where the position vector $\mathbf{r}$ is the position of the dipole and $\mathbf{B}$ the magnetic induction. If $\mathbf{B}$ is uniform and so does not depend on $\mathbf{r}$, then the gradient is zero and so the force is also zero, in accord with experiment.

Around 1820, Oersted discovered experimentally that electric currents could exert forces similar to those exerted by permanent magnets, for example on a compass needle. Figure 13.9 shows a simple current loop located in the xz plane and carrying a steady current I.

If the loop is flat, then the dipole is perpendicular to the plane, and if we consider points on the axis far away from the loop, then it turns out that the magnitude of $\mathbf{p}_m$ is

$$p_m = IA$$

where A is the area of the loop (which need not be circular).

It is easy to demonstrate from classical physics that a particle with mass M, charge Q and angular momentum $\mathbf{l}$ is a magnetic dipole

$$\mathbf{p}_m = \frac{Q}{2M}\mathbf{l}$$ (13.33)

According to Bohr's model of the hydrogen atom, the allowed electron orbits each had an angular momentum that was an integral multiple of $h/2\pi$. Since magnetic dipoles are linked to angular momentum, the possibility arises that if we could

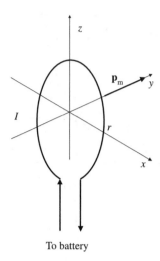

Figure 13.9   Current loop

measure the magnetic dipole moments of individual atoms, then we could investigate the quantization of angular momentum. In the presence of an external magnetic field, the bar magnet will take up one of its possible alignments with the axis of quantization (the direction of the magnetic field lines).

O. Stern and W. Gerlach [45] performed the first and most famous experiment designed to investigate the quantization of angular momentum. Whilst the force on a magnetic dipole is zero in a uniform magnetic field, the force is not zero for a *non-uniform* field. It is difficult to hang a single atom between the poles of a magnet, so a beam of atoms was passed through such a field and the deflection of the beam measured. A schematic diagram of the Stern–Gerlach equipment is shown in Figure 13.10.

A beam of atoms is produced in an oven, and passes through a collimating slit and often a velocity selector in order to form a monoenergetic beam. The beam then travels in the *x*-direction through a long, powerful electromagnet whose pole pieces are deliberately shaped to give a non-uniform magnetic field in the *z*-direction.

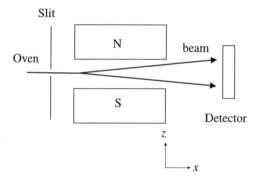

Figure 13.10   Stern–Gerlach experiment

Detailed analysis shows that the force on the magnetic dipoles in the z-direction is given by a force proportional to the z-component of the magnetic dipole moment (and hence the z-component of the angular momentum vector)

$$F_z = \mathbf{p}_{m,z} \frac{\partial B_z}{\partial z} \tag{13.34}$$

It is a difficult experiment to perform satisfactorily, for the atoms collide with each other in the beam and so distort the result. An essential part of such an *atomic beam* apparatus is that there should be a good vacuum system, and very low beam intensities have to be used. Despite the experimental difficulties, the findings were positive; the atom beam was shown to split on passage through the magnetic field and this confirmed the quantization of angular momentum.

13.9 Electron Spin

The first experiments were made with silver atoms. The surprising result of the Stern–Gerlach experiment was not that the beam split, but that it split into two components, with a different separation than that expected. Similar results were obtained for copper and gold, and in later work for the alkali metals and for hydrogen atoms. The point is that an atomic dipole moment characterized by quantum number l should show $2l + 1$ orientations with the magnetic field. Since the quantum number takes values $0, 1, 2, \ldots$ there should always be an odd number of orientations and hence the atom beam should split into an odd number of components.

The explanation came in 1925 when S. Goudsmit and G. Uhlenbeck [46] analysed the splittings of spectral lines occurring when atoms are subjected to an external magnetic field (the *Zeeman effect*). They showed that the measured splittings could be explained if electrons were postulated to have an intrinsic magnetic moment, in addition to the one they acquired by motion about the atomic nucleus. In order to explain the experimental results, they assumed the existence of a spin angular momentum vector **s**, which had similar properties to **l** above. In order to account for the splitting into two beams, they postulated a spin quantum number for electrons s of $\frac{1}{2}$ and so a spin magnetic quantum number m_s of $\pm\frac{1}{2}$. The spin wavefunctions are usually written α (corresponding to the $+\frac{1}{2}$ spin quantum number) and β (corresponding to the $m_s = -\frac{1}{2}$ quantum number). So we write, just as for ordinary angular momentum

$$\hat{s}^2\alpha = \frac{1}{2}\left(\frac{1}{2} + 1\right)\frac{h^2}{4\pi^2}\alpha$$

$$\hat{s}^2\beta = \frac{1}{2}\left(\frac{1}{2} + 1\right)\frac{h^2}{4\pi^2}\beta$$

$$\hat{s}_z\alpha = \frac{1}{2}\frac{h}{2\pi}\alpha \tag{13.35}$$

$$\hat{s}_z\beta = -\frac{1}{2}\frac{h}{2\pi}\beta$$

A problem appeared once the splittings were analysed; according to classical theory, the spin magnetic moment of an electron should be

$$\mathbf{P}_{\text{spin}} = \frac{-e}{2m_{\text{e}}}\mathbf{s} \tag{13.36}$$

whilst to get agreement with the splittings, an extra factor of (almost) 2 was needed. The solution to the problem was the introduction of the g-factor, an experimentally determined quantity that made Equation (13.36) above, correct. Thus we write

$$\mathbf{P}_{\text{spin}} = -g_{\text{e}}\frac{e}{2m_{\text{e}}}\mathbf{s}$$

The electronic g_{e} factor is another of physical science's very accurately known constants, having a value of

$$g_{\text{e}} = 2.0023193043787 \pm (82 \times 10^{-12})$$

13.10 Total Angular Momentum

When orbital angular momentum and spin angular momentum both exist in the same atom, the magnetic moments that result from these two angular momenta interact to cause a splitting of the energy level. The interaction that leads to this splitting is called *spin–orbit coupling* and it couples the two into a resultant *total angular momentum*. A simple vector model that is very similar to the model used to describe orbital angular momentum can describe this. According to this model, the total angular momentum of an electron is characterized by a quantum number j. For any given non-zero value of l the possible values of j are given by

$$j = l \pm s$$

(The use of j as an atomic quantum number is not to be confused with the use of j for the square root of -1.) The rule is that the quantum number j must always be positive, so if $l = 1$, then $j = \frac{3}{2}$ and $\frac{1}{2}$, but if $l = 0$, then we only have $j = \frac{1}{2}$.

I emphasized earlier the interplay between spectroscopic data and theoretical developments; even before Schrödinger's time many highly precise spectroscopic data had been obtained experimentally. The traditional reference source for such data are the three volumes of tables in the *Atomic Energy Levels* series [47]. In the past few years, the National Institute of Standards and Technology (NIST) atomic spectroscopists have made available a unified comprehensive Atomic Spectra Database on the World Wide Web, which contains spectral reference data for 91 000 wavelengths (http://www.nist.gov). The sample in Table 13.5 is taken from the database.

Table 13.5 Internet search for hydrogen term values

Configuration	Term symbol	j	Term value (cm^{-1})
$1s^1$	^{2}S	$\frac{1}{2}$	0
$2p^1$	^{2}P	$\frac{1}{2}$	82 258.9206
		$\frac{3}{2}$	82 259.2865
$2s^1$	^{2}S	$\frac{1}{2}$	82 258.9559
$3p^1$	^{2}P	$\frac{1}{2}$	97 492.2130
		$\frac{3}{2}$	97 492.3214
$3s^1$	^{2}S	$\frac{1}{2}$	97 492.2235
$3d^1$	^{2}D	$\frac{3}{2}$	97 492.3212
		$\frac{5}{2}$	97 492.3574
$4p^1$	^{2}P	$\frac{1}{2}$	102 823.8505
		$\frac{3}{2}$	102 823.8962
$4s^1$	^{2}S	$\frac{1}{2}$	102 823.8549
$4d^1$	^{2}D	$\frac{3}{2}$	102 823.8961
		$\frac{5}{2}$	102 823.9114
$4f^1$	^{2}F	$\frac{5}{2}$	102 823.9113
		$\frac{7}{2}$	102 823.9190

Note that the data contain *spectroscopic term symbols* for each level, which are discussed in all elementary physical chemistry undergraduate texts. Spectroscopists traditionally deal with *term values* rather than energies; these are just ε/hc_0.

13.11 Dirac Theory of the Electron

There is no mention of electron spin from the Schrödinger equation, and certainly no clue as to why the classical equation for magnetic moments is in error by (roughly) a factor of 2 when applied to electron spin, Equation (13.36). If we consider the time-dependent Schrödinger equation for a free electron

$$-\frac{h^2}{8\pi^2 m_e}\left(\frac{\partial^2}{\partial x^2} + \frac{\partial^2}{\partial y^2} + \frac{\partial^2}{\partial z^2}\right)\Psi(\mathbf{r},t) = j\frac{h}{2\pi}\frac{\partial}{\partial t}\Psi(\mathbf{r},t) \qquad (13.37)$$

(where j is the square root of -1), it is seen to be a second-order partial differential equation with respect to the spatial coordinates and a first-order partial differential equation with respect to time. It therefore is not consistent with the Special Theory of Relativity, which requires that time and space should enter such equations on an equal footing. If the equation is second order in space, it should also be second order in time.

Erwin Schrödinger decided that the way ahead was to abandon the classical energy expression and start again with the relativistically correct equation

$$\varepsilon^2 = m_e^2 c_0^4 + c_0^2 \mathbf{p}^2 \tag{13.38}$$

He then made the operator substitutions

$$\varepsilon \rightarrow j\frac{h}{2\pi}\frac{\partial}{\partial t}; \qquad p_x \rightarrow -j\frac{h}{2\pi}\frac{\partial}{\partial x}, \quad \text{etc.}$$

to arrive at the Klein–Gordan equation

$$\left(\frac{\partial^2}{\partial x^2} + \frac{\partial^2}{\partial y^2} + \frac{\partial^2}{\partial z^2} - \frac{1}{c_0^2}\frac{\partial^2}{\partial t^2} - \frac{4\pi^2 m_e^2 c_0^2}{h^2}\right)\Psi(\mathbf{r}, t) = 0 \tag{13.39}$$

In discussing relativistic matters, it is usual to write equations such as this in *four-vector notation*. We define a four-vector with components

$$\begin{pmatrix} x_1 \\ x_2 \\ x_3 \\ x_4 \end{pmatrix} = \begin{pmatrix} x \\ y \\ z \\ jc_0 t \end{pmatrix}$$

so that Equation (13.39) becomes

$$\left(\frac{\partial^2}{\partial x_1^2} + \frac{\partial^2}{\partial x_2^2} + \frac{\partial^2}{\partial x_3^2} + \frac{\partial^2}{\partial x_4^2} - \frac{4\pi^2 m_e^2 c_0^2}{h^2}\right)\Psi(x_1, x_2, x_3, x_4) = 0 \tag{13.40}$$

The Klein–Gordan equation is more satisfactory in that it has the desirable symmetry but it turns out that it cannot describe electron spin. In the limit of low energy, it is equivalent to the familiar Schrödinger equation.

Paul Dirac had the ingenious idea of working with a relativistic equation that was linear in the vector components. He wrote

$$\left(\gamma_1 \frac{\partial}{\partial x_1} + \gamma_2 \frac{\partial}{\partial x_2} + \gamma_3 \frac{\partial}{\partial x_3} + \gamma_4 \frac{\partial}{\partial x_4} - \frac{2\pi m_e c_0}{h}\right)\Psi(x_1, x_2, x_3, x_4) = 0 \tag{13.41}$$

where the multipliers γ_i have to be determined. This equation is called the *Dirac equation*. Both the Schrödinger and the Klein–Gordan equation are second order, and it is usual to manipulate the Dirac equation in order to give a corresponding second-order equation. This can be done by operating on Equation (13.41) with the operator

$$\gamma_1 \frac{\partial}{\partial x_1} + \gamma_2 \frac{\partial}{\partial x_2} + \gamma_3 \frac{\partial}{\partial x_3} + \gamma_4 \frac{\partial}{\partial x_4} + \frac{2\pi m_e c_0}{h}$$

A little operator algebra shows that the multipliers have to satisfy

$$\gamma_i \gamma_j + \gamma_j \gamma_i = \begin{cases} 2 & \text{if } i = j \\ 0 & \text{if } i \neq j \end{cases}$$

It can be shown that any particle whose wavefunction satisfies the Dirac equation must be a spin-$\frac{1}{2}$ particle. Not only that, the Dirac treatment gives the correct value for the magnetic moment in that it gives

$$\mathbf{p}_{\text{spin}} = -2\frac{e}{2m_e}\mathbf{s}$$

13.12 Measurement in the Quantum World

The process of measurement in quantum mechanics is subtler than in classical mechanics. As discussed in the Appendix, the possible results of measurements depend on the eigenvalues of the appropriate operators. Also, if we wish to make simultaneous measurements of two observables (such as the linear momentum and position, or two components of an angular momentum vector), we have to take account of Heisenberg's uncertainty principle. Certain pairs of observables can be measured simultaneously to arbitrary precision, certain other pairs cannot.

The word 'measurement' gets used in two different ways in quantum mechanics. Suppose we have a one-electron atom as discussed above; we know that the energies are found from Schrödinger's equation, which I will write in Hamiltonian form as

$$\hat{H}\psi_i = \varepsilon_i\psi_i$$

If the atom is in state ψ_i, then repeated energy measurements on the same atom will always yield the same result, ε_i (we say that the system is in a *pure state*).

If, on the other hand, we pass a beam of electrons through a Stern–Gerlach apparatus, then the magnet separates the beam into two components that correspond to the two spin eigenvalues $m_s = +\frac{1}{2}$ and $m_s = -\frac{1}{2}$. This kind of measurement is referred to as *state preparation*, for if we pass the $m_s = +\frac{1}{2}$ beam through a further Stern–Gerlach apparatus oriented in the same way as the first, we simply observe the one beam. I have shown this schematically in Figure 13.11.

However, electrons in the incident beam are not in pure states, and their spin wavefunction can be written as linear combinations of the two spin functions α and β

$$\psi = a\alpha + b\beta$$

They are said to be in *mixed states*. Here, a and b are scalar (complex) constants. The question is: How do we interpret the measurement process? For example, does a given electron in some sense know which spin state it should be in, before passage through the apparatus? Similar questions puzzled Schrödinger and his contemporaries, and led Schrödinger to state the famous *cat paradox*. In a version of this

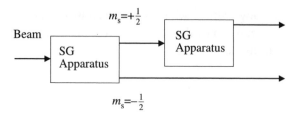

Figure 13.11   Repeated Stern–Gerlach measurements

thought experiment, a cat is placed in a closed box together with, (for example), a radioactive atom that releases a deadly poison when it decays. If the wavefunction of the living cat is ψ_L and that of a dead cat ψ_D, then the state of the cat at time t can be described by

$$\psi_{\text{cat}}(t) = c_L(t)\psi_L + c_D(t)\psi_D$$

The coefficients are time dependent, as is the wavefunction. Before the box is closed, the cat is alive. Once the box is closed, the cat is apparently neither dead nor alive; it is in a mixed state. On opening the box at a certain time, the cat is either dead or alive, but is it the act of measurement that has forced the cat into one state or the other?

There are whole rafts of similar questions that can be asked. Most of the difficulties can be resolved by recognizing that we are asking a statistical question, not one about a specific cat. If we prepare a large number of identical experiments, with identical cats and amounts of radioactive isotopes, close the lids on the boxes and then examine a statistically significant number of boxes, we will find that a fraction $|c_D(t)|^2$ will have died and a fraction $|c_L(t)|^2$ will be alive, but no prediction can be made as to the state of a given cat in a given box.

14 The Orbital Model

Our next step is to consider a many-electron atom, such as that illustrated in Figure 14.1. I am going to make the 'infinite nuclear mass' approximation, and only consider the electronic problem. The atom is therefore fixed in space with the nucleus at the centre of the coordinate system. The n electrons are at position vectors $\mathbf{r}_1$, $\mathbf{r}_2, \ldots, \mathbf{r}_n$ and the scalar distance between (say) electrons 1 and 2 is r_{12} in an obvious notation. The electronic wavefunction will depend on the coordinates of all the electrons, and I will write it $\Psi(\mathbf{r}_1, \mathbf{r}_2, \ldots, \mathbf{r}_n)$. If the nuclear charge is Ze, then we have to consider three contributions to the electronic energy: the kinetic energy of each electron, the Coulomb attraction between the nucleus and the electrons, and finally the Coulomb repulsion between pairs of electrons. We therefore write the Hamiltonian operator as

$$\hat{H} = \sum_{i=1}^{n} \left(-\frac{h^2}{8\pi^2 m_e} \left(\frac{\partial^2}{\partial x_i^2} + \frac{\partial^2}{\partial y_i^2} + \frac{\partial^2}{\partial z_i^2} \right) - \frac{Ze^2}{4\pi\epsilon_0 r_i} \right)$$
$$+ \sum_{i=1}^{n-1} \sum_{j=i+1}^{n} \frac{e^2}{4\pi\epsilon_0 r_{ij}} \tag{14.1}$$

14.1 One- and Two-Electron Operators

In order to stop the notation becoming unwieldy, I will group these terms as follows. Each term in the first bracket refers to the coordinates of the individual electrons, and gives the kinetic energy of each electron together with its attraction to the nucleus. I will call such terms *one-electron operators* and authors normally write them as

$$\hat{h}^{(1)}(\mathbf{r}_i) = -\frac{h^2}{8\pi^2 m_e} \left(\frac{\partial^2}{\partial x_i^2} + \frac{\partial^2}{\partial y_i^2} + \frac{\partial^2}{\partial z_i^2} \right) - \frac{Ze^2}{4\pi\epsilon_0 r_i} \tag{14.2}$$

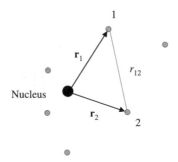

Figure 14.1   Many-electron atom

Each term in the second double summation gives the Coulomb repulsion of a pair of electrons; I will refer to them as *two-electron operators* and write

$$\hat{g}(\mathbf{r}_i, \mathbf{r}_j) = \frac{e^2}{4\pi\epsilon_0 r_{ij}} \tag{14.3}$$

The Hamiltonian is then, in our compact notation

$$\hat{H} = \sum_{i=1}^{n} \hat{h}^{(1)}(\mathbf{r}_i) + \sum_{i=1}^{n-1} \sum_{j=i+1}^{n} \hat{g}(\mathbf{r}_i, \mathbf{r}_j) \tag{14.4}$$

and we wish to investigate the solutions of the Schrödinger equation

$$\hat{H}\Psi(\mathbf{r}_1, \mathbf{r}_2, \dots, \mathbf{r}_n) = \varepsilon\Psi(\mathbf{r}_1, \mathbf{r}_2, \dots, \mathbf{r}_n) \tag{14.5}$$

14.2 The Many-Body Problem

The many-electron atom is an example of a so-called *many-body problem*. These are not unique to quantum theory; a familiar example is planetary motion. Newton's equations of motion can be solved exactly for the motion of any one of the planets around the sun individually, but the planets also attract each other. During the eighteenth and nineteenth centuries a great deal of effort was expended trying to find an exact solution to planetary motion, but all efforts failed and it is generally accepted that exact solutions do not exist, even for just three bodies. Astronomers are lucky in the sense that the gravitational force depends on the product of two masses and the mass of the sun is much greater than the masses of the individual planets. Ingenious techniques were developed to treat the inter-planetary attractions as small perturbations, and the planetary problem can be solved numerically to any accuracy required.

To continue with the planetary motion analogy, chemists are less lucky in one sense because the electrostatic force depends on the product of two charges, and in the case of an electron and a proton these forces are roughly equal in magnitude. It will therefore come as no surprise when I tell you that an exact solution of the many-electron atomic Schrödinger problem seems to be impossible, because of the electron–electron repulsions, and that we are apparently in a weaker position than the astronomers. On the positive side, I have stressed the Born interpretation of quantum mechanics. Here we do not focus attention on the trajectories of the individual particles, but rather we ask about the probability that a region of space is occupied by any one of the particles. The quantum mechanical problem therefore seems to be more hopeful than the astronomical one.

14.3 The Orbital Model

Let me now investigate the Schrödinger equation that would result from an atom consisting of electrons that did not repel each other. We can think of this as some 'zero-order' approximation to a true many-electron atom, just as the astronomers might have investigated their simple model of planetary motion. We therefore write

$$\hat{H}\Psi(\mathbf{r}_1, \mathbf{r}_2, \ldots, \mathbf{r}_n) = \varepsilon\Psi(\mathbf{r}_1, \mathbf{r}_2, \ldots, \mathbf{r}_n)$$

$$\left(\sum_{i=1}^{n} \hat{h}^{(1)}(\mathbf{r}_i)\right)\Psi(\mathbf{r}_1, \mathbf{r}_2, \ldots, \mathbf{r}_n) = \varepsilon\Psi(\mathbf{r}_1, \mathbf{r}_2, \ldots, \mathbf{r}_n)$$

and this appears to be a candidate for separation of variables; I write

$$\Psi(\mathbf{r}_1, \mathbf{r}_1, \ldots, \mathbf{r}_1) = \psi_1(\mathbf{r}_1)\psi_2(\mathbf{r}_2) \cdots \psi_n(\mathbf{r}_n) \tag{14.6}$$

Substitution and separation gives n identical one-electron atom Schrödinger equations, and so the total wavefunction is a product of the familiar 1s, 2s and 2p atomic orbitals discussed in Chapter 13 (with nuclear charge Ze). The energy is given by the sum of the orbital energies.

We then have to take account of electron spin and the Pauli Principle, as discussed in Chapter 13. I can remind you of the principles by writing down some of the lowest energy solutions for helium, in particular those that formally involve the 1s and 2s orbitals. I will label the atomic orbitals 1s and 2s for obvious reasons, and I will adopt the habit of writing s for the spin variable (sorry about the double use of the same symbol s) as in Chapter 13.

The allowed wavefunctions are given in Table 14.1, where Ψ_1 describes the spectroscopic ground state, giving energy $2\varepsilon_{1s}$. The remaining wavefunctions describe excited states. Ψ_2 is the first excited singlet state whilst Ψ_3, Ψ_4 and Ψ_5 are the three components of the first triplet state. We refer to Ψ_2 through Ψ_5 as *singly excited*

Table 14.1 Some electronic states for helium

State	Spatial part	Spin part	Energy, ε
Ψ_1	$1s(\mathbf{r}_1)1s(\mathbf{r}_2)$	$\frac{1}{\sqrt{2}}(\alpha(s_1)\beta(s_2) - \alpha(s_2)\beta(s_1))$	$2\varepsilon_{1s}$
Ψ_2	$\frac{1}{\sqrt{2}}(1s(\mathbf{r}_1)2s(\mathbf{r}_2) + 1s(\mathbf{r}_2)2s(\mathbf{r}_1))$	$\frac{1}{\sqrt{2}}(\alpha(s_1)\beta(s_2) - \alpha(s_2)\beta(s_1))$	$\varepsilon_{1s} + \varepsilon_{2s}$
Ψ_3		$\alpha(s_1)\alpha(s_2)$	$\varepsilon_{1s} + \varepsilon_{2s}$
Ψ_4	$\frac{1}{\sqrt{2}}(1s(\mathbf{r}_1)2s(\mathbf{r}_2) - 1s(\mathbf{r}_2)2s(\mathbf{r}_1))$	$\frac{1}{\sqrt{2}}(\alpha(s_1)\beta(s_2) + \alpha(s_2)\beta(s_1))$	$\varepsilon_{1s} + \varepsilon_{2s}$
Ψ_5		$\beta(s_1)\beta(s_2)$	$\varepsilon_{1s} + \varepsilon_{2s}$
Ψ_6	$2s(\mathbf{r}_1)2s(\mathbf{r}_2)$	$\frac{1}{\sqrt{2}}(\alpha(s_1)\beta(s_2) - \alpha(s_2)\beta(s_1))$	$2\varepsilon_{2s}$

wavefunctions because they have been formally produced from the ground state wavefunction by exciting a single electron. Owing to my neglect of electron repulsion, the energies of Ψ_2 through Ψ_5 are the same. Ψ_6 is a *doubly excited wavefunction* and so on.

Our zero-order model predicts that the singlet and triplet excited states derived from a $1s^1\,2s^1$ orbital configuration will have energy

$$\varepsilon_{2s} - \varepsilon_{1s} = \frac{Z^2 m_e e^4}{8h^2 \epsilon_0^2} \left(\frac{1}{1^2} - \frac{1}{2^2} \right)$$

above the ground state. Since $Z = 2$ for helium we calculate a wavenumber of $329\,212\,\text{cm}^{-1}$ Experimental data can be found at the NBS/NIST website http://www.nist.gov as in Table 14.2.

I should explain that the J quantum number is similar to the j quantum number we met in our study of one-electron atoms in Chapter 13; for light atoms such as helium it is determined by combining the individual electron orbital and spin angular momentum quantum numbers, according to a set of well-known rules called the *Russell–Saunders* scheme. We combine the l quantum numbers for the two electrons; since $l = 0$ for an s electron, the allowed resultant L is also 0. The electronic states are therefore S states. We also combine the spin quantum numbers s. Since $s = \frac{1}{2}$ for an electron, the allowed values of the resultant S are $\frac{1}{2} + \frac{1}{2}$ and $\frac{1}{2} - \frac{1}{2}$ and the spin multiplicities $2S + 1$ are 1 and 3. We then combine the L's and the S's in the same way to get J.

The zero-order model is not even qualitatively correct; it overestimates the energy difference between the ground state and the excited states, and has nothing at all to

Table 14.2 Experimental data for helium

Configuration	Term	J	Term value (cm^{-1})
$1s^2$	^1S	0	0
$1s^12s^1$	^3S	1	159 856.07760
$1s^12s^1$	^1S	0	166 277.542

say about the experimentally interesting difference between the singlet and the triplet excited states. This poor agreement with experiment is mostly due to our neglect of electron repulsion but is in part due to the fact that each electron shields the other electron from the Coulomb attraction due to the nucleus.

14.4 Perturbation Theory

There are very few physically interesting problems that we can solve exactly by the standard methods of quantum theory. The great majority of problems, including those of atomic and molecular structure, must therefore be tackled by approximate methods. Suppose that our problem is to solve (for example), the helium atom electronic Schrödinger equation

$$\hat{H}_i \Psi_i(\mathbf{r}_1, \mathbf{r}_2) = \varepsilon_i \Psi_i(\mathbf{r}_1, \mathbf{r}_2) \tag{14.7}$$

We might suspect that this problem is similar to that of two superimposed hydrogen atoms, for which we can find exact solutions to the *zero-order* Schrödinger equation

$$\hat{H}^{(0)} \Psi_i^{(0)}(\mathbf{r}_1, \mathbf{r}_2) = \varepsilon_i^{(0)} \Psi_i^{(0)}(\mathbf{r}_1, \mathbf{r}_2) \tag{14.8}$$

The aim of perturbation theory is to relate the solutions of problem (14.7) to the exact zero-order solutions of (14.8). To simplify the notation, I will drop all references to two electrons; perturbation theory is a general technique, not one that is specific to helium. It is also general in that it can be applied to every solution not just the lowest energy one. There are two technical points; first I am going to assume that the state of interest is not degenerate. There is a special version of perturbation theory that is applicable to degenerate states, and if you are interested I can refer you to the classic texts such as Eyring, Walter and Kimball. Second, I am going to assume that the wavefunctions are real rather than complex. It makes the equations look a bit easier on the eye.

We proceed as follows: first we write the Hamiltonian as

$$\hat{H} = \hat{H}^{(0)} + \lambda \hat{H}^{(1)} \tag{14.9}$$

where λ is called the *perturbation parameter*. The second term in (14.9) is called the *perturbation*. We assume that the energies and wavefunctions for our problem can be expanded in terms of the zero-order problem as

$$\Psi_i = \Psi_i^{(0)} + \lambda \Psi_i^{(1)} + \lambda^2 \Psi_i^{(2)} + \cdots$$
$$\varepsilon_i = \varepsilon_i^{(0)} + \lambda \varepsilon_i^{(1)} + \lambda^2 \varepsilon_i^{(2)} + \cdots \tag{14.10}$$

The superscript $^{(k)}$ refers to the order of perturbation theory, and the equation should demonstrate why the perturbation parameter λ is added; it is a formal device used to keep track of the 'orders' of the perturbation. It might physically correspond to an applied electric field (as in the Stark effect) or an applied magnetic induction (as in the Zeeman effect, in which case we need to use complex wavefunctions). If we substitute the expansions into our problem we find

$$\hat{H}^{(0)}\Psi_i^{(0)} + \lambda(\hat{H}^{(1)}\Psi_i^{(0)} + \hat{H}^{(0)}\Psi_i^{(1)}) + \lambda^2(\hat{H}^{(1)}\Psi_i^{(1)} + \hat{H}^{(0)}\Psi_i^{(2)}) + \cdots$$
$$= \varepsilon_i^{(0)}\Psi_i^{(0)} + \lambda(\varepsilon_i^{(1)}\Psi_i^{(0)} + \varepsilon_i^{(0)}\Psi_i^{(1)})$$
$$+ \lambda^2(\varepsilon_i^{(2)}\Psi_i^{(0)} + \varepsilon_i^{(1)}\Psi_i^{(1)} + \varepsilon_i^{(0)}\Psi_i^{(2)}) + \cdots \tag{14.11}$$

In order that Equation (14.11) may be true for all values of λ, the coefficients of λ on either side of the equation must be equal. Equating and rearranging we find

$$\hat{H}^{(0)}\Psi_i^{(0)} = \varepsilon_i^{(0)}\Psi_i^{(0)}$$
$$(\hat{H}^{(0)} - \varepsilon_i^{(0)})\Psi_i^{(1)} = (\varepsilon_i^{(1)} - \hat{H}^{(1)})\Psi_i^{(0)} \tag{14.12}$$
$$(\hat{H}^{(0)} - \varepsilon_i^{(0)})\Psi_i^{(2)} = \varepsilon_i^{(2)}\Psi_i^{(0)} + \varepsilon_i^{(1)}\Psi_i^{(1)} - \hat{H}^{(1)}\Psi_i^{(1)}$$

Solution of these equations gives

$$\varepsilon_i = \varepsilon_i^{(0)} + \lambda H_{ii}^{(1)} + \lambda^2 \sum_{j \neq i} \frac{H_{ij}^{(1)} H_{ji}^{(1)}}{\varepsilon_i^{(0)} - \varepsilon_j^{(0)}} + \cdots$$

$$\Psi_i = \Psi_i^{(0)} + \lambda \sum_{j \neq i} \frac{H_{ji}^{(1)}}{\varepsilon_i^{(0)} - \varepsilon_j^{(0)}} \Psi_j^{(0)} + \cdots \tag{14.13}$$

I have used the following shorthand for the integrals

$$H_{ij}^{(1)} = \int \Psi_i^{(0)} \hat{H}^{(1)} \Psi_j^{(0)} \mathrm{d}\tau \tag{14.14}$$

The first-order correction to the energy can therefore be calculated from the unperturbed wavefunction and the perturbing Hamiltonian

$$\varepsilon_i^{(1)} = H_{ii}^{(1)}$$
$$= \int \Psi_i^{(0)} \hat{H}^{(1)} \Psi_i^{(0)} \mathrm{d}\tau$$

On the other hand, the second-order correction to the energy requires knowledge of the remaining states

$$\varepsilon_i^{(2)} = \sum_{j \neq i} \frac{\left(\int \Psi_i^{(0)} \hat{H}^{(1)} \Psi_j^{(0)} \mathrm{d}\tau \right)^2}{\varepsilon_i^{(0)} - \varepsilon_j^{(0)}} \tag{14.15}$$

If we return to the helium atom where the zero-order problem is two non-interacting atomic electrons, the zero-order wavefunctions and energies are shown in Table 14.1. The perturbation is the Coulomb repulsion between the two electrons

$$\hat{H}^{(1)} = \frac{e^2}{4\pi\epsilon_0} \frac{1}{r_{12}}$$

and the first-order correction to the ground state energy is

$$\varepsilon^{(1)} = \int \Psi_1 \left(\frac{e^2}{4\pi\epsilon_0} \frac{1}{r_{12}} \right) \Psi_1 d\tau$$

where Ψ_1 is given in Table 14.1.

Evaluation of the integral is far from easy, since it involves the coordinates of both electrons (it is therefore a six-dimensional integral), and it has a singularity (it tends to infinity as the two electrons approach each other). It is shown in the classic texts such as Eyring, Walter and Kimball that

$$\varepsilon^{(1)} = \frac{5Z}{8} \frac{e^2}{4\pi\epsilon_0 a_0}$$
$$\frac{\varepsilon^{(1)}}{hc_0} = \frac{5Z}{8hc_0} \frac{e^2}{4\pi\epsilon_0 a_0}$$
$$= 274\,343\,\text{cm}^{-1}$$

and my revised estimate of the first ionization energy is now $164\,606\,\text{cm}^{-1}$, in better agreement with experiment.

14.5 The Variation Method

Another, completely different method of finding approximate solutions to the Schrödinger equation is based on the following theorem:

Theorem 14.1   *If the lowest eigenvalue of a Hamiltonian $\hat{H}$ is ε and Φ is a function with the correct boundary conditions, then*

$$\frac{\int \Phi^* \hat{H} \Phi d\tau}{\int \Phi^* \Phi d\tau} \geq \varepsilon$$

Once again the proof is given in all the classical texts such as Eyring, Walter and Kimball. I can illustrate the use of this technique by reference to the helium atom.

The spatial part of the zero-order wavefunction is

$$\Psi(\mathbf{r}_1, \mathbf{r}_2) = \frac{Z^3}{\pi a_0^3} \exp\left(-Z\frac{(r_1 + r_2)}{a_0}\right)$$

where Z is the atomic number (2 in this case). We now allow for the possibility that one electron partially screens the nucleus and so the second electron sees a reduced nuclear charge Z'. This suggests that we use a trial wavefunction that comprises a product of modified helium 1s orbitals with effective nuclear charge $Z'e$

$$\Phi(\mathbf{r}_1, \mathbf{r}_2) = \frac{(Z')^3}{\pi a_0^3} \exp\left(-Z'\frac{(r_1 + r_2)}{a_0}\right)$$

We would expect Z' to lie between 1 and 2 and we look for the value of Z' that makes the *variational integral*

$$\varepsilon' = \frac{\int \Phi(\mathbf{r}_1, \mathbf{r}_2)(\hat{h}^{(1)}(\mathbf{r}_1) + \hat{h}^{(1)}(\mathbf{r}_2) + \hat{g}(\mathbf{r}_1, \mathbf{r}_2))\Phi(\mathbf{r}_1, \mathbf{r}_2)\mathrm{d}\tau_1 \mathrm{d}\tau_2}{\int \Phi^2(\mathbf{r}_1, \mathbf{r}_2)\mathrm{d}\tau_1 \mathrm{d}\tau_2} \tag{14.16}$$

a minimum. I have dropped the complex conjugate signs * because of the real nature of the wavefunction. The denominator is

$$\int \Phi^2(\mathbf{r}_1, \mathbf{r}_2)\mathrm{d}\tau_1 \mathrm{d}\tau_2 = \int 1s^2(\mathbf{r}_1)\mathrm{d}\tau_1 \int 1s^2(\mathbf{r}_2)\mathrm{d}\tau_2$$

$$= 1$$

The numerator is

$$\varepsilon' = \int \Phi(\mathbf{r}_1, \mathbf{r}_2)\hat{h}^{(1)}(\mathbf{r}_1)\Phi(\mathbf{r}_1, \mathbf{r}_2)\,\mathrm{d}\tau_1 \mathrm{d}\tau_2 + \int \Phi(\mathbf{r}_1, \mathbf{r}_2)\hat{h}^{(1)}(\mathbf{r}_2)\Phi(\mathbf{r}_1, \mathbf{r}_2)\,\mathrm{d}\tau_1 \mathrm{d}\tau_2$$

$$+ \int \Phi(\mathbf{r}_1, \mathbf{r}_2)\hat{g}(\mathbf{r}_1, \mathbf{r}_2)\Phi(\mathbf{r}_1, \mathbf{r}_2)\,\mathrm{d}\tau_1 \mathrm{d}\tau_2$$

$$\varepsilon' = \int 1s(\mathbf{r}_1)\hat{h}^{(1)}(\mathbf{r}_1)1s(\mathbf{r}_1)\mathrm{d}\tau_1 \int 1s^2(\mathbf{r}_2)\mathrm{d}\tau_2 + \int 1s(\mathbf{r}_2)\hat{h}^{(1)}(\mathbf{r}_2)1s(\mathbf{r}_2)\mathrm{d}\tau_2 \int 1s^2(\mathbf{r}_1)\mathrm{d}\tau_1$$

$$+ \frac{e^2}{4\pi\epsilon_0}\int 1s(\mathbf{r}_1)1s(\mathbf{r}_2)\frac{1}{r_{12}}1s(\mathbf{r}_1)1s(\mathbf{r}_2)\,\mathrm{d}\tau_1 \mathrm{d}\tau_2 \tag{14.17}$$

The first two integrals are related to the energy of a 1s orbital He$^+$; each integral is equal because of the indistinguishability of the two electrons, and I can write each one as

$$
\int 1s(\mathbf{r}_1)\left(-\frac{h^2}{8\pi^2 m_e}\left(\frac{\partial^2}{\partial x_1^2}+\frac{\partial^2}{\partial y_1^2}+\frac{\partial^2}{\partial z_1^2}\right)-\frac{Z'e^2}{4\pi\epsilon_0 r_1}\right)1s(\mathbf{r}_1)\,d\tau_1
$$

$$
+\int 1s(\mathbf{r}_1)\left(\frac{(Z'-Z)e^2}{4\pi\epsilon_0 r_1}\right)1s(\mathbf{r}_1)\,d\tau_1
$$

$$
=-\tfrac{1}{2}(Z')^2 E_h+\int 1s(\mathbf{r}_1)\left(\frac{(Z'-Z)e^2}{4\pi\epsilon_0 r_1}\right)1s(\mathbf{r}_1)\,d\tau_1
$$

$$
=-\tfrac{1}{2}(Z')^2 E_h+Z'(Z'-Z)E_h
$$

Adding the two-electron integral from above we find a variational energy of

$$
\varepsilon = ((Z')^2 - 2ZZ' + \tfrac{5}{8}Z')E_h \tag{14.18}
$$

We obtain the best approximation to the true energy by giving Z' the value that will make the energy a minimum. This means

$$
\frac{d\varepsilon}{dZ'} = (2Z' - 2Z + \tfrac{5}{8})E_h
$$

$$
= 0
$$

and so the best energy results when we take

$$
Z' = Z - \tfrac{5}{16}
$$

This leads to an improved ground state energy of

$$
\varepsilon = -\left(\frac{27}{16}\right)^2 E_h
$$

By introducing more and more parameters like Z' into the wavefunction Φ we can approach more and more closely the experimental result. E. Hylleraas [48] experimented with wavefunctions that were explicit functions of the inter-electron coordinate r_{12}. His first attempt was to write

$$
\Phi = A(\exp(-Z'(r_1 + r_2))(1 + cr_{12})) \tag{14.19}
$$

where A is the normalization constant and Z' and c are adjustable parameters. In later work he made use of expansions such as

$$
\Phi = A(\exp(-Z'(r_1 + r_2))(\text{polynomial in } r_1, r_2, r_{12})) \tag{14.20}
$$

and was able to demonstrate impressive agreement with experiment. We refer to such wavefunctions as *correlated* wavefunctions; as we will soon see, the most sophisticated orbital models average over the electron interactions whilst correlated wavefunctions allow for the 'instantaneous' interaction between electrons.

Table 14.3   Hylleraas's variational results for helium

Approximate wavefunction	ε/E_h
$\exp(-k(r_1 + r_2))$	-2.8478
$\exp(-k(r_1 + r_2))\exp(-c_1 r_{12})$	-2.8896
$\exp(-k(r_1 + r_2))\exp(-c_1 r_{12})\cosh{(c(r_1 - r_2))}$	-2.8994
$\exp(-(\frac{r_1+r_2}{2}))(c_0 + c_1 r_2 + c_2(r_1 - r_2)^2 + c_3(r_1 + r_2) + c_4(r_1 + r_2)^2 + c_5 r_{12}^2)$	-2.9032
'Exact'	-2.90372437703

Some of Hylleraas's results are shown in Table 14.3; the most accurate theoretical value known is that of K. Frankowski and C. L. Pekeris [49]. Hylleraas's approach, whereby we write the interelectron distances explicitly in the wavefunction (and so abandon the orbital model), gives by far the most accurate treatment of atomic systems, but like most attractive propositions there is a catch. I glossed over calculation of the two-electron integral in the discussion above, but this causes a major problem for molecular systems. Any attempt to use hydrogenic orbitals for a molecular system leads to two-electron integrals that are impossibly hard to evaluate. Despite many attempts, Hylleraas's method has never been successfully applied to a large polyatomic molecule.

14.6 The Linear Variation Method

The variation principle as stated above applies only to the lowest energy solution of any given symmetry (spatial and spin). A special kind of variation function widely used in molecular structure theory is the *linear variation function*. For the sake of argument, suppose we try to improve the helium atom ground state wavefunction Ψ_1 by adding Ψ_2 through Ψ_6 (given in Table 14.1) and so write

$$\Phi = c_1\Psi_1 + c_2\Psi_2 + \cdots + c_6\Psi_6$$

We keep the effective nuclear charge Z' constant throughout the calculation (since it is a non-linear parameter), but seek the values of the linear parameters c_1 through c_6 that minimize the variational integral

$$\varepsilon = \frac{\int \Phi^* \hat{H} \Phi \, d\tau}{\int \Phi^* \Phi \, d\tau}$$

For neatness, I will recast the problem in matrix notation and treat the general case with n rather than 6 Ψs. We write

$$\Phi = (\Psi_1 \quad \Psi_2 \cdots \Psi_n)\begin{pmatrix} c_1 \\ c_2 \\ \cdots \\ c_n \end{pmatrix} = \Psi\mathbf{c} \qquad (14.21)$$

I will also collect integrals such as

$$H_{ij} = \int \Psi_i^* \hat{H} \Psi_j \, d\tau \quad \text{and} \quad S_{ij} = \int \Psi_i^* \Psi_j \, d\tau$$

into the $n \times n$ matrices $\mathbf{H}$ and $\mathbf{S}$. I will generally work with real rather than complex wavefunctions, so we can drop the complex conjugate sign * and the variational integral becomes

$$\varepsilon = \frac{\mathbf{c}^T \mathbf{H} \mathbf{c}}{\mathbf{c}^T \mathbf{S} \mathbf{c}}$$

We now let $\mathbf{c}$ change by a small amount $\delta\mathbf{c}$ and find the change in ε (which we will then set to zero for a stationary point). We have

$$\begin{aligned}
\varepsilon + \delta\varepsilon &= \frac{(\mathbf{c} + \delta\mathbf{c})^T \mathbf{H} (\mathbf{c} + \delta\mathbf{c})}{(\mathbf{c} + \delta\mathbf{c})^T \mathbf{S} (\mathbf{c} + \delta\mathbf{c})} \\
&= \frac{\mathbf{c}^T\mathbf{H}\mathbf{c} + (\delta\mathbf{c})^T\mathbf{H}\mathbf{c} + \mathbf{c}^T\mathbf{H}\delta\mathbf{c} + (\delta\mathbf{c})^T\mathbf{H}\delta\mathbf{c}}{\mathbf{c}^T\mathbf{S}\mathbf{c} + (\delta\mathbf{c})^T\mathbf{S}\mathbf{c} + \mathbf{c}^T\mathbf{S}\delta\mathbf{c}(\delta\mathbf{c})^T\mathbf{S}\delta\mathbf{c}} \\
&= \frac{\mathbf{c}^T\mathbf{H}\mathbf{c} + (\delta\mathbf{c})^T\mathbf{H}\mathbf{c} + \mathbf{c}^T\mathbf{H}\delta\mathbf{c} + (\delta\mathbf{c})^T\mathbf{H}\delta\mathbf{c}}{\mathbf{c}^T\mathbf{S}\mathbf{c}\left(1 + \frac{(\delta\mathbf{c})^T\mathbf{S}\mathbf{c}}{\mathbf{c}^T\mathbf{S}\mathbf{c}} + \frac{\mathbf{c}^T\mathbf{S}\delta\mathbf{c}}{\mathbf{c}^T\mathbf{S}\mathbf{c}} + \frac{(\delta\mathbf{c})^T\mathbf{S}\delta\mathbf{c}}{\mathbf{c}^T\mathbf{S}\mathbf{c}}\right)}
\end{aligned}$$

Expanding the denominator by the binomial theorem and retaining only the first order in $\delta\mathbf{c}$ we find

$$\varepsilon + \delta\varepsilon = \frac{(\mathbf{c}^T\mathbf{H}\mathbf{c} + (\delta\mathbf{c})^T\mathbf{H}\mathbf{c} + \mathbf{c}^T\mathbf{H}\delta\mathbf{c} + \cdots)}{\mathbf{c}^T\mathbf{S}\mathbf{c}}\left(1 - \frac{(\delta\mathbf{c})^T\mathbf{S}\mathbf{c} + \mathbf{c}^T\mathbf{S}\delta\mathbf{c}}{\mathbf{c}^T\mathbf{S}\mathbf{c}} + \cdots\right) \quad (14.22)$$

and after multiplying out

$$(\varepsilon + \delta\varepsilon)\mathbf{c}^T\mathbf{S}\mathbf{c} = (\delta\mathbf{c})^T(\mathbf{H}\mathbf{c} - \varepsilon\mathbf{S}\mathbf{c}) + (\mathbf{c}^T\mathbf{H} - \varepsilon\mathbf{c}^T\mathbf{S})\delta\mathbf{c} + \cdots \quad (14.23)$$

The two terms on the right-hand side are simply matrix transposes of each other, they carry the same information. I didn't place any requirements on $\delta\mathbf{c}$; it is quite arbitrary and so for a stationary point we must have

$$\mathbf{H}\mathbf{c} = \varepsilon\mathbf{S}\mathbf{c}$$

This is called a *generalized matrix eigenvalue problem* and there are exactly n possible solutions $\mathbf{c}_1, \varepsilon_1; \mathbf{c}_2, \varepsilon_2; \ldots; \mathbf{c}_n, \varepsilon_n$. Each of the n solutions is an upper bound to n electronic states (including the ground state), and so the linear variation method has the added bonus of giving approximations to each of n states simultaneously. Not only that, but if we add further Ψs and repeat the calculation, each of the energies will at worst stay the same or possibly get closer to the true energy. I have illustrated

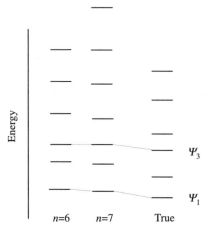

Figure 14.2   MacDonald's Theorem

MacDonald's Theorem for the case $n = 6$ in Figure 14.2. The addition of Ψ_7 has had no effect on the $n = 6$ approximation for Ψ_3 but has (for example) lowered the energy of the Ψ_1 approximation.

I have stressed the special property of atoms; they are spherical and their electronic Hamiltonian commutes with the square of the angular momentum operator, together with the z-component. The Schrödinger equation makes no mention of electron spin and so the Hamiltonian also commutes with the square of the spin angular momentum operator and the z-component. In Table 14.4, I have summarized the relevant quantum numbers for the six states of helium considered.

It can be shown (see, for example, Eyring, Walter and Kimball) that if Ψ_i and Ψ_j are eigenfunctions of an operator $\hat{A}$ that commutes with the Hamiltonian $\hat{H}$, and that the eigenvalues are different

$$\hat{A}\Psi_i = a_i \Psi_i$$
$$\hat{A}\Psi_j = a_j \Psi_j$$
$$a_i \neq a_j$$

Table 14.4   Angular momentum quantum numbers for helium states

	L	M_L	S	M_S
Ψ_1	0	0	0	0
Ψ_2	0	0	0	0
Ψ_3	0	0	1	1
Ψ_4	0	0	1	0
Ψ_5	0	0	1	-1
Ψ_6	0	0	0	0

then

$$\int \Psi_i \hat{H} \Psi_j \, d\tau = 0$$

This means that the Hamiltonian matrix will have a simple form

$$\begin{pmatrix} H_{11} & H_{12} & 0 & 0 & 0 & H_{16} \\ H_{21} & H_{22} & 0 & 0 & 0 & H_{26} \\ 0 & 0 & H_{33} & 0 & 0 & 0 \\ 0 & 0 & 0 & H_{44} & 0 & 0 \\ 0 & 0 & 0 & 0 & H_{55} & 0 \\ H_{61} & H_{62} & 0 & 0 & 0 & H_{66} \end{pmatrix}$$

so instead of a 6×6 matrix eigenvalue problem, we have a 3×3 and 3 at 1×1.

There is nothing particularly hard about solving a 6×6 matrix eigenvalue problem, but this simple example demonstrates how angular momentum can be used to help with atomic calculations. In the case of polyatomic molecules, things are not so easy. Angular momentum operators do not generally commute with molecular Hamiltonians and so molecular problems are much harder than atomic ones. Spin operators commute with molecular Hamiltonians, as do symmetry operators.

14.7 Slater Determinants

Solving the appropriate electronic Schrödinger equation is only one aspect of a calculation; we also have to take account of electron spin and, because electrons are fermions, the electronic wavefunction has to satisfy the Pauli Principle. Neither electron spin nor the Pauli Principle appears from the Schrödinger treatment. As I mentioned in Chapters 12 and 13, Pauli's Principle can be stated in a number of different ways; I am going to restate it as

Electronic wavefunctions must be antisymmetric to exchange of electron names.

I produced the helium orbital wavefunctions in Table 14.1 by a somewhat ad hoc method; I constructed suitable spatial parts and spin parts, which I combined in such a way that the Pauli Principle was satisfied. A more systematic method for constructing antisymmetric orbital wavefunctions is needed.

Electron spin can be conveniently treated by combining spatial orbitals with the spin functions α and β; for a given spatial orbital $\psi(\mathbf{r})$ we work with two space and spin wavefunctions that we write $\psi(\mathbf{r})\alpha(s)$ and $\psi(\mathbf{r})\beta(s)$; these are usually called *spinorbitals* and electrons are allocated to spinorbitals. A spinorbital can hold a maximum of one electron. Suppose then that we have four electrons that we wish to allocate to the four spinorbitals $\psi_A(\mathbf{r})\alpha(s)$, $\psi_B(\mathbf{r})\alpha(s)$, $\psi_A(\mathbf{r})\beta(s)$ and $\psi_B(\mathbf{r})\beta(s)$. One possible allocation is

$$\psi_A(\mathbf{r}_1)\alpha(s_1)\psi_A(\mathbf{r}_2)\beta(s_2)\psi_B(\mathbf{r}_3)\alpha(s_3)\psi_B(\mathbf{r}_4)\beta(s_4)$$

but we must now allow for the indistinguishability of the electrons and take account of all the remaining $4! - 1$ permutations of electrons through the spinorbitals, for example

$$\psi_A(\mathbf{r}_2)\alpha(s_2)\psi_A(\mathbf{r}_1)\beta(s_1)\psi_B(\mathbf{r}_3)\alpha(s_3)\psi_B(\mathbf{r}_4)\beta(s_4)$$

which has been obtained from the original allocation by permuting the names of electrons 1 and 2. All the $4!$ permutations have to appear with equal weight in the total wavefunction, and in order to satisfy the Pauli Principle we multiply each of them by the sign of the permutation. This is -1 if we permute an even number of electrons, and $+1$ if we permute an odd number. A total orbital wavefunction that satisfies the Pauli Principle will therefore be

$$\Psi = \psi_A(\mathbf{r}_1)\alpha(s_1)\psi_A(\mathbf{r}_2)\beta(s_2)\psi_B(\mathbf{r}_3)\alpha(s_3)\psi_B(\mathbf{r}_4)\beta(s_4)$$
$$- \psi_A(\mathbf{r}_2)\alpha(s_2)\psi_A(\mathbf{r}_1)\beta(s_1)\psi_B(\mathbf{r}_3)\alpha(s_3)\psi_B(\mathbf{r}_4)\beta(s_4) + \cdots$$

John C. Slater is credited with having noticed that these terms could be written as a determinant (of order 4, in this case), which we construct as

$$\Psi = \begin{vmatrix} \psi_A(\mathbf{r}_1)\alpha(s_1) & \psi_A(\mathbf{r}_1)\beta(s_1) & \psi_B(\mathbf{r}_1)\alpha(s_1) & \psi_B(\mathbf{r}_1)\beta(s_1) \\ \psi_A(\mathbf{r}_2)\alpha(s_2) & \psi_A(\mathbf{r}_2)\beta(s_2) & \psi_B(\mathbf{r}_2)\alpha(s_2) & \psi_B(\mathbf{r}_2)\beta(s_2) \\ \psi_A(\mathbf{r}_3)\alpha(s_3) & \psi_A(\mathbf{r}_3)\beta(s_3) & \psi_B(\mathbf{r}_3)\alpha(s_3) & \psi_B(\mathbf{r}_3)\beta(s_3) \\ \psi_A(\mathbf{r}_4)\alpha(s_4) & \psi_A(\mathbf{r}_4)\beta(s_4) & \psi_B(\mathbf{r}_4)\alpha(s_4) & \psi_B(\mathbf{r}_4)\beta(s_4) \end{vmatrix} \qquad (14.24)$$

Some authors write the determinants with rows and columns interchanged, which of course leaves the value of the determinant unchanged

$$\Psi = \begin{vmatrix} \psi_A(\mathbf{r}_1)\alpha(s_1) & \psi_A(\mathbf{r}_2)\alpha(s_2) & \psi_A(\mathbf{r}_3)\alpha(s_3) & \psi_A(\mathbf{r}_4)\alpha(s_4) \\ \psi_A(\mathbf{r}_1)\beta(s_1) & \psi_A(\mathbf{r}_2)\beta(s_2) & \psi_A(\mathbf{r}_3)\beta(s_2) & \psi_A(\mathbf{r}_4)\beta(s_2) \\ \psi_B(\mathbf{r}_1)\alpha(s_1) & \psi_B(\mathbf{r}_2)\alpha(s_2) & \psi_B(\mathbf{r}_3)\alpha(s_3) & \psi_B(\mathbf{r}_3)\alpha(s_3) \\ \psi_B(\mathbf{r}_1)\beta(s_1) & \psi_B(\mathbf{r}_2)\beta(s_2) & \psi_B(\mathbf{r}_3)\beta(s_3) & \psi_B(\mathbf{r}_4)\beta(s_4) \end{vmatrix}$$

It is an attractive property of determinants that they change sign if we interchange two rows (or columns), and this is formally equivalent to interchanging the name of two of the electrons. Also, if two columns are the same, then the determinant is zero, which is formally equivalent to letting two electrons occupy the same spinorbital.

Not every electronic state of every atom or molecule can be written as a single Slater determinant and linear combinations are then needed. For example, of the wavefunctions shown in Table 14.1, we see by inspection that Ψ_1, Ψ_3, Ψ_5 and Ψ_6 can be written as single Slater determinants, for example

$$\Psi_1 = \sqrt{\frac{1}{2}} \begin{vmatrix} 1s(\mathbf{r}_1)\alpha(s_1) & 1s(\mathbf{r}_1)\beta(s_1) \\ 1s(\mathbf{r}_2)\alpha(s_2) & 1s(\mathbf{r}_2)\beta(s_2) \end{vmatrix}; \qquad \Psi_3 = \sqrt{\frac{1}{2}} \begin{vmatrix} 1s(\mathbf{r}_1)\alpha(s_1) & 2s(\mathbf{r}_1)\alpha(s_1) \\ 1s(\mathbf{r}_2)\alpha(s_2) & 2s(\mathbf{r}_2)\alpha(s_2) \end{vmatrix}$$

but Ψ_2 and Ψ_4 have to be written as a sum of two determinants, for example

$$\Psi_2 = \frac{1}{2} \left\{ \begin{vmatrix} 1s(\mathbf{r}_1)\alpha(s_1) & 2s(\mathbf{r}_1)\beta(s_1) \\ 1s(\mathbf{r}_2)\alpha(s_2) & 2s(\mathbf{r}_2)\beta(s_2) \end{vmatrix} - \begin{vmatrix} 1s(\mathbf{r}_1)\beta(s_1) & 2s(\mathbf{r}_1)\alpha(s_1) \\ 1s(\mathbf{r}_2)\beta(s_2) & 2s(\mathbf{r}_2)\alpha(s_2) \end{vmatrix} \right\}$$

14.8 The Slater–Condon–Shortley Rules

Slater determinants are compact summaries of all possible permutations of electrons and spin orbitals, but the way I have written them down is unwieldy, and many authors adopt simplified notations. Suppose, for example, we have a many-electron system whose electronic configuration can be written

$$(\psi_A)^2 (\psi_B)^2 \cdots (\psi_M)^2$$

This is chemical shorthand for 2M spinorbitals $\psi_A\alpha \; \psi_A\beta \; \psi_B\alpha \; \psi_B\beta \cdots \psi_M\alpha \; \psi_M\beta$ occupied by 2M electrons. One convention is to write A for $\psi_A\alpha_A$ and $\bar{A}$ for $\psi_A\beta_A$ with the Slater determinant represented by

$$D = |A\bar{A}B\bar{B} \cdots M\bar{M}|$$

From time to time we need to know the expectation values of sums of certain one-electron operators and certain two-electron operators. Suppose that there are n electrons; these are indistinguishable and so any sum of operators must include all of them on an equal footing. Expectation values are typically the electronic contribution to the molecular electric dipole moment

$$-e \int \Psi \left(\sum_{i=1}^{n} \mathbf{r}_i \right) \Psi d\tau$$

and the electron repulsion in a polyatomic system with n electrons is

$$\frac{e^2}{4\pi\epsilon_0} \int \int \Psi \left(\sum_{i=1}^{n-1} \sum_{j=i+1}^{n} \frac{1}{r_{ij}} \right) \Psi d\tau_1 d\tau_2$$

Ψ has to be a linear combination of Slater determinants $D_1, D_2, \ldots$ so we need a systematic set of rules for working out such expectation values between single Slater determinants that I will call D_1 and D_2 which are, in the simplified notation

$$D_1 = |UVW \cdots Z|$$
$$D_2 = |U'V'W' \cdots Z'|$$

The first step is to rearrange one of the determinants to make as many of the spin-orbitals equal as possible. This may introduce a sign change. The algebra is easier if we assume that the individual orbitals are normalized and orthogonal. Consider the overlap integral

$$\int D_1 D_2 d\tau$$

where the integration is over the space and spin coordinates of all the electrons. Each determinant expands into $n!$ terms, and the product has $(n!)^2$ terms. On integration, orthonormality of the individual spinorbitals means that there will be just $n!$ non-zero terms each equal to 1 with all remaining terms zero. If at least one of the spinorbitals is different, say $U \neq U'$, then the complete overlap integral is zero. Thus

$$\int D_1 D_2 d\tau = \begin{pmatrix} n! & \text{if } D_1 = D_2 \\ 0 & \text{otherwise} \end{pmatrix}$$

This gives a Slater determinant normalizing factor of $1/\sqrt{n!}$.

The rules for sums of one- and two-electron operators can be found in more advanced texts such as Eyring, Walter and Kimball; all we need to note are the following results:

1. If two or more spin orbitals are zero, then the expectation value of a sum of one-electron operators is zero.

2. If three or more spin orbitals are different, then the expectation value of a sum of two-electron operators is zero.

14.9 The Hartree Model

We now return to the problem of the helium atom. We have established that the electronic wavefunction would be exactly a product of hydrogenic orbitals in the absence of electron repulsion. We have also seen that neglect of electron repulsion leads to impossibly poor agreement with experiment. The orbital model is an extremely attractive one, so the question is how can we both allow for electron repulsion in some average way, whilst retaining the orbital picture. D. R. Hartree's solution to the problem was to allow each electron to come under the influence of an *average* potential due to the other electron and the nucleus. Suppose, for the sake of argument,

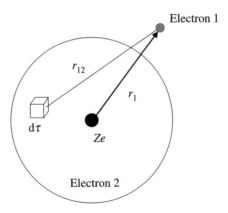

Figure 14.3 The Hartree model

that electrons 1 and 2 occupy orbitals ψ_A and ψ_B (which might be the same, but are to be determined). We could have a guess that the ψs might be hydrogenic, to get the calculation started.

According to the familiar Born interpretation, $\psi_B{}^2\mathrm{d}\tau$ is a probability and so we can regard the second electron as a charge distribution, with density $-e\psi_B{}^2$. Electron 1 therefore sees a potential due to the nucleus and the smeared out electron 2, as shown in Figure 14.3.

The electronic Schrödinger equation for electron 1 is

$$\left(-\frac{h^2}{8\pi^2 m_e}\nabla^2 - \frac{Ze^2}{4\pi\epsilon_0 r_1} + \frac{e^2}{4\pi\epsilon_0}\int\frac{\psi_B^2}{r_{12}}\mathrm{d}\tau_B\right)\psi_A = \varepsilon_A\psi_A \qquad (14.25)$$

That gives us ψ_A. We now focus on electron 2, for which

$$\left(-\frac{h^2}{8\pi^2 m_e}\nabla^2 - \frac{Ze^2}{4\pi\epsilon_0 r_2} + \frac{e^2}{4\pi\epsilon_0}\int\frac{\psi_A^2}{r_{12}}\mathrm{d}\tau_A\right)\psi_B = \varepsilon_B\psi_B \qquad (14.26)$$

and calculate ψ_B then back to electron 1 and so on. The calculation is an iterative one and we stop once the change between iterations is sufficiently small. Each electron experiences a field due to the remaining electrons, and at the end of the calculation the average electron density derived from the field must be the same as the field and so D. R. Hartree coined the phrase *self consistent field* (SCF for short) in 1927 [50]. William Hartree and his son Douglas R. Hartree did much of the early work and so we speak of the *Hartree self consistent field method*.

The theory of atomic structure is dominated by angular momentum considerations, since the square of the orbital angular momentum operator and its z-component commute both with each other and the electronic Hamiltonian. This simplifies the problem considerably and the Hartrees wrote each atomic orbital as

$$\psi(\mathbf{r}) = \frac{1}{r}P(nl; r)Y_{l,m_l}(\theta, \phi) \qquad (14.27)$$

where $P(nl; r)$ is a radial function and Y_{lm} a spherical harmonic. Their notation for P should be clear; each shell has a different radial function. From now on I am going to make the notation more consistent with previous chapters and write

$$\psi(\mathbf{r}) = \frac{1}{r} P_{nl}(r) Y_{l,m_l}(\theta, \phi)$$

Thus, for fluorine we would expect three different radial functions, $P_{1s}(r)$, $P_{2s}(r)$ and $P_{2p}(r)$. Because of the spherical symmetry of atoms, all the 2p solutions have the same radial part.

Details of the method are given in D. R. Hartree's 1957 book *The Calculation of Atomic Structures* [51] and essentially the radial functions are determined from the variation principle. That is to say, they are chosen so as to minimize the variational energy

$$\frac{\int \Phi^* \hat{H} \Phi \, d\tau}{\int \Phi^* \Phi \, d\tau}$$

14.10 The Hartree–Fock Model

Hartree's calculations were done numerically. It soon became apparent that these early calculations gave energies that were in poor agreement with experiment; V. Fock [52] pointed out that Hartree had not included the Pauli principle in his method. Essentially, the *Hartree model* considered a simple orbital product such as

$$\Psi_{\text{Hartree}} = \psi_A(\mathbf{r}_1)\alpha(s_1)\psi_A(\mathbf{r}_2)\beta(s_2)\psi_B(\mathbf{r}_3)\alpha(s_3)\psi_B(\mathbf{r}_4)\beta(s_4) \tag{14.28}$$

whilst the *Hartree–Fock (HF)* model uses a fully antisymmetrized wavefunction such as

$$\Psi_{\text{Hartree-Fock}} = \begin{vmatrix} \psi_A(\mathbf{r}_1)\alpha(s_1) & \psi_A(\mathbf{r}_1)\beta(s_1) & \psi_B(\mathbf{r}_1)\alpha(s_1) & \psi_B(\mathbf{r}_1)\beta(s_1) \\ \psi_A(\mathbf{r}_2)\alpha(s_2) & \psi_A(\mathbf{r}_2)\beta(s_2) & \psi_B(\mathbf{r}_2)\alpha(s_2) & \psi_B(\mathbf{r}_2)\beta(s_2) \\ \psi_A(\mathbf{r}_3)\alpha(s_3) & \psi_A(\mathbf{r}_3)\beta(s_3) & \psi_B(\mathbf{r}_3)\alpha(s_3) & \psi_B(\mathbf{r}_3)\beta(s_3) \\ \psi_A(\mathbf{r}_4)\alpha(s_4) & \psi_A(\mathbf{r}_4)\beta(s_4) & \psi_B(\mathbf{r}_4)\alpha(s_4) & \psi_B(\mathbf{r}_4)\beta(s_4) \end{vmatrix} \tag{14.29}$$

In the simplest version, HF theory concentrates on electronic states that can be represented as a single Slater determinant. We find an extra term in the energy expression called the *exchange energy*, discussed earlier in Chapter 12. Evaluation of the energy needs a more complicated numerical procedure then the simpler Hartree theory. Inclusion of electron exchange by the numerical methods used in their day

Table 14.5   Selection of atoms treated in D. R. Hartree's book

Atom	Atom	Atom	Atom
H^-	Ne^{2+}	Cl^-	Fe^{16+}
He^+	Na	Cl	Zn^{2+}
Li^+	Na^+	Ca	Zr^{4+}
Be	Mg	Ti^{2+}	Mo^+
B	Al^{2+}	V^+	In^{3+}
C	Al^{3+}	Mn	Sb^{3+}
O^{+6}	S^-	Mn^+	Au^+
F^-	S	Fe^{13+}	Tl^+
			Tl^{2+}

proved more and more difficult for atoms towards the bottom right-hand corner of the Periodic Table. The Hartrees were able to study a wide range of atoms in different electronic states, together with their ions. To give a flavour of systems studied, I have reproduced in Table 14.5 the systems that D. R. Hartree reports in his book, which cover the period 1948 through 1957. The author mentions that all up to Mn^{2+} and also Fe^{16+} and Zn^{2+} have exchange terms included in the calculations.

14.11 Atomic Shielding Constants

Hartree–Fock wavefunctions are the best wavefunctions that are possible within the orbital model, as shown in Figure 14.4. Wavefunctions A and B are simple orbital wavefunctions, perhaps hydrogenic or improved hydrogenic. The difference between the HF energy and experiment is called the *correlation energy*.

Output from an atomic HF program consists of the radial function, together with data for each shell such as that shown in Table 14.6.

It is usual to work with normalized radial functions, and this determines the functions apart from their sign (for if $P_{nl}(r)$ is a solution, so is $-P_{nl}(r)$). Hartree used a

Table 14.6   Properties output from an atomic HF study

ε	The orbital energy
A	The initial slope $P_{nl}(r)/r^{l+1}, r \to 0$
s	The screening parameter
$\langle 1/R^3 \rangle$	Expectation value of $1/R^3$
$\langle 1/R \rangle$	Expectation value of $1/R$
$\langle R \rangle$	Expectation value of R
$\langle R^2 \rangle$	Expectation value of R^2
Virial ratio	$\langle \text{Potential} \rangle / \langle \text{Kinetic} \rangle$
Spin–orbit coupling	ζ_{nl}
Orbit–orbit coupling	$M^k\,(nl, nl)$

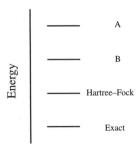

Figure 14.4 Hartree–Fock energy

convention that the radial function should be positive near the nucleus, whilst other authors use a convention that it should be positive for large r.

Several authors pointed out that it was often desirable to have simple approximations to the wavefunctions and energy levels of atoms and ions. For an atom of n electrons, there are $3n$ independent variables in the Schrödinger equation (plus spin). Douglas R. Hartree expressed the need most dramatically as follows:

One way of representing a solution quantitatively would be by a table of its numerical values, but an example will illustrate that such a table would be far too large ever to evaluate, or to use if it were evaluated. Consider, for example, the tabulation of a solution for one stationary state of Fe. Tabulation has to be at discrete values of the variables, and 10 values of each variable would provide only a very coarse tabulation; but even this would require 10^{78} entries to cover the whole field; and even though this might be reduced to, say, $5^{78} \cong 10^{53}$ by use of the symmetry properties of the solution, the whole solar system does not contain enough matter to print such a table. And, even if it could be printed, such a table would be far too bulky to use. And all this is for a single stage of ionization of a single atom.

14.11.1 Zener's wavefunctions

In Section 14.5 we addressed ways of improving the analytical 1s orbital for He, especially by treating the effective nuclear charge Z' as a variational parameter. In his 1930 keynote paper entitled 'Analytic Atomic Wave Functions', C. Zener [53] extended this simple treatment to first row atoms. As usual, I will let the author tell the story in his own words through the Abstract:

The wave functions for the atoms Be, B, C, N, O, F and Ne are written as simple analytic expressions with several parameters. The best values of these parameters are then determined by the variation method. In the final wave functions the effective quantum number is very nearly two, the radial node is so small as to have little effect upon the charge distribution, the coefficient in the exponential is related to an empirical 'mean effective charge'.

14.11.2 Slater's rules

Finally we turn to the work of J. C. Slater [54], and once again you might like to read his 1930 Abstract:

> In analogy with the method of Zener for the atoms Li to F, simple rules are set up giving approximate analytical atomic wavefunctions for all the atoms, in any stage of ionization. These are applied to X-ray levels, sizes of atoms and ions, diamagnetic susceptibility etc. In connection with ferromagnetism, it is shown that if this really depends on the existence of incomplete shells within the atoms, rather far apart in the crystal, then the metals most likely to show it would be Fe, Co, Ni and alloys of Mn and Cu (Heuser alloys).

Slater extended Zener's shielding constants for Li to F to the other atoms by adjusting the values until he got agreement with experimental results of stripped atom and X-ray levels, atom sizes and the other quantities mentioned. He noticed that Zener's wavefunctions had radial nodes but argued that they were unimportant since they come much closer to the nucleus than for hydrogen. Consequently, he decided to ignore them altogether and wrote a radial part as

$$ r^{n^*-1} \exp\left(-\left(\frac{Z-s}{n^*}\right)r\right) \tag{14.30} $$

where n^* is an effective quantum number and s the shielding constant. n^* and s are found by simple rules (that have become known as Slater's rules) as shown in Table 14.7 and below.

To determine $Z - s$, the electrons are divided into the following groups, each having a different shielding constant: 1s; 2s, 2p; 3s, 3p; 3d; 4s, 4p; 4d; 4f; 5s, 5p; etc. That is, the s and p of a given n are grouped together but the d and f are separated. The shells are considered to be arranged from inside out in the order named.

The shielding constant s is formed, for any group of electrons, from the following contributions:

1. Nothing from any shell outside the one considered.

2. An amount 0.35 from each other electron in the group considered (except the 1s group, where 0.30 is used instead).

3. If the shell considered is an s, p shell, then an amount 0.85 from each electron with total quantum number less by one, and an amount 1.00 from every electron still farther in. But if the shell is a d or f, then an amount 1.00 from every electron inside it.

Slater gives three examples, reproduced in Table 14.8, which are worth quoting; C,

Table 14.7   Slater n^* values

n	n^*
1	1
2	2
3	3
4	3.7
5	4.0
6	4.2

Table 14.8   Slater $Z - s$ values

	C	Fe	Fe$^+$ (1s^1)
1s	$5.70 = 6 - 0.30$	$25.70 = 26 - 0.30$	26.00
2s, 2p	$3.25 = 6 - 3(0.35) - 2(0.85)$	$21.85 = 26 - 7(0.35) - 2(0.85)$	22.70
3s, 3p		$14.75 = 26 - 7(0.35) - 8(0.85) - 2(1.00)$	15.75
3d		$6.25 = 26 - 5(0.35) - 18(1.00)$	7.25
4s		$3.75 = 26 - 1(0.35) - 14(0.85) - 18(0.85)$	4.75
		$- 10(1.00)$	

Fe and Fe cations lacking a K-electron so that there is only one 1s electron. Slater's rules are still widely quoted in atomic and molecular structure theory.

14.12 Koopman's Theorem

Consider the ionization process

$$Ne(1s^2 2s^2 2p^6) \rightarrow Ne^+(1s^2 2s^1 2p^6)$$

where I have ionized a 2s electron from neon (Figure 14.5). Suppose that the energy level diagram represents a HF calculation on neon before ionization, and that the orbitals do not relax in any way after ionization. That is, the neutral atom and the ionized cation have the same HF orbitals. According to Koopmans' theorem, the ionization energy for the process is the negative of the HF orbital energy. The theorem holds for all HF orbitals.

Koopmans' theorem

Ionization from HF orbital ψ_i

Ionization energy = −orbital energy

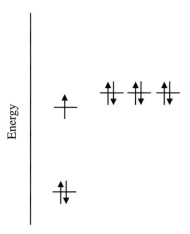

Figure 14.5 Ionized neon

Ionization energies can be measured accurately by modern-day versions of the photoelectric effect. For valence shell ionization we use UltraViolet PhotoElectron Spectroscopy (UVPES). As the name suggests, ultraviolet photons are the ionizing source. Inner shell electrons need X-rays for ionization.

15 Simple Molecules

It is time to progress to the quantum theory of molecules, and it should come as no surprise when I tell you that the orbital model is a good starting point for many of the calculations we professionals do today.

Molecules are more complicated than atoms because:

1. Whilst angular momentum considerations dominate the theory of atomic structure (it is amazing what can be done with pencil and paper before any nasty integrals have to be evaluated) the $\hat{L}^2$ and $\hat{L}_z$ operators do not commute with molecular Hamiltonians (except for linear molecules; if the molecular axis is the z-axis, then L_z does commute). To a certain extent, molecular symmetry operators help us since they commute with molecular Hamiltonians, but most molecules of any real chemical interest have no symmetry apart from the identity operation. So we are stuck with the Hamiltonian, and nothing to help us simplify the eigenvalue problem apart from electron spin. The Schrödinger equation for a free molecule in the absence of an applied field does not contain spin and so both $\hat{S}^2$ and $\hat{S}_z$ inevitably commute with molecular Hamiltonians.

2. Molecules aren't spherical, and the great simplifying feature of atomic Hartree–Fock (HF) theory

$$\psi(\mathbf{r}) = \frac{1}{r} P_{nl}(r) Y_{l,m_l}(\theta, \phi)$$

is no longer appropriate. This statement follows from the one above.

3. The HF limit is attainable for atoms by direct numerical integration of the HF equations. This limit is unattainable for molecules except in a few simple cases of high symmetry, and numerical integration techniques that are fine for atoms are inapplicable for molecules. In any case, we have to concern ourselves with calculations that are beyond the HF limit if we want to study chemical reactions.

15.1 The Hydrogen Molecule Ion H_2^+

The simplest molecule is the hydrogen molecule ion H_2^+ shown in Figure 15.1. It is formed by passing an electric discharge through dihydrogen, and it has been well studied experimentally. Its dissociation energy is known to be $D_e = 269.6\,\text{kJ mol}^{-1}$ and it has an equilibrium bond length of 106 pm. There is only one well-established electronic state, namely the ground state.

First of all, in quantum mechanics just as in classical mechanics, we can rigorously separate off the translational motion of the molecule. That leaves us to concentrate on the two nuclei (each of mass $m_p = 1.673 \times 10^{-27}\,\text{kg}$) and the electron (of mass $m_e = 9.109 \times 10^{-31}\,\text{kg}$) about the centre of mass. The wavefunction therefore depends on the coordinates of the electron ($\mathbf{r}$) and the two nuclei ($\mathbf{R}_A$ and $\mathbf{R}_B$)

$$\Psi_{\text{tot}} = \Psi_{\text{tot}}(\mathbf{R}_A, \mathbf{R}_B, \mathbf{r})$$

If we regard such a system from the viewpoint of classical mechanics, we would be tempted to try to separate the motions of the nuclei and the electron, because of the great difference in their masses (a factor of 1 : 1836). This doesn't mean that the motions are truly separable like the molecular translational motion of the centre of mass and the relative motion of the nuclei and electrons within the molecule, more that the separation can be done with only a small error.

M. Born and J. R. Oppenheimer [55] first investigated this possibility and wrote

$$\Psi_{\text{tot}}(\mathbf{R}_A, \mathbf{R}_B, \mathbf{r}) = \Psi_{\text{nuc}}(\mathbf{R}_A, \mathbf{R}_B)\psi_e(\mathbf{R}_A, \mathbf{R}_B, \mathbf{r})$$

I am using the convention that lower case ψ refers to one electron, whilst upper case Ψ refers to many electrons. In this case there is only one electron and so the electronic wavefunction is technically an *orbital*; in this case a *molecular orbital* (MO). Born and Oppenheimer showed that the approximation was good to almost (but not exactly) the ratio of the particle masses and so we normally glue the nuclei to fixed

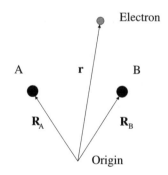

Figure 15.1 Hydrogen molecule ion

positions in space and concentrate on the electron(s) when investigating many problems in molecular electronic structure theory.

We treat the nuclei separately; if we want to know about vibrational and rotational motion, then we have to solve the relevant nuclear vibrational Schrödinger equation and the relevant nuclear rotational Schrödinger equation. This is not the same as doing a molecular mechanics (MM) calculation for the nuclei because there are proper quantum mechanical vibrational and rotational Schrödinger equations as we saw in Chapter 11.

The separation of electronic and nuclear motions allows us to specify a molecular geometry; the electrons experience an electrostatic potential energy due to the nuclei, which are fixed at positions in space for the purpose of calculating the electronic wavefunction. Just as for MM, molecular geometries can be found by investigating stationary points on the potential energy surface.

The electronic wavefunction is given as the solution of an electronic Schrödinger equation

$$-\frac{h^2}{8\pi^2 m_e}\left(\frac{\partial^2}{\partial x^2}+\frac{\partial^2}{\partial x^2}+\frac{\partial^2}{\partial x^2}\right)\psi_e(\mathbf{R}_A, \mathbf{R}_B, \mathbf{r}) + U\psi_e(\mathbf{R}_A, \mathbf{R}_B, \mathbf{r}) = \varepsilon_e\psi_e(\mathbf{R}_A, \mathbf{R}_B, \mathbf{r})$$

$$(15.1)$$

where the electrostatic potential energy is

$$U = \frac{-e^2}{4\pi\epsilon_0}\left(\frac{1}{|\mathbf{r}-\mathbf{R}_A|}+\frac{1}{|\mathbf{r}-\mathbf{R}_B|}\right)$$

We solve the electronic equation and the total energy is given by adding on the fixed nuclear repulsion

$$\varepsilon_{\text{tot}} = \varepsilon_e + \frac{e^2}{4\pi\epsilon_0|\mathbf{R}_A-\mathbf{R}_B|} \qquad (15.2)$$

The hydrogen molecule ion is unique amongst molecules in that we can solve the electronic Schrödinger equation exactly (by numerical methods) to any required accuracy, within the Born–Oppenheimer approximation. The first step is to make a change of variable to so-called *elliptic coordinates* that are defined with reference to Figure 15.2

$$\mu = \frac{r_A + r_B}{R_{AB}}; \qquad \nu = \frac{r_A - r_B}{R_{AB}}; \qquad \phi$$

In this system of coordinates the electronic Schrödinger equation is

$$-\frac{h^2}{8\pi^2 m_e}\frac{4}{R_{AB}^2(\mu^2-\nu^2)}\left(\frac{\partial}{\partial\mu}\left[(\mu^2-1)\frac{\partial}{\partial\mu}\right]+\frac{\partial}{\partial\nu}\left[(1-\nu^2)\frac{\partial}{\partial\nu}\right]\right.$$

$$\left.+\frac{\mu^2-\nu^2}{(\mu^2-1)(1-\nu^2)}\frac{\partial^2}{\partial\phi^2}\right)\psi_e + U\psi_e = \varepsilon_e\psi_e \qquad (15.3)$$

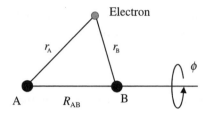

Figure 15.2   Elliptic coordinates

and we look for a separation of variables solution

$$\psi_e(\mu, \nu, \phi) = M(\mu)N(\nu)\Phi(\phi)$$

Since ϕ only enters the equation in the $\frac{\partial^2}{\partial \phi^2}$ term, it is at once apparent that we can factor off the ϕ term. If we call the separation parameter $-\lambda^2$ we have

$$\frac{d^2\Phi}{d\phi^2} = -\lambda^2\Phi$$

and so

$$\Phi(\phi) = \frac{1}{\sqrt{2\pi}}\exp\left(j\lambda\phi\right) \tag{15.4}$$

where λ can take on positive and negative values, and j is the square root of -1. In the limit as $R_{AB} \rightarrow 0$, the quantum number λ becomes equivalent to the atomic quantum number m_l.

Separation into a μ and a ν equation also proves possible, and the equations have been solved by E. Teller [56], by O. Burrau [57] and by others leading to results in complete agreement with experiment. Solution of the differential equations is far from easy and the best references are D. R. Bates et al. [58], H. Wind [59] and of course Eyring, Walter and Kimball (EWK).

15.2 The LCAO Model

The hydrogen molecule ion is unusual amongst molecules in that we can solve the electronic problem exactly (by numerical methods). Once we consider polyelectron systems, we have to seek approximate methods. Any chemist would argue that molecules are built from atoms and so we should capitalize on this chemical knowledge by attempting to build molecular wavefunctions from atomic ones.

Suppose we build the hydrogen molecular ion starting from a hydrogen atom and a proton initially separated by a large distance. The electronic wavefunction will be a hydrogen 1s orbital until the proton is close enough to make any significant perturbation and so we might guess that the molecular wavefunction should resemble an atomic 1s orbital, at least near the appropriate nucleus.

We therefore guess that the low energy molecular orbitals of H_2^+ might be represented

$$\psi = c_A 1s_A + c_B 1s_B$$

where the coefficients c_A and c_B have to be determined. This technique is called the *linear combination of atomic orbitals* (LCAO) and I have used the shorthand that $1s_A$ is a hydrogen 1s orbital centred on nucleus A. In this particular case we can deduce the coefficients from symmetry. According to the Born interpretation, $\psi^2 d\tau$ gives the chance that an electron can be found in the volume element $d\tau$. We have

$$\psi^2 d\tau = (c_A^2 1s_A^2 + 2c_A c_B 1s_A 1s_B + c_B^2 1s_B^2) \, d\tau$$

Electron densities around the two H atoms have to be the same by symmetry, and the nuclear labels A and B can be used interchangeably, which means that

$$c_A^2 = c_B^2; \qquad c_A = \pm c_B$$

This gives two possibilities, that I will label ψ_+ and ψ_-, and we usually write them in normalized form as

$$\psi_+ = \frac{1}{\sqrt{2(1+S)}} (1s_A + 1s_B)$$

$$\psi_- = \frac{1}{\sqrt{2(1-S)}} (1s_A - 1s_B) \qquad (15.5)$$

where

$$S = \int 1s_A 1s_B d\tau$$

$$= \exp(-R_{AB}/a_0) \left(1 + \frac{R_{AB}}{a_0} + \frac{R_{AB}^2}{3a_0^2} \right)$$

We can test these approximate wavefunctions by calculating the variational energies, which can be written

$$\varepsilon_\pm = \varepsilon_H(1s) + \frac{e^2}{4\pi\epsilon_0 R_{AB}} - \frac{\varepsilon_{AA} \pm \varepsilon_{AB}}{1 \pm S}$$

$$\varepsilon_{AA} = \frac{e^2}{4\pi\epsilon_0} \int \frac{1s_A^2}{r_A} d\tau$$

$$= \frac{e^2}{4\pi\epsilon_0 a_0} \frac{a_0}{R_{AB}} \left(1 - \left(1 + \frac{R_{AB}}{a_0} \right) \exp\left(-2\frac{R_{AB}}{a_0} \right) \right)$$

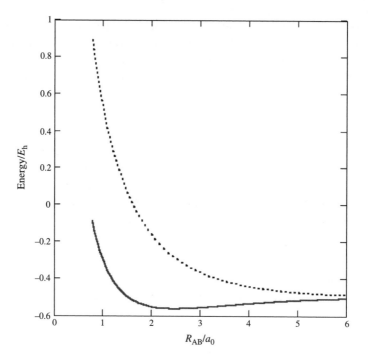

Figure 15.3 Potential energy curves for lowest states of H_2^+

$$\varepsilon_{AB} = \frac{e^2}{4\pi\epsilon_0} \int \frac{1s_A^2}{r_B} \, d\tau$$

$$= \frac{e^2}{4\pi\epsilon_0 a_0} \left(1 + \frac{R_{AB}}{a_0}\right) \exp\left(-\frac{R_{AB}}{a_0}\right) \qquad (15.6)$$

The analytical solution of these integrals together with the overlap integral S (in Equation (15.5) is covered by the classic texts such as EWK; I have simply quoted the results. This gives us the two potential energy curves shown in many elementary quantum chemistry textbooks (Figure 15.3). There is a subtle point; they are often called molecular potential energy curves because the nuclei experience just this potential.

The ground state energy of a hydrogen atom is $-\frac{1}{2}E_h$ and the curves tend asymptotically to the correct limit. The upper curve describes an excited state whilst the lower curve describes the ground state. The calculated binding energy is in poor agreement with experiment (Table 15.1), whilst the equilibrium bond length is in modest agreement with experiment.

A hydrogenic 1s orbital has the form

$$1s = \sqrt{\frac{\zeta^3}{\pi a_0^3}} \exp\left(-\frac{\zeta r}{a_0}\right) \qquad (15.7)$$

where the orbital exponent $\zeta = 1$ for a hydrogen atom, 2 for He^+ and so on. The next step is to find the best value of ζ that is appropriate for a hydrogen atom within a

Table 15.1 Hydrogen molecule ion results

	D_e (eV)	R_e (pm)
Experiment	2.791	106
Simple LCAO with $\zeta = 1$	1.76	132.3
Simple LCAO, best $\zeta = 1.238$	2.25	106
James, elliptic	2.772	106

molecule. The best value turns out to be 1.238, showing that the 1s orbital contracts a little on molecule formation. Table 15.1 shows better agreement with experiment.

15.3 Elliptic Orbitals

The best orbitals that have been obtained for H_2^+ were found by a different approach; the natural coordinates to use are the elliptic coordinates μ, ν and ϕ, and H. M. James [60] found that a good approximation to the lowest orbital is

$$\psi = \exp(-\delta\mu)(1 + c\nu^2) \qquad (15.8)$$

His results are shown in Table 15.1.

The hydrogen molecular ion is interesting in that it doesn't have an electron pair, and yet it is stable. The obvious question is: Can we give a hand-waving explanation for this stability, one that doesn't rely on the presence of electron pairs?

The obvious place to look is the electron density. The electron density plots by themselves are not particularly informative; the ground state wavefunction shows a region of apparently enhanced electron density between the nuclei, whilst the excited state wavefunction has less electron density in this region. It is more interesting to look at the difference between the electron density in the molecule and two ground state hydrogen atoms, each of which contains half an electron. Such plots are called *density differences*, and they are shown in Figure 15.4. Positive contours correspond to a gain of electron density (in units of e) compared with two overlapped half hydrogens; negative contours correspond to a loss. The molecule lies along the horizontal axis.

The density difference plot for the ground state illustrates an electron density enhancement in the region between the nuclei and so gives a stable molecule. Electron density has to be redistributed, it can't appear from nowhere and we see that the bond region gains at the expense of regions beyond the nuclei. The excited state wavefunction shows the opposite effect (see Figure 15.5); there is a depletion of electron density in the bond region and a gain of electron density beyond the nuclei. This explains the instability of this state, so we can give a simple electrostatic explanation for the stability of the molecule in its ground state, and its instability in the excited state.

It is a general principle that, once we have calculated an electron density from the laws of quantum mechanics, we can analyse it using the laws of classical electrostatics.

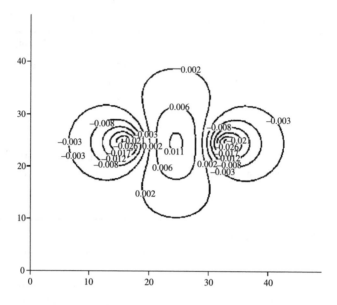

Figure 15.4   Density difference for the ground state

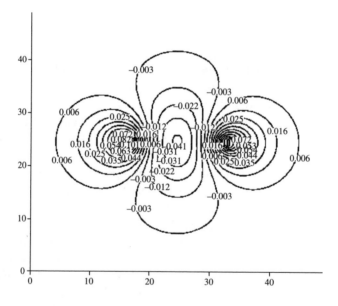

Figure 15.5   Density difference for the excited state

15.4 The Heitler–London Treatment of Dihydrogen

The simplest molecule of any true chemical importance is dihydrogen (see Table 15.2). The methods discussed above for the hydrogen molecule ion are not applicable to dihydrogen, because of the extra electron and the electron–electron repulsion. The

Table 15.2 Dihydrogen elementary valence bond (Heitler–London) calculation

Ψ	Combination	S	M_S	Comment
Ψ_1	$D_1 - D_2$	0	0	Covalent ground state
Ψ_2	$D_1 + D_2$	1	0	Excited triplet state
Ψ_3	D_3	1	1	Excited triplet state
Ψ_4	D_4	1	-1	Excited triplet state
Ψ_5	D_5	0	0	Ionic term
Ψ_6	D_6	0	0	Ionic term

first successful treatment of dihydrogen was that of W. Heitler and F. London [61]. They argued as follows: consider a 'reaction'

$$H_A + H_B \rightarrow H_2$$

where the left-hand side hydrogen atoms labelled A and B are initially at infinite separation. Each H atom is in its lowest electronic state and there are two spin orbitals per atom, giving four in total: $1s_A\alpha$, $1s_A\beta$, $1s_B\alpha$ and $1s_B\beta$. As we bring the atoms closer and closer together, eventually their orbitals will overlap significantly and we have to take account of the Pauli Principle. We need to make sure that the total wavefunction is antisymmetric to exchange of electron names. There are $(4!/2!) = 12$ possible ways of distributing two electrons amongst four atomic spin orbitals but not all such distributions satisfy the Pauli Principle. Analysis along the lines given for helium in Chapter 14 gives the following unnormalized building blocks (Slater determinants)

$$D_1 = \begin{vmatrix} 1s_A(\mathbf{r}_1)\alpha(s_1) & 1s_B(\mathbf{r}_1)\beta(s_1) \\ 1s_A(\mathbf{r}_2)\alpha(s_2) & 1s_B(\mathbf{r}_2)\beta(s_2) \end{vmatrix}; \quad D_2 = \begin{vmatrix} 1s_A(\mathbf{r}_1)\beta(s_1) & 1s_B(\mathbf{r}_1)\alpha(s_1) \\ 1s_A(\mathbf{r}_2)\beta(s_2) & 1s_B(\mathbf{r}_2)\alpha(s_2) \end{vmatrix}$$

$$D_3 = \begin{vmatrix} 1s_A(\mathbf{r}_1)\alpha(s_1) & 1s_B(\mathbf{r}_1)\alpha(s_1) \\ 1s_A(\mathbf{r}_2)\alpha(s_2) & 1s_B(\mathbf{r}_2)\alpha(s_2) \end{vmatrix}; \quad D_4 = \begin{vmatrix} 1s_A(\mathbf{r}_1)\beta(s_1) & 1s_B(\mathbf{r}_1)\beta(s_1) \\ 1s_A(\mathbf{r}_2)\beta(s_2) & 1s_B(\mathbf{r}_2)\beta(s_2) \end{vmatrix}$$

$$D_5 = \begin{vmatrix} 1s_A(\mathbf{r}_1)\alpha(s_1) & 1s_A(\mathbf{r}_1)\beta(s_1) \\ 1s_A(\mathbf{r}_2)\alpha(s_2) & 1s_A(\mathbf{r}_2)\beta(s_2) \end{vmatrix}; \quad D_6 = \begin{vmatrix} 1s_B(\mathbf{r}_1)\alpha(s_1) & 1s_B(\mathbf{r}_1)\beta(s_1) \\ 1s_B(\mathbf{r}_2)\alpha(s_2) & 1s_B(\mathbf{r}_2)\beta(s_2) \end{vmatrix}$$

It is useful to combine the Ds into spin eigenfunctions, since the spin operators commute with the molecular Hamiltonian. The advantage is that we only need take combinations of those wavefunctions having the same spin quantum numbers when seeking to improve our description of the electronic states. The combination $\Psi_1 = D_1 - D_2$ is a singlet spin state and is said to represent the covalent bond, since it gives an equal sharing to the two equivalent 1s orbitals by the two electrons. Ψ_2 through Ψ_4 correspond to the first excited state, which is a triplet spin state. They have the same energy in the absence of an external magnetic field. Ψ_5 and Ψ_6 are called ionic terms, because they represent an electron density distribution in which both electrons are associated with the same nucleus. Heitler and London included Ψ_1 (for the electronic ground state) and Ψ_2 through Ψ_4 (for the excited triplet state) in their original calculation. The necessary integrals needed for a variational calculation are given in Heitler and London's paper, and in the paper by Y. Sugiura [62]. We often refer to the Heitler–London approach as the *valence bond* (VB) method and I will use the two descriptors interchangeably. It was the first successful treatment of an electron pair bond.

The energy corresponding to Ψ_1 is sometimes written

$$\varepsilon = \frac{J + K}{1 + S^2}$$

$$J = \int 1s_A(\mathbf{r}_1)1s_B(\mathbf{r}_2)\hat{H}1s_A(\mathbf{r}_1)1s_B(\mathbf{r}_2)\,d\tau_1 d\tau_2$$

$$K = \int 1s_A(\mathbf{r}_1)1s_B(\mathbf{r}_2)\hat{H}1s_A(\mathbf{r}_2)1s_B(\mathbf{r}_1)\,d\tau_1 d\tau_2$$

$$S = \int 1s_A(\mathbf{r}_1)1s_B(\mathbf{r}_1)\,d\tau \qquad (15.9)$$

and early 'explanations' of chemical bonding focused on the *Coulomb* $(=J/1 + S^2)$, and *Exchange* $(=K/1 + S^2)$ contributions to molecular energies. Several improvements were made to the simple VB treatment of dihydrogen, for example treating the orbital exponent as a variational parameter, and inclusion of the ionic terms once again correctly weighted by use of the variation principle. This latter procedure is referred to as *configuration interaction* (CI).

15.5 The Dihydrogen MO Treatment

The molecular orbital treatment was given by H. Hellmann [64] amongst others. He took the lowest energy MO as a combination of hydrogen 1s orbitals, as for H_2^+

$$\psi_+ = \frac{1}{\sqrt{2(1 + S)}}(1s_A + 1s_B)$$

and wrote

$$D_7 = \sqrt{\frac{1}{2!}}\begin{vmatrix} \psi_+(\mathbf{r}_1)\alpha(s_1) & \psi_+(\mathbf{r}_1)\beta(s_1) \\ \psi_+(\mathbf{r}_2)\alpha(s_2) & \psi_+(\mathbf{r}_2)\beta(s_2) \end{vmatrix}$$

The calculated bond length and dissociation energy are in poorer agreement with experiment than those obtained from the simple VB treatment (Table 15.3), and this puzzled many people at the time. It also led them to believe that the VB method was the correct way forward for the description of molecular structure; in the event,

Table 15.3 Dihydrogen calculations

Comment	D_e (eV)	R_e/a_0
Experiment	4.72	1.40
Simple valence bond	3.14	1.64
Simple MO	2.65	1.60
James and Coolidge [63]	4.698	1.40

advances in computer technology and numerical linear algebra have proved them wrong, but that's for a later chapter.

But why is the simple MO treatment of dihydrogen so poor? If we expand D_7 in terms of the atomic orbitals $1s_A$ and $1s_B$ we find (apart from the normalizing constant)

$$D_7 = (1s_A(\mathbf{r}_1)1s_B(\mathbf{r}_2) + 1s_B(\mathbf{r}_1)1s_A(\mathbf{r}_2) + 1s_A(\mathbf{r}_1)1s_A(\mathbf{r}_2) + 1s_B(\mathbf{r}_1)1s_B(\mathbf{r}_2))$$
$$\times (\alpha(s_1)\beta(s_2) - \alpha(s_2)\beta(s_1))$$

which is equal to the simple VB wavefunction but with ionic terms included and weighted equally to the covalent terms. The solution to the problem is to include excited states in the wavefunction, just as we did for helium, but with a variable weight. This process is CI. In the limit, once all the refinements are made, the two refined treatments (VB with CI and MO with CI) give exactly the same results and so

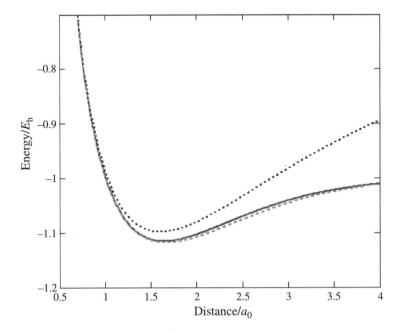

Figure 15.6 Simple dihydrogen calculations

there is no particular reason to prefer one rather than the other. The starting points are different, but not the refined end results.

To summarize, Figure 15.6 shows the potential energy curves for the VB, LCAO and the CI treatments. The full curve is the VB calculation, the dotted curve the MO and the dashed curve is a CI treatment.

15.6 The James and Coolidge Treatment

Just as for H_2^+, it is again found that an accurate value of the binding energy can be obtained by writing the wavefunction in terms of elliptic coordinates. H. James and M. Coolidge [63] wrote such a wavefunction that included the interelectron distance r_{12} explicitly

$$\psi = \exp(-\delta(\mu_1 + \mu_2)) \sum_{klmnp} c_{klmnp} (\mu_1^k \mu_2^l \nu_1^m \nu_2^n u^p + \mu_1^l \mu_2^k \nu_1^n \nu_2^m u^p)$$

$$\mu_1 = \frac{r_{A,1} + r_{B,1}}{R_{AB}}; \qquad \nu_1 = \frac{r_{A,1} - r_{B,1}}{R_{AB}}; \qquad u = \frac{2r_{12}}{R_{AB}} \qquad (15.10)$$

Here k, l, m, n and p are integers and the form of the function is that it is symmetric to the interchange of electron names. δ is the orbital exponent. In order to make the wavefunction symmetric in the nuclear coordinates, the authors included only those terms having $(m + n)$ as an even integer. They found that a 13-term function gave essentially complete agreement with experiment.

15.7 Population Analysis

Our simple treatment of the hydrogen molecule ion was based on the bonding orbital

$$\psi_+ = \frac{1}{\sqrt{2(1 + S)}} (1s_A + 1s_B)$$

which corresponds to a charge distribution of density $-e\psi_+^2$. We normally use the symbol ρ for volume charge densities and to emphasize that it depends on positions in space we write

$$\rho(\mathbf{r}) = -e(\psi_+(\mathbf{r}))^2$$

Substituting and expanding we have

$$\rho(\mathbf{r}) = -\frac{e}{2(1 + S)} ((1s_A(\mathbf{r}))^2 + (1s_B(\mathbf{r}))^2 + 2 \times 1s_A(\mathbf{r}) \times 1s_B(\mathbf{r}))$$

which we can write formally in matrix language as

$$\rho(\mathbf{r}) = -e(1s_A(\mathbf{r}) \quad 1s_B(\mathbf{r})) \begin{pmatrix} \dfrac{1}{2(1+S)} & \dfrac{1}{2(1+S)} \\ \dfrac{1}{2(1+S)} & \dfrac{1}{2(1+S)} \end{pmatrix} \begin{pmatrix} 1s_A(\mathbf{r}) \\ 1s_B(\mathbf{r}) \end{pmatrix} \qquad (15.11)$$

Chemists often forget about the $-e$ since they tend to think in positive numbers of electrons, and they write the middle symmetric matrix $\mathbf{P}$. It is referred to as the *charge density matrix*, or the *matrix of charges and bond orders*. In this case we have

$$\mathbf{P} = \begin{pmatrix} \dfrac{1}{2(1+S)} & \dfrac{1}{2(1+S)} \\ \dfrac{1}{2(1+S)} & \dfrac{1}{2(1+S)} \end{pmatrix}$$

From the early days of molecular quantum theory, authors have tried to divide up the electron density in chemically appealing ways by allocating parts of the density to atom regions and parts to bond regions. If we integrate the electron density

$$\int \rho(\mathbf{r}) \, d\tau = -\frac{e}{2(1+S)} \left(\int (1s_A(\mathbf{r}))^2 \, d\tau + \int (1s_B(\mathbf{r}))^2 \, d\tau \right.$$
$$\left. + 2 \int 1s_A(\mathbf{r}) \times 1s_B(\mathbf{r}) \, d\tau \right)$$

which must come to the number of electrons (1 in the case of hydrogen molecule ion) times $-e$. We therefore have, on doing the integrals and keeping the terms in order

$$\int \rho(\mathbf{r}) \, d\tau = -\frac{e}{2(1+S)}(1 + 1 + 2S)$$

We interpret this by saying that the electron charge fractions

$$\frac{-e}{2(1+S)}, \quad \frac{-e}{2(1+S)} \quad \text{and} \quad \frac{-2Se}{2(1+S)}$$

are associated with atom A, with atom B and with the bond region between atoms A and B. These fractions are referred to as the *net atom and bond* (or *overlap*) *populations*. Robert S. Mulliken introduced the idea of population analysis in his 1955 paper [65], but the ideas had already been around for a very long time. He had the idea to divide each overlap population into equal parts and allocate part to each atom formally contributing to the bond. Here is the Abstract of his famous 1955 paper, entitled 'Electronic Population Analysis on LCAO–MO Molecular Wave Functions I':

With increasing availability of good all-electron LCAO–MO wavefunctions for molecules, a systematic procedure for obtaining maximum insight from such data has become desirable. An analysis in quantitative form is given here in terms of breakdowns of the electronic population into partial and total 'gross atomic populations' or into partial and total 'net atomic populations' together with 'overlap populations'. 'Gross atomic populations' distribute the electrons almost perfectly among the various AOs of the various atoms in the molecule. From these numbers, a definite figure is obtained for the amount of promotion (e.g., from 2s to 2p) in each atom; and also for the gross charge Q on each atom if the bonds are polar. The total overlap population for any pair of atoms in a molecule is in general made up of positive and negative contributions.

Mulliken's aim was to give rough-and-ready indices that characterize a molecular charge distribution and that can be used for comparisons between molecules. The Mulliken gross atom populations in our case are

$$\frac{1}{2(1+S)} + \frac{S}{2(1+S)} = \frac{1}{2}$$

15.7.1 Extension to many-electron systems

A more careful analysis is needed when dealing with multi-electron wavefunctions. Suppose Ψ is an m-electron wavefunction, then it will depend on spatial coordinates and spin coordinates

$$\Psi = \Psi(\mathbf{r}_1, s_1, \mathbf{r}_2, s_2, \ldots, \mathbf{r}_m s_m)$$

If I use the convention that $d\tau$ is a spatial differential volume element (equal to dx, dy, dz in Cartesian coordinates) and ds refers to the spin variable, then the Born interpretation is that

$$|\Psi(\mathbf{r}_1, s_1, \mathbf{r}_2, s_2, \ldots, \mathbf{r}_m s_m)|^2 \, d\tau_1 ds_1 d\tau_2 ds_2 \cdots d\tau_m ds_m$$

gives the probability of finding simultaneously electron 1 in $d\tau_1 ds_1$, electron 2 in $d\tau_2 ds_2$, ... electron m in $d\tau_m ds_m$. Many simple physical properties such as the electric dipole moment depend only on the occupancy of an arbitrary space–spin differential element $d\tau\, ds$ by any electron. We can find this by averaging over all electrons except electron number 1. Indeed, the electric dipole moment doesn't depend on electron spin and so we can also average over the electron spin. I can illustrate with reference to dihydrogen LCAO

$$D_7 = \sqrt{\frac{1}{2!}} \begin{vmatrix} \psi_+(\mathbf{r}_1)\alpha(s_1) & \psi_+(\mathbf{r}_1)\beta(s_1) \\ \psi_+(\mathbf{r}_2)\alpha(s_2) & \psi_+(\mathbf{r}_2)\beta(s_2) \end{vmatrix}$$

$$= \sqrt{\tfrac{1}{2}}\psi_+(\mathbf{r}_1)\psi_+(\mathbf{r}_2)\{\alpha(s_1)\beta(s_2) - \beta(s_1)\alpha(s_2)\}$$

$$D_7^2 = \tfrac{1}{2}\{\psi_+(\mathbf{r}_1)\psi_+(\mathbf{r}_2)\}^2\{\alpha^2(s_1)\beta^2(s_2) + \beta^2(s_1)\alpha^2(s_2) - 2\alpha(s_1)\alpha(s_2)\beta(s_1)\beta(s_2)\}$$

If we average over electron 2 (i.e. integrate with respect to its space and spin co-ordinates) we get

$$\int D_7^2 \, d\tau_2 ds_2 = \tfrac{1}{2}\psi_+(\mathbf{r}_1)^2\{\alpha^2(s_1) + \beta^2(s_1)\}$$

and if we are not particularly interested in the spin variable we find, then on integration over the spin of electron 1

$$\int D_7^2 \, d\tau_2 ds_2 ds_1 = \{\psi_+(\mathbf{r}_1)\}^2 \qquad (15.12)$$

I have of course assumed that all the wavefunctions are real rather than complex; it wouldn't have made any difference to my argument.

This latter quantity, Equation (15.12) times $d\tau_1$, gives the probability of finding electron 1 in $d\tau_1$ with either spin, and the other electrons anywhere, again with either spin. Since there are two indistinguishable electrons in dihydrogen, the total electron density must be twice my result

$$\rho(\mathbf{r}) = -2e\{\psi_+(\mathbf{r})\}^2$$

The charges and bond orders matrix is therefore just twice what we found for the hydrogen molecular ion

$$\mathbf{P}^{\text{LCAO}} = \begin{pmatrix} \dfrac{1}{1+S} & \dfrac{1}{1+S} \\ \dfrac{1}{1+S} & \dfrac{1}{1+S} \end{pmatrix} \qquad (15.13)$$

A corresponding analysis for the VB function gives

$$\mathbf{P}^{\text{VB}} = \begin{pmatrix} \dfrac{1}{1+S^2} & \dfrac{S}{1+S^2} \\ \dfrac{S}{1+S^2} & \dfrac{1}{1+S^2} \end{pmatrix} \qquad (15.14)$$

and the gross Mulliken population for each hydrogen nucleus is once again $\tfrac{1}{2}$, just as it should be.

There is a nice way to remember the result, provided you are happy with matrices. If we write an *overlap matrix* $\mathbf{S}$ for the two atomic orbitals $1s_A$ and $1s_B$ as

$$\mathbf{S} = \begin{pmatrix} \int 1s_A 1s_A \, d\tau & \int 1s_A 1s_B \, d\tau \\ \int 1s_B 1s_A \, d\tau & \int 1s_B 1s_B \, d\tau \end{pmatrix}$$

then a little analysis shows that

$$\sum \sum P_{ij} S_{ij} = \text{number of electrons}$$

This can be restated in terms of the trace of the matrix product

$$\text{tr}(\mathbf{PS}) = \text{number of electrons} \qquad (15.15)$$

16 The HF−LCAO Model

In this chapter I want to explain how we progress the Hartree−Fock (HF) orbital model to more complicated molecules. The HF model encompasses the Born−Oppenheimer approximation and so we glue the nuclei into place before trying to calculate the electronic wavefunction. Each electron experiences an average field due to the remaining electrons, together with the field due to the fixed nuclei. If we are interested in geometry optimization, then we have to perform HF calculations at each point on the molecular potential energy surface.

We can solve the relevant HF equation numerically for atoms to whatever accuracy we require, because of their high symmetry. In the case of a molecule, we have to resort to the LCAO approximation, as discussed in Chapter 15.

In the case of dihydrogen and the hydrogen molecular ion, we were able to deduce simple forms for the lowest energy molecular orbitals

$$\psi_\pm = \frac{1}{\sqrt{2(1 \pm S)}}(1s_A \pm 1s_B)$$

just on the basis of symmetry considerations, assuming that only the hydrogenic 1s orbitals had a role to play. We have to use very many atomic orbitals (*basis functions*, to use the correct jargon) for polyatomic molecules and it is not possible to deduce the LCAO coefficients by hand waving symmetry considerations. The 'form' of the resulting orbitals, by which I mean the LCAO coefficients, is determined by the variation principle. We therefore have to find a variational expression for the electronic energy.

There is a third issue, namely which basis functions should we use for a polyatomic molecule; we have already met hydrogenic and Slater orbitals, but it turns out that neither of these is suitable by itself for molecular calculations. The reason is one of pragmatism, and has to do with the difficulty of calculating certain two-electron integrals that appear in our orbital model. There will be more on this later in the chapter.

16.1 Roothaan's Landmark Paper

Our next landmark paper is 'New Developments in Molecular Orbital Theory' by C. C. J. Roothaan [66]. It's an old paper (1951), and in those days they didn't always have a synopsis. We can learn a great deal from the first paragraph of the Introduction, as follows:

> *For dealing with the problems of molecular quantum mechanics, two methods of approximation have been developed which are capable of handling many-electron systems. The Heitler–London–Pauling–Slater or valence bond (VB) method originated from a chemical point of view. The atoms are considered as the material from which the molecule is built; accordingly, the molecular wave function is constructed from the wave functions of the individual atoms. The Hund–Mulliken or molecular orbital (MO) method is an extension of the Bohr theory of electron configurations from atoms to molecules. Each electron is assigned to a one-electron wave function or molecular orbital, which is the quantum mechanical analog of an electron orbit... It is the purpose of this paper to build a rigorous mathematical framework for the MO method.*

Within the Born–Oppenheimer approximation we consider a molecule as a set of N point charges of magnitudes $eZ_1, eZ_2, \ldots, eZ_N$ at fixed position vectors $\mathbf{R}_1$, $\mathbf{R}_2, \ldots, \mathbf{R}_N$. Their mutual potential energy is

$$U_{\text{nuc}} = \frac{e^2}{4\pi\epsilon_0} \sum_{i=1}^{N-1} \sum_{j=i+1}^{N} \frac{Z_i Z_j}{R_{ij}} \tag{16.1}$$

Assume for the moment that we have chosen n basis functions, which could be the Slater orbitals from Chapter 13. Basis functions usually are real quantities in the mathematical sense, but complex basis functions have to be used in difficult cases such as when we have to deal with molecular magnetic properties. I will assume that we have chosen a set of real basis functions written $\chi_1(\mathbf{r}), \chi_2(\mathbf{r}), \ldots, \chi_n(\mathbf{r})$. The HF–LCAO method seeks to express each orbital $\psi_1(\mathbf{r}), \psi_2(\mathbf{r}), \ldots, \psi_M(\mathbf{r})$ as a linear combination of the basis functions

$$\psi_i = c_{i,1}\chi_1 + c_{i,2}\chi_2 + \cdots + c_{i,n}\chi_n \tag{16.2}$$

and the process gives n LCAO–MO orbitals in total.

Roothaan's original treatment only applies to molecular electronic states where each HF–LCAO is doubly occupied (so-called 'closed shell states'). This covers the case of the vast majority of organic molecules in their electronic ground state, and we think of the electronic configuration as

$$(\psi_1)^2 (\psi_2)^2 \cdots (\psi_M)^2$$

Figure 16.1   Electronic closed shell with $2M$ electrons

as shown in Figure 16.1. Also, the treatment is restricted to the lowest energy state of each allowed symmetry type; for most organic molecules, which have no symmetry, this means that only the electronic ground state can be treated. The HF–LCAO wavefunction is a single Slater determinant, and we can assume without any loss of generality that the HF–LCAO orbitals are normalized and orthogonal. It is rarely the case that the *basis functions* are orthogonal, but this can be dealt with very simply. Each orbital is doubly occupied and there are therefore $m = 2M$ electrons.

We now need to find the variational energy, which for a real wavefunction is

$$\varepsilon_{el} = \frac{\int \Psi \hat{H} \Psi \, d\tau}{\int \Psi^2 \, d\tau}$$

It is neatest if I express the Hamiltonian as the sum of one-electron and two-electron operators as in Chapter 14

$$\hat{H} = \sum_{i=1}^{m} \hat{h}^{(1)}(\mathbf{r}_i) + \sum_{i=1}^{m-1} \sum_{j=i+1}^{m} \hat{g}(\mathbf{r}_i, \mathbf{r}_j) \tag{16.3}$$

I also discussed the Slater–Condon–Shortley rules in Chapter 14. Application of these rules to our wavefunction gives

$$\varepsilon_{el} = 2 \sum_{i=1}^{M} \int \psi_i(\mathbf{r}_1)\hat{h}^{(1)}(\mathbf{r}_1)\psi_i(\mathbf{r}_1) \, d\tau_1$$

$$+ \sum_{i=1}^{M} \sum_{j=1}^{M} 2 \iint \psi_i(\mathbf{r}_1)\psi_i(\mathbf{r}_1)\hat{g}(\mathbf{r}_1, \mathbf{r}_2)\psi_j(\mathbf{r}_2)\psi_j(\mathbf{r}_2) \, d\tau_1 d\tau_2$$

$$- \sum_{i=1}^{M} \sum_{j=1}^{M} \iint \psi_i(\mathbf{r}_1)\psi_j(\mathbf{r}_1)\hat{g}(\mathbf{r}_1, \mathbf{r}_2)\psi_i(\mathbf{r}_2)\psi_j(\mathbf{r}_2) \, d\tau_1 d\tau_2 \tag{16.4}$$

The one-electron term represents the kinetic energy of the $2M$ electrons and the mutual electrostatic potential energy of the electrons and the nuclei. The first of the two-electron terms is the Coulomb term; it represents the mutual electrostatic repulsion of a pair of electrons whose charge densities are $-e\psi_R^2$ and $-e\psi_S^2$. The second of the two-electron terms is the exchange term, which arises because of the fermion nature of the electrons. These equations apply for any single determinant closed shell, they do not yet contain any reference to the LCAO procedure.

16.2 The $\hat{J}$ and $\hat{K}$ Operators

It is sometimes useful to recast the equation as the expectation value of a sum of one-electron and certain *pseudo one-electron operators*

$$\varepsilon_e = 2 \int \sum_{i=1}^{M} \psi_i(\mathbf{r_1})(\hat{h}^{(1)}(\mathbf{r}_1) + \hat{J}(\mathbf{r}_1) - \tfrac{1}{2}\hat{K}(\mathbf{r}_1))\psi_i(\mathbf{r}_1)\, d\tau_1 \qquad (16.5)$$

The operator $\hat{h}^{(1)}$ represents the kinetic energy of an electron and the nuclear attraction. The operators $\hat{J}$ and $\hat{K}$ are called the *Coulomb* and the *exchange operators*. They can be defined through their expectation values as follows

$$\int \psi_R(\mathbf{r}_1)\hat{J}(\mathbf{r}_1)\psi_R(\mathbf{r}_1)\, d\tau_1 = \sum_{i=1}^{M} \iint \psi_R^2(\mathbf{r}_1)\hat{g}(\mathbf{r}_1,\mathbf{r}_2)\psi_i^2(\mathbf{r}_2)\, d\tau_1 d\tau_2 \qquad (16.6)$$

and

$$\int \psi_R(\mathbf{r}_1)\hat{K}(\mathbf{r}_1)\psi_R(\mathbf{r}_1)\, d\tau_1 = \sum_{i=1}^{M} \iint \psi_R(\mathbf{r}_1)\psi_i(\mathbf{r}_1)\hat{g}(\mathbf{r}_1,\mathbf{r}_2)\psi_R(\mathbf{r}_2)\psi_i(\mathbf{r}_2)\, d\tau_1 d\tau_2$$

$$(16.7)$$

The HF Hamiltonian is a one-electron operator, defined by

$$\hat{h}^{F}(\mathbf{r}) = \hat{h}^{(1)}(\mathbf{r}) + \hat{J}(\mathbf{r}) - \tfrac{1}{2}\hat{K}(\mathbf{r}) \qquad (16.8)$$

where the coordinates $\mathbf{r}$ refer to an arbitrary electron. HF orbitals are solutions of the eigenvalue equation

$$\hat{h}^{F}(\mathbf{r})\psi(\mathbf{r}) = \varepsilon\psi(\mathbf{r})$$

16.3 The HF–LCAO Equations

I am going to make use of matrices and matrix algebra for many of the derived equations, for the simple reason that they look neater than they would otherwise do. The n basis functions are usually real and usually overlap each other and so they are not necessarily orthogonal. Following the arguments of Chapters 13 and 14, we collect together the basis function overlap integrals into an $n \times n$ real symmetric matrix **S** that has typically an i, j element

$$S_{i,j} = \int \chi_i(\mathbf{r}_1)\chi_j(\mathbf{r}_1)\, d\tau_1$$

It is convenient to store the LCAO coefficients in an $n \times n$ matrix

$$\mathbf{U} = \begin{pmatrix} c_{1,1} & c_{2,1} & \cdots & c_{n,1} \\ c_{1,2} & c_{2,2} & \cdots & c_{n,2} \\ \cdots & \cdots & \cdots & \cdots \\ c_{1,n} & c_{2,n} & \cdots & c_{n,n} \end{pmatrix} \tag{16.9}$$

so that the first column collects the coefficient of the first occupied HF–LCAO orbital and so on. If we collect together the m occupied LCAO orbitals into an $n \times m$ matrix $\mathbf{U}^{\mathrm{occ}}$

$$\mathbf{U}^{\mathrm{occ}} = \begin{pmatrix} c_{1,1} & c_{2,1} & \cdots & c_{m,1} \\ c_{1,2} & c_{2,2} & \cdots & c_{m,2} \\ \cdots & \cdots & \cdots & \cdots \\ c_{1,n} & c_{2,n} & \cdots & c_{m,n} \end{pmatrix} \tag{16.10}$$

then the matrix $2\mathbf{U}^{\mathrm{occ}}\,(\mathbf{U}^{\mathrm{occ}})^{\mathrm{T}}$ gives the charges and bond orders matrix **P** discussed earlier. The next step is to express the energy in terms of the basis functions and the matrix **P**.

The one-electron contribution is

$$2\sum_{R=1}^{M} \int \psi_R(\mathbf{r}_1)\hat{h}^{(1)}(\mathbf{r}_1)\psi_R(\mathbf{r}_1)\, d\tau = 2\sum_{R=1}^{M}\sum_{i=1}^{n}\sum_{j=1}^{n} c_{R,i}c_{R,j} \int \chi_i(\mathbf{r}_1)\hat{h}^{(1)}(\mathbf{r}_1)\chi_j(\mathbf{r}_1)\, d\tau$$

If we switch the summation signs on the right-hand side we recognize elements of the charges and bond orders matrix **P**

$$2\sum_{R=1}^{M}\sum_{i=1}^{n}\sum_{j=1}^{n} c_{R,i}c_{R,j} \int \chi_i(\mathbf{r}_1)\hat{h}^{(1)}(\mathbf{r}_1)\chi_j(\mathbf{r}_1)\, d\tau$$

$$= \sum_{i=1}^{n}\sum_{j=1}^{n}\left\{ 2\sum_{R=1}^{M} c_{R,i}c_{R,j} \int \chi_i(\mathbf{r}_1)\hat{h}^{(1)}(\mathbf{r}_1)\chi_j(\mathbf{r}_1)\, d\tau \right\}$$

$$= \sum_{i=1}^{n}\sum_{j=1}^{n}\left\{ P_{i,j} \int \chi_i(\mathbf{r}_1)\hat{h}^{(1)}(\mathbf{r}_1)\chi_j(\mathbf{r}_1)\, d\tau \right\}$$

Finally, on collecting together the one-electron integrals over the basis functions into an $n \times n$ matrix $\mathbf{h}_1$ whose i,jth element is

$$(\mathbf{h}_1)_{ij} = \int \chi_i(\mathbf{r}_1)\hat{h}^{(1)}(\mathbf{r}_1)\chi_i(\mathbf{r}_1)\,d\tau_1$$

then the one-electron term emerges as the trace of the matrix product $\mathbf{Ph}_1$

$$\sum_{i=1}^{n}\sum_{j=1}^{n} P_{ij}(h^{(1)})_{ij} = \mathrm{tr}(\mathbf{Ph}_1) \tag{16.11}$$

A corresponding analysis shows that the two-electron terms can be written as $\frac{1}{2}\mathrm{tr}(\mathbf{P}(\mathbf{J} - \frac{1}{2}\mathbf{K}))$, where the elements of the matrices $\mathbf{J}$ and $\mathbf{K}$ depend on those of $\mathbf{P}$ in a more complicated way

$$J_{ij} = \sum_{k=1}^{n}\sum_{l=1}^{n} P_{kl} \iint \chi_i(\mathbf{r}_1)\chi_j(\mathbf{r}_1)\hat{g}(\mathbf{r}_1,\mathbf{r}_2)\chi_k(\mathbf{r}_2)\chi_l(\mathbf{r}_2)\,d\tau_1 d\tau_2$$

$$K_{ij} = \sum_{k=1}^{n}\sum_{l=1}^{n} P_{kl} \iint \chi_i(\mathbf{r}_1)\chi_k(\mathbf{r}_1)\hat{g}(\mathbf{r}_1,\mathbf{r}_2)\chi_j(\mathbf{r}_2)\chi_l(\mathbf{r}_2)\,d\tau_1 d\tau_2 \tag{16.12}$$

Many authors collect together these Coulomb and exchange matrices into a composite called the *electron repulsion matrix*

$$\mathbf{G} = \mathbf{J} - \tfrac{1}{2}\mathbf{K}$$

The electronic energy therefore comes out as

$$\begin{aligned}\varepsilon_{el} &= \mathrm{tr}(\mathbf{Ph}_1) + \tfrac{1}{2}\mathrm{tr}(\mathbf{PJ}) - \tfrac{1}{4}\mathrm{tr}(\mathbf{PK})\\ &= \mathrm{tr}(\mathbf{Ph}_1) + \tfrac{1}{2}\mathrm{tr}(\mathbf{PG})\end{aligned} \tag{16.13}$$

We now examine how the electronic energy changes when $\mathbf{P}$ changes by a small amount $\delta\mathbf{P}$ (equivalent to asking how ε_{el} changes as the LCAO coefficients change). We let

$$\mathbf{P} \rightarrow \mathbf{P} + \delta\mathbf{P} \tag{16.14}$$

and after a little manipulation find the first-order change in the electronic energy

$$\begin{aligned}\varepsilon_{el} &= \mathrm{tr}(\mathbf{h}_1\delta\mathbf{P}) + \mathrm{tr}(\mathbf{G}\delta\mathbf{P})\\ &= \mathrm{tr}(\mathbf{h}^{F}\delta\mathbf{P})\end{aligned} \tag{16.15}$$

where I have defined the *Hartree–Fock Hamiltonian matrix*

$$\mathbf{h}^{F} = \mathbf{h}_1 + \mathbf{G} \tag{16.16}$$

We want to find $\delta \mathbf{P}$ so that $\delta \varepsilon_{\mathrm{el}}$ is zero; the trivial solution is when $\delta \mathbf{P}$ equals the zero matrix, but obviously there have to be restrictions on the allowed form of $\delta \mathbf{P}$. I noted above the requirement that the HF–LCAO orbitals be normalized and orthogonal. If we collect together all the overlap integrals over basis functions into the matrix $\mathbf{S}$, then a little matrix manipulation will establish that

$$\mathbf{U}^{\mathrm{T}}\mathbf{S}\mathbf{U} = 1$$

$$\mathbf{P}\mathbf{S}\mathbf{P} = 4\mathbf{P}$$

The condition that the HF–LCAO orbitals are normalized and orthogonal is equivalent to the matrix equation $\mathbf{P}\mathbf{S}\mathbf{P} = 4\mathbf{P}$ and the modified $\mathbf{P}$ must also satisfy this condition

$$(\mathbf{P} + \delta \mathbf{P})\mathbf{S}(\mathbf{P} + \delta \mathbf{P}) = 4(\mathbf{P} + \delta \mathbf{P})$$

A little manipulation shows that, at the energy minimum

$$\mathbf{h}^{F}\mathbf{P} = \mathbf{P}\mathbf{h}^{F} \qquad (16.17)$$

and whilst this doesn't actually help find the electron density at the minimum, it gives a condition that has to be satisfied.

Roothaan solved the problem in a different but equivalent way; he let the HF–LCAO coefficients vary subject only to the condition that the HF–LCAO orbitals remained normalized and orthogonal. He demonstrated that the coefficients $\mathbf{c}_i$ are given by the generalized matrix eigenvalue problem

$$\mathbf{h}^{F}\mathbf{c}_i = \varepsilon_i \mathbf{S}\mathbf{c}_i \qquad (16.18)$$

The HF matrix is $n \times n$ and there are exactly n solutions to the eigenvalue problem. The lowest energy m solutions correspond to the occupied HF–LCAO orbitals. The energy ε is called the HF orbital energy, and for each value of ε there is a column vector of HF–LCAO coefficients. Once again, this doesn't help us to find the coefficients, because they are contained within $\mathbf{h}^{F}$. So, an iterative procedure is necessary.

16.3.1 The HF–LCAO equations

For the record, since I will need to refer to the HF Hamiltonian many times, here it is (for the closed shell system as shown in Figure 16.1, and assuming real basis functions)

$$
\begin{aligned}
h_{ij}^{\mathrm{F}} = & \int \chi_i(\mathbf{r}_1)\hat{h}^{(1)}(\mathbf{r}_1)\chi_j(\mathbf{r}_1)\,\mathrm{d}\tau \\
& + \sum_{k=1}^{n}\sum_{l=1}^{n} P_{kl} \iint \chi_i(\mathbf{r}_1)\chi_j(\mathbf{r}_1)\hat{g}(\mathbf{r}_1,\mathbf{r}_2)\chi_k(\mathbf{r}_2)\chi_l(\mathbf{r}_2)\,\mathrm{d}\tau_1\mathrm{d}\tau_2 \\
& - \frac{1}{2}\sum_{k=1}^{n}\sum_{l=1}^{n} P_{kl} \iint \chi_i(\mathbf{r}_1)\chi_k(\mathbf{r}_1)\hat{g}(\mathbf{r}_1,\mathbf{r}_2)\chi_j(\mathbf{r}_2)\chi_l(\mathbf{r}_2)\,\mathrm{d}\tau_1\mathrm{d}\tau_2 \qquad (16.19)
\end{aligned}
$$

Over the years, many workers have devised iterative procedures for solving the problem. The simplest procedure is as follows:

- Choose a molecular geometry.

- Choose a suitable basis set.

- Calculate all integrals over the basis functions and store them.

- Make an educated guess at the HF–LCAO coefficients $\mathbf{U}^{occ}$.

- Calculate $\mathbf{P}$ and $\mathbf{h}^F$ (the time-consuming step).

- Solve the matrix eigenvalue problem $\mathbf{h}^F \mathbf{c} = \varepsilon \mathbf{S} \mathbf{c}$ to give the new $\mathbf{U}^{occ}$.

- Check for convergence (test ε_{el} and/or $\mathbf{P}$)

- Exit or go back three steps.

There are other procedures. I'll give you a numerical example later and explain some of the methods that people use in order to speed up their calculation. Naturally, having progressed this far in the book you will know that the HF–LCAO calculation simply gives one point on a molecular potential energy surface, as defined by the Born–Oppenheimer approximation. If your interest in life is molecular geometry optimization, then you will have to follow the same kind of procedures as with molecular mechanics (MM) in Chapter 5; there is a very big difference in that MM energies can be calculated in almost no time at all, whilst HF–LCAO energies consume considerable resource.

16.4 The Electronic Energy

As noted in Equation (16.4), the HF–LCAO electronic energy is given by

$$\varepsilon_{el} = \mathrm{tr}(\mathbf{P} \mathbf{h}_1) + \tfrac{1}{2}\mathrm{tr}(\mathbf{P}\mathbf{G})$$
$$= \mathrm{tr}(\mathbf{P} \mathbf{h}^F) - \tfrac{1}{2}\mathrm{tr}(\mathbf{P}\mathbf{G})$$

The HF–LCAO matrix eigenvalue equation is $\mathbf{h}^F \mathbf{c} = \varepsilon \mathbf{S} \mathbf{c}$, and the lowest energy m solutions determine the electronic ground state of a closed shell molecule. The sum of orbital energies ε_{orb} is therefore

$$\varepsilon_{orb} = 2(\varepsilon_A + \varepsilon_B + \cdots + \varepsilon_M)$$

If

$$\mathbf{h}^F \mathbf{c}_A = \varepsilon_A \mathbf{S} \mathbf{c}_A$$
$$(\mathbf{c}_A)^T \mathbf{h}^F \mathbf{c}_A = \varepsilon_A (\mathbf{c}_A)^T \mathbf{S} \mathbf{c}_A$$

$$\varepsilon_A = \frac{(\mathbf{c}_A)^T \mathbf{h}^F \mathbf{c}_A}{(\mathbf{c}_A)^T \mathbf{S} \mathbf{c}_A}$$
$$= (\mathbf{c}_A)^T \mathbf{h}^F \mathbf{c}_A$$

then a little manipulation shows that

$$\varepsilon_{el} = \varepsilon_{orb} - \tfrac{1}{2} \mathrm{tr}(\mathbf{PG}) \qquad (16.20)$$

Orbital energies do not sum to give the total electronic energy. I have written the formula in matrix terms for convenience when working in the LCAO approximation. The result is true for all HF wavefunctions, whether they are numerical atomic ones (and therefore at the HF limit) or just rough-and-ready ones that have been through the HF treatment but are nowhere near the HF limit.

16.5 Koopmans' Theorem

Koopmans' theorem relates ionization energies to HF orbital energies, as discussed already for the atomic HF case. It is valid for any closed shell HF wavefunction, no matter how good or bad. It is subject to the following small print: the orbitals of the parent molecule and cation must remain unchanged on ionization. The fact is that the electron density invariably reorganizes on ionization, but Koopmans' theorem calculations have been widely used to interpret photoelectron spectra.

16.6 Open Shell Systems

The vast majority of organic molecules have electronic configurations of the type

$$(\psi_1)^2 (\psi_2)^2 \cdots (\psi_M)^2$$

that can be described by the closed shell version of the HF–LCAO procedure detailed above. A good deal of chemistry is concerned with excited triplet states, free radicals, cations and anions where there is at least one unpaired electron, and the procedure outlined above is not appropriate because the energy formula is incorrect.

The HF–LCAO procedure is a chemically attractive one since it gives a clear orbital picture and it turns out that several classes of open shell molecules can still be treated using modified forms of the basic theory. The simplest case is shown in Figure 16.2. We have a number $n_1 = M$ of doubly occupied orbitals and a number $n_2 = P - (M + 1) N$ of singly occupied orbitals, all with parallel spin electrons. We refer to this as the *restricted open shell (ROHF)* model. The closed shell energy formula, Equation (16.4), has to be modified, and for the sake of neatness I will introduce the *occupation numbers* ν_1 ($= 2$) and ν_2 ($= 1$) for the two shells. If I use i

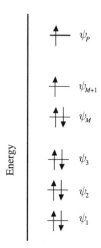

Figure 16.2   Restricted open shell HF

and j as indices for the doubly occupied orbitals and u and v for the singly occupied orbitals we find

$$
\varepsilon_{\mathrm{el}} = \nu_1 \left(\begin{array}{l}
\displaystyle\sum_{i=1}^{M} \int \psi_i(\mathbf{r}_1)\hat{h}^{(1)}(\mathbf{r}_1)\psi_i(\mathbf{r}_1)\,\mathrm{d}\tau_1 \\[1em]
\displaystyle +\frac{1}{2}\nu_1 \sum_{i=1}^{M}\sum_{j=1}^{M}\left(\iint \psi_i^2(\mathbf{r}_1)\hat{g}(\mathbf{r}_1,\mathbf{r}_2)\psi_j^2(\mathbf{r}_2)\,\mathrm{d}\tau_1\mathrm{d}\tau_2 \right. \\[1em]
\displaystyle \left. -\frac{1}{2}\iint \psi_i(\mathbf{r}_1)\psi_j(\mathbf{r}_1)\hat{g}(\mathbf{r}_1,\mathbf{r}_2)\psi_i(\mathbf{r}_2)\psi_j(\mathbf{r}_2)\,\mathrm{d}\tau_1\mathrm{d}\tau_2 \right)
\end{array} \right)
$$

$$
+ \nu_2 \left(\begin{array}{l}
\displaystyle\sum_{u=M+1}^{P} \int \psi_u(\mathbf{r}_1)\hat{h}^{(1)}(\mathbf{r}_1)\psi_u(\mathbf{r}_1)\,\mathrm{d}\tau_1 \\[1em]
\displaystyle +\frac{1}{2}\nu_2 \sum_{u=M+1}^{P}\sum_{v=M+1}^{P}\left(\iint \psi_u^2(\mathbf{r}_1)\hat{g}(\mathbf{r}_1,\mathbf{r}_2)\psi_v^2(\mathbf{r}_2)\,\mathrm{d}\tau_1\mathrm{d}\tau_2 \right. \\[1em]
\displaystyle \left. -\iint \psi_u(\mathbf{r}_1)\psi_v(\mathbf{r}_1)\hat{g}(\mathbf{r}_1,\mathbf{r}_2)\psi_u(\mathbf{r}_2)\psi_v(\mathbf{r}_2)\,\mathrm{d}\tau_1\mathrm{d}\tau_2 \right)
\end{array} \right)
$$

$$
+ \nu_1\nu_2 \left(\sum_{i=1}^{M}\sum_{u=M+1}^{P}\left(\iint \psi_i^2(\mathbf{r}_1)\hat{g}(\mathbf{r}_1,\mathbf{r}_2)\psi_u^2(\mathbf{r}_2)\,\mathrm{d}\tau_1\mathrm{d}\tau_2 \right. \right.
$$

$$
\left. \left. -\frac{1}{2}\iint \psi_i(\mathbf{r}_1)\psi_u(\mathbf{r}_1)\hat{g}(\mathbf{r}_1,\mathbf{r}_2)\psi_i(\mathbf{r}_2)\psi_u(\mathbf{r}_2)\,\mathrm{d}\tau_1\mathrm{d}\tau_2 \right) \right)
$$

$$(16.21)$$

In the LCAO variation we introduce n basis functions. We can collect the HF–LCAO coefficients of the n_1 doubly occupied orbitals into columns of a matrix

U_1 ($n \times n_1$) and the coefficients of the n_2 singly occupied orbitals into columns of U_2 ($n \times n_2$) and define the $n \times n$ density matrices

$$\mathbf{R}_1 = \mathbf{U}_1\mathbf{U}_1^T; \qquad \mathbf{R}_2 = \mathbf{U}_2\mathbf{U}_2^T$$

Repeating the analysis given for closed shell states gives

$$\varepsilon_{el} = \nu_1\text{tr}\left(\mathbf{R}_1\left(\mathbf{h}_1 + \frac{1}{2}\mathbf{G}_1\right)\right) + \nu_2\text{tr}\left(\mathbf{R}_2\left(\mathbf{h}_1 + \frac{1}{2}\mathbf{G}_2\right)\right) \qquad (16.22)$$

where the two **G** matrices contain Coulomb and exchange contributions similar to those defined by Equation (16.12) for the closed shell case. We then allow $\mathbf{R}_1$ and $\mathbf{R}_2$ to vary, subject to orthonormality, just as in the closed shell case. Once again, C. C. J. Roothaan [67] showed, in his 1960 paper, how to write a Hamiltonian matrix whose eigenvectors give the columns $\mathbf{U}_1$ and $\mathbf{U}_2$ above.

16.7 The Unrestricted Hartree–Fock Model

The methyl radical CH_3 can be easily produced by breaking the C–C bond in ethane $C_2H_6 \rightarrow 2\ CH_3$

It has nine electrons and there are two ways we can progress, as shown in Figure 16.3. First we can insist that the lowest energy molecular orbitals are doubly occupied, with the extra electron occupying the highest energy orbital (ROHF). On the other hand, we can let the α and β spin electrons occupy different spatial orbitals (the *unrestricted Hartree–Fock* model, known by the acronym UHF).

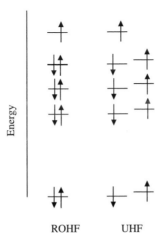

Figure 16.3 Methyl radical

The treatment will be familiar to you by now. We write a single Slater determinant but allow the p α and the q β spin electrons to have different spatial orbitals. Application of the Slater–Condon–Shortley rules gives an energy expression in terms of the HF orbitals as

$$
\begin{aligned}
\varepsilon_{\text{el}} = & \sum_{i=1}^{P} \int \psi_i^{\alpha}(\mathbf{r}_1)\hat{h}^{(1)}(\mathbf{r}_1)\psi_i^{\alpha}(\mathbf{r}_1)\, d\tau_1 \sum_{j=1}^{Q} \int \psi_j^{\beta}(\mathbf{r}_1)\hat{h}^{(1)}(\mathbf{r}_1)\psi_j^{\beta}(\mathbf{r}_1)\, d\tau_1 \\
& + \sum_{i=1}^{P}\sum_{j=1}^{P} \iint \psi_i^{\alpha}(\mathbf{r}_1)\psi_i^{\alpha}(\mathbf{r}_1)\hat{g}(\mathbf{r}_1,\mathbf{r}_2)\psi_j^{\alpha}(\mathbf{r}_2)\psi_j^{\alpha}(\mathbf{r}_2)\, d\tau_1 d\tau_2 \\
& - \frac{1}{2}\sum_{i=1}^{P}\sum_{j=1}^{P} \iint \psi_i^{\alpha}(\mathbf{r}_1)\psi_j^{\alpha}(\mathbf{r}_1)\hat{g}(\mathbf{r}_1,\mathbf{r}_2)\psi_i^{\alpha}(\mathbf{r}_2)\psi_j^{\alpha}(\mathbf{r}_2)\, d\tau_1 d\tau_2 \\
& + \sum_{i=1}^{Q}\sum_{j=1}^{Q} \iint \psi_i^{\beta}(\mathbf{r}_1)\psi_i^{\beta}(\mathbf{r}_1)\hat{g}(\mathbf{r}_1,\mathbf{r}_2)\psi_j^{\beta}(\mathbf{r}_2)\psi_j^{\beta}(\mathbf{r}_2)\, d\tau_1 d\tau_2 \\
& - \frac{1}{2}\sum_{i=1}^{P}\sum_{j=1}^{P} \iint \psi_i^{\beta}(\mathbf{r}_1)\psi_j^{\beta}(\mathbf{r}_1)\hat{g}(\mathbf{r}_1,\mathbf{r}_2)\psi_i^{\beta}(\mathbf{r}_2)\psi_j^{\beta}(\mathbf{r}_2)\, d\tau_1 d\tau_2 \qquad (16.23)
\end{aligned}
$$

There are no cross terms between the two sets of orbitals because of the orthogonality of the spin functions. We now introduce the LCAO concept; we expand each set of HF orbitals in terms of the same basis set $\chi_1, \chi_2, \ldots, \chi_n$ and form two density matrices, one for the α-spin electrons $\mathbf{P}^{\alpha}$ and one for the β spin electrons $\mathbf{P}^{\beta}$ in the obvious way. We finally arrive at two linked HF–LCAO Hamiltonian matrices; the α-spin matrix has elements

$$
\begin{aligned}
h_{ij}^{F,\alpha} = & \int \chi_i(\mathbf{r}_1)\hat{h}^{(1)}(\mathbf{r}_1)\chi_j(\mathbf{r}_1)\, d\tau \\
& + \sum_{k=1}^{n}\sum_{l=1}^{n} (P_{kl}^{\alpha} + P_{kl}^{\beta}) \iint \chi_i(\mathbf{r}_1)\chi_j(\mathbf{r}_1)\hat{g}(\mathbf{r}_1,\mathbf{r}_2)\chi_k(\mathbf{r}_2)\chi_l(\mathbf{r}_2)\, d\tau_1 d\tau_2 \\
& - \sum_{k=1}^{n}\sum_{l=1}^{n} P_{kl}^{\alpha} \iint \chi_i(\mathbf{r}_1)\chi_k(\mathbf{r}_1)\hat{g}(\mathbf{r}_1,\mathbf{r}_2)\chi_j(\mathbf{r}_2)\chi_l(\mathbf{r}_2)\, d\tau_1 d\tau_2 \qquad (16.24)
\end{aligned}
$$

with a similar expression for the β electrons. The UHF–LCAO orbitals are found by any of the standard techniques already discussed, for example repeated construction of the density matrices, the two Hamiltonians, matrix diagonalization and so on until consistency is attained. The orbital energies can be associated with ionization energies using an extension of Koopmans' theorem.

16.7.1 Three technical points

There are three technical points that you should be aware of.

1. I chose the example CH_3 with some care to make sure that the electronic ground state could be written as a single Slater determinant. There are some electronic states of very simple molecules where this is not possible, and more advanced considerations apply; for example, the first excited singlet state of dihydrogen cannot be represented as a single Slater determinant.

2. The UHF method does not generally give wavefunctions that are eigenfunctions of the spin operator $\hat{S}^2$. The methyl radical UHF wavefunction will actually be a mixture of various spin components ranging from the (correct) doublet state (spin quantum number $s = \frac{1}{2}$) through in steps of 1 to the spin state with $s = \frac{9}{2}$

$$\Psi_{UHF} = c_2 \Psi_{s=1/2} + c_4 \Psi_{s=3/2} + \cdots \tag{16.25}$$

and very often it happens that the 'correct' spin state dominates ($c_2 \approx 1$ in this case, a doublet) with the largest impurity coming from the next highest spin state (c_4 in this case, a quartet). There are two strategies; the easier option is to run the UHF calculation and eliminate the largest contaminant after the UHF wavefunction has been found, whilst the more correct but incredibly more complicated procedure is to eliminate all spin contaminants from the wavefunction before performing the UHF calculation. The second option leads to a linear combination of Slater determinants, such as that given in Equation (16.25), where we have to find the HF–LCAO coefficients and also the expansion coefficients c_2, c_4 and so on. Such *extended Hartree–Fock* (EHF) calculations are very expensive in computer resource, and I will return to this idea in Chapter 19.

3. The 'charges and bond orders matrix' concept needs to be generalized to cover the case of open shells, because some orbitals are singly occupied (in the ROHF model) and because the spatial parts of the spinorbitals can be different (in the case of the UHF model). We collect the LCAO coefficients for the occupied α and β spin electrons separately into two matrices that I will call $\mathbf{U}^\alpha$ and $\mathbf{U}^\beta$. Each matrix product $\mathbf{U}^\alpha(\mathbf{U}^\alpha)^T$ and $\mathbf{U}^\beta(\mathbf{U}^\beta)^T$ gives us the charges and bond orders matrices for the individual α- and β-spin electrons. These matrices are sometimes referred to as $\mathbf{P}^\alpha$ and $\mathbf{P}^\beta$. The usual charge density $\mathbf{P}$ is the sum of these, and the difference is called the spin density $\mathbf{Q}$.

16.8 Basis Sets

The stumbling block for molecular HF applications was the very large number of difficult integrals, especially those of the two-electron kind. A typical two-electron integral over basis functions

$$\frac{e^2}{4\pi\epsilon_0} \int\int \chi_i(\mathbf{r}_1)\chi_j(\mathbf{r}_1) \frac{1}{r_{12}} \chi_k(\mathbf{r}_2)\chi_l(\mathbf{r}_2)\, d\tau_1 d\tau_2$$

might involve four different basis functions, all of which might be on different atomic centres, or not even on atomic centres. Generally there will be a lot of them; there is no obvious choice of coordinate origin, the integral has a singularity (that is to say, it becomes infinite as the two electrons approach each other) and it is six-dimensional. Authors therefore spoke of the *integral bottleneck*, and the bottleneck was not broken until the early 1950s.

16.8.1 Clementi and Raimondi

Slater's atomic orbital rules were the first step to a correct treatment of basis functions that could be used in molecules. Clementi and Raimondi [68] refined Slater's empirical ideas in 1963 by performing atomic HF–LCAO calculations on atoms with atomic number 2 through 36, in order to calculate the variationally correct orbital exponents. The synopsis is worth reading:

> *The self-consistent field function for atoms with 2 to 36 electrons are computed with a minimum basis set of Slater-type orbitals. The orbital exponents of the atomic orbitals are optimized as to ensure the energy minimum. The analysis of the optimized orbital exponents allows us to obtain simple and accurate rules for the 1s, 2s, 3s, 4s, 2p, 3p, 4p and 3d electronic screening constants. These rules are compared with those proposed by Slater and reveal the need for the screening due to the outside electrons. The analysis of the screening constants (and orbital exponents) is extended to the excited states of the ground state configuration and the positive ions.*

What they did, starting from Slater's ideas of 1s, 2s, 2p... atomic orbitals with modified orbital exponents (effective nuclear charges), was as follows for each atom in their set (see Table 16.1):

- Decide on the electronic ground state configuration and choose starting values of the orbital exponents from Slater's rules.

- Optimize each orbital exponent individually by the HF–LCAO procedure. At the end of each optimization, the earlier optimized values will have changed and so...

- Check for self consistency amongst the orbital exponents and either exit or go back one step.

Table 16.1 Comparison of Slater's exponents with those of Clementi and Raimondi (CR)

Atom	CR 1s exponent	Slater value for 1s	CR 2s exponent	CR 2p exponent	Slater value for 2s/2p
H	1	1			
He	1.6875	1.70			
Li	2.6906	2.70	0.6396		0.650
Be	3.6848	3.70	0.9560		0.975
B	4.6795	4.70	1.2881	1.2107	1.300
C	5.6727	5.70	1.6083	1.5679	1.625
N	6.6651	6.70	1.9237	1.9170	1.950
O	7.6579	7.70	2.2458	2.2266	2.275
F	8.6501	8.70	2.5638	2.5500	2.600
Ne	9.6241	9.70	2.8792	2.8792	2.925

There are better optimization procedures, as you will know from reading earlier chapters, but that is how the early workers did things.

We call such basis sets *single zeta* or *minimal* because they use exactly the same number of atomic orbitals as in descriptive chemistry. For each atom there is just one 1s orbital, one 2s, three 2p and so on.

16.8.2 Extension to second-row atoms

Clementi extended this treatment to the second row in his 1964 paper [69], and he wrote the following in his Abstract:

> *The self-consistent field functions for the ground state of the first and second row atoms (from He to Ar) are computed with a basis set in which two Slater-type orbitals (STO) are chosen for each atomic orbital. The reported STOs have carefully optimized orbital exponents. The total energy is not far from the accurate Hartree–Fock energy given by Clementi, Roothaan and Yoshimine for the first row atoms and unpublished data for the second-row atoms. The obtained basis sets have sufficient flexibility to be a most useful starting set for molecular computations, as noted by Richardson. With the addition of 3d and 4f functions, the reported atomic basis sets provide a molecular basis set which duplicates quantitatively most of the chemical information derivable by the more extended basis sets needed to obtain accurate Hartree–Fock molecular functions.*

Clementi thus doubled the number of atomic orbitals and so used two slightly different 1s orbitals, two slightly different 2s orbitals and so on. Once again, he optimized the orbital exponents by systematic variations and a small sample from his results is shown in Table 16.2. We refer to such a basis set as a *double zeta* basis set. Where the single zeta basis set for atomic lithium has a 1s exponent of 2.6906, the double zeta basis set has two 1s orbitals with exponents 2.4331 and 4.5177 (the inner and outer 1s orbitals).

Table 16.2 A selection from Clementi's double zeta basis set

	1s exponents	2s exponents	2p exponents
He	1.4461		
	2.8622		
Li	2.4331	0.6714	
	4.5177	1.9781	
Be	3.3370	0.6040	
	5.5063	1.0118	
B	4.3048	0.8814	1.0037
	6.8469	1.4070	2.2086
C	5.2309	1.1678	1.2557
	7.9690	1.8203	2.7263
N	6.1186	1.3933	1.5059
	8.9384	2.2216	3.2674
O	7.0623	1.6271	1.6537
	10.1085	2.6216	3.6813
F	7.9179	1.9467	1.8454
	11.0110	3.0960	4.1710
Ne	8.9141	2.1839	2.0514
	12.3454	3.4921	4.6748

The lowest energy HF–LCAO atomic orbital for a Li atom will be a combination of the four s-type basis functions, and we call this combination 'the' atomic 1s orbital.

16.8.3 Polarization functions

There is a second point to note from Clementi's paper, where he speaks about '*the addition of 3d and 4f functions*...' with reference to first- and second- row atoms, respectively. Any chemist would write the fluorine electronic ground state configuration

$$\text{F:}\quad 1s^2\,2s^2\,2p^5$$

and so the d and f orbitals are unoccupied. When atoms combine to form molecules the atomic charge density may well distort from spherical symmetry and such *polarization functions* are needed to describe this distortion accurately.

16.9 Gaussian Orbitals

None of these considerations actually broke the integrals bottleneck. There are two considerations. First of all, the sheer number of integrals that have to be processed.

If we run a HF–LCAO calculation with n basis functions, then we have to calculate $p = \frac{1}{2}n(n+1)$ one-electron integrals of each type (overlap, kinetic energy and electron-nuclear attraction). If $n = 100$, then we have to calculate 15 150 one-electron integrals. It is the two-electron integrals that cause the problem, for we have to calculate at most $q = \frac{1}{2}p(p+1)$ of them; if $n = 100$, then $q = 1.25 \times 10^7$. There are approximately $\frac{1}{8}n^4$ integrals (for large n), although this is to be seen as an upper limit. Many of the integrals turn out to be so tiny that they can be ignored for practical calculations on large molecules.

The second consideration is the mathematical intractability of the integrals, as mentioned above. The real breakthrough came in the 1950s with the introduction of (*Cartesian*) *Gaussian basis functions*; these are similar to Slater orbitals but they have an exponential factor that goes as the *square* of the distance between the electron and the orbital centre

$$\chi_G = N_G x^l y^m z^n \exp\left(-\alpha \frac{r^2}{a_0^2}\right) \tag{16.26}$$

Here l, m and n can have any integral values and the orbital exponent α is positive. If all of l, m and n are zero, then we talk about a 1s-type *Gaussian-type orbital* (GTO). If one of l, m or n is equal to 1, then we have a 2p-type GTO. Expressions for normalized 1s and $2p_x$ GTO are

$$G_{1s} = \left(\frac{2\alpha}{\pi a_0^2}\right)^{3/4} \exp\left(-\frac{\alpha r^2}{a_0^2}\right)$$

$$G_{2p_x} = \left(\frac{128\alpha^5}{\pi^3 a_0^{10}}\right)^{1/4} x \exp\left(-\frac{\alpha r^2}{a_0^2}\right)$$

When dealing with d-type (where $l+m+n = 2$) we note that there are six possibilities, xx, xy, xz, yy, yz and zz, rather than the five combinations we normally encounter for STO. The combination $(x^2 + y^2 + z^2)$ actually gives a 3s GTO, but we normally include all six Cartesian Gaussians in calculations. Similar considerations apply to the f, g . . . orbitals.

Why use Gaussians? Let me illustrate the answer by considering the two one-dimensional normal distribution curves (which are one-dimensional Gaussians) shown in Figure 16.4

$$G_A(x) = \exp\left(-\alpha_A(x - x_A)^2\right)$$

$$G_B(x) = \exp\left(-\alpha_B(x - x_B)^2\right)$$

I have taken $\alpha_A = 0.1$, $x_A = 1$ (the full curve) and $\alpha_B = 0.3$ and $x_B = -2$ (the dotted curve). It is easily shown that the product $G_A(x) \times G_B(x)$ is another Gaussian G_C

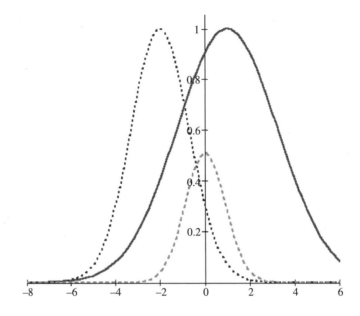

Figure 16.4   Product of two one-dimensional Gaussians

whose centre lies between G_A and G_B

$$G_C(x) = \exp\left(-\frac{\alpha_A \alpha_B}{\alpha_A + \alpha_B}\{x_A - x_B\}^2\right)\exp\left(-(\alpha_A + \alpha_B)(x - x_C)^2\right)$$

$$x_C = \frac{\alpha_A x_A + \alpha_B x_B}{\alpha_A + \alpha_A}$$

(16.27)

which is shown as the dashed curve. This is a general property of Gaussians.

Credit for the introduction of Gaussian basis functions is usually given to S. F. Boys [70] for his 1950 paper, and here is the famous synopsis:

This communication deals with the general theory of obtaining numerical electronic wavefunctions for the stationary states of atoms and molecules. It is shown that by taking Gaussian functions, and functions derived from these by differentiation with respect to the parameters, complete systems of functions can be constructed appropriate to any molecular problem, and that all the necessary integrals can be explicitly evaluated. These can be used in connection with the molecular orbital treatment, or localized bond method, or the general method of linear combinations of many Slater determinants by the variation procedure. This general method of obtaining a sequence of solutions converging to the accurate solution is examined. It is shown that the only obstacle to the evaluation of wavefunctions of any required degree of accuracy is the labour of computation. A modification of the general method applicable to atoms is discussed and considered to be extremely practicable

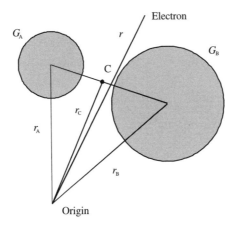

Figure 16.5 Overlap integral between two GTOs

GTOs have one great advantage over Slater-type orbitals (STOs); the nasty integrals (especially the two-electron integrals) we need for molecular quantum mechanics are relatively straightforward because they can be reduced from at most a four-centre integral to a one-centre integral by repeated application of the principle above. Consider, for example, an overlap integral between the two s-type GTOs shown in Figure 16.5. This is a three-dimensional extension to the one-dimensional problem discussed above.

Gaussian A has exponent α_A and is centred at $\mathbf{r}_A$; Gaussian B has exponent α_B and is centred at $\mathbf{r}_B$. If the position vector of the electron is $\mathbf{r}$, then its position vector relative to the centre of G_A is $\mathbf{r} - \mathbf{r}_A$, with a similar expression for G_B. The overlap integral (apart from the normalizing constants N and N') is

$$S_{AB} = N \int \exp\left(-\alpha_A \frac{|\mathbf{r} - \mathbf{r}_A|^2}{a_0^2}\right) \exp\left(-\alpha_B \frac{|\mathbf{r} - \mathbf{r}_B|^2}{a_0^2}\right) d\tau$$

$$= N' \int \exp\left(-\alpha_C \frac{|\mathbf{r} - \mathbf{r}_C|^2}{a_0^2}\right) d\tau \tag{16.28}$$

The product GTO G_C has exponent α_C and centre $\mathbf{r}_C$ given by

$$\alpha_C = \alpha_A + \alpha_B$$

$$\mathbf{r}_C = \frac{1}{\alpha_A + \alpha_B}(\alpha_A \mathbf{r}_A + \alpha_B \mathbf{r}_B)$$

The remaining integral is a product of three standard integrals

$$\int \exp\left(-\alpha_C \frac{|\mathbf{r} - \mathbf{r}_C|^2}{a_0^2}\right) d\tau = \int \exp\left(-\alpha_C \frac{(x - x_C)^2}{a_0^2}\right) dx$$

$$\int \exp\left(-\alpha_C \frac{(y - y_C)^2}{a_0^2}\right) dy \int \exp\left(-\alpha_C \frac{(z - z_C)^2}{a_0^2}\right) dz$$

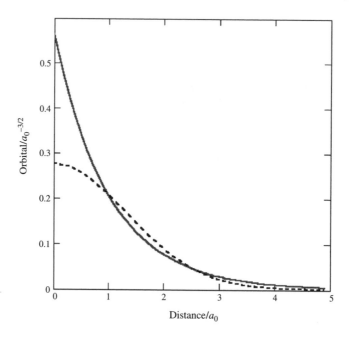

Figure 16.6 GTO vs. STO for a hydrogen atom

One major problem that concerned early workers is that GTOs do not give terribly good energies. If we try a variational calculation on a hydrogen atom with a single s-type GTO

$$G(\alpha) = \left(\frac{\alpha}{\pi a_0^3}\right)^{3/4} \exp\left(-\alpha \frac{r^2}{a_0^2}\right)$$

and calculate the optimal Gaussian exponent, then we find $\alpha_{opt} = 0.283$ with a variational energy of $-0.424\,E_h$. The experimental energy is $-\frac{1}{2}E_h$ and the error is some 15%. A second problem is that the shapes of GTOs and STOs are quite different. Figure 16.6 shows the dependences of the STO (with exponent $\zeta = 1$) and the GTO (with exponent $\alpha = 0.283$) on distance for a hydrogen atom. The plot is of wavefunction$/a_0^{-3/2}$ vs. distance from the nucleus, r/a_0.

The full curve is the STO, the dashed curve the best GTO. GTOs show the wrong behaviour at the nucleus, where they should have a cusp because the mutual potential energy of the electron and the nucleus becomes infinite as the distance becomes zero. GTOs also fall off far too quickly with the distance from the nucleus.

16.9.1 STO/nG

The next step was to address the long-distance behaviour, and Hehre, Stewart and Pople proposed the idea of fitting a fixed linear combination of n GTOs to a given

STO. The GTOs are not explicitly included in a HF–LCAO calculation, they are just used to give a good fit to an STO for integral evaluation. The resulting HF–LCAO orbitals can be thought of as minimal basis STOs. The GTOs are called *primitive* GTOs, and we say that the resulting atomic (STO) orbital is *contracted*. So, for example, we would use least squares fitting techniques to find the best three primitive GTO exponents α_i and contraction coefficients d_i in the STO/3G fit to a 1s STO orbital with exponent 1.

$$STO(\zeta = 1) = d_1 GTO(\alpha_1) + d_2 GTO(\alpha_2) + d_3 GTO(\alpha_3) \qquad (16.29)$$

The next keynote paper is 'Use of Gaussian Expansions of Slater-Type Atomic Orbitals' by W. J. Hehre *et al.* [71]. As usual I will let the authors explain their ideas:

Least Squares representations of Slater-type atomic orbitals as a sum of Gaussian-type orbitals are presented. These have the special feature that common Gaussian exponents are shared between Slater-type 2s and 2p functions. Use of these atomic orbitals in self-consistent molecular-orbital calculations is shown to lead to values of atomisation energies, atomic populations, and electric dipole moments which converge rapidly (with increasing size of the Gaussian expansion) to the values appropriate for pure Slater-type orbitals. The ζ exponents (or scale factors) for the atomic orbitals which are optimized for a number of molecules are also shown to be nearly independent of the number of Gaussian functions. A standard set of ζ values for use in molecular calculations is suggested on the basis of this study and is shown to be adequate for the calculation of total and atomisation energies, but less appropriate for studies of the charge distribution.

As we increase the number of primitive GTOs in the expansion, the resultant looks more and more like an STO, except at the nucleus where it can never attain the correct shape (the cusp). I have shown the comparison in Figure 16.7 for the STO/3G basis set. We therefore regard a minimal molecular basis set as comprised of STOs, except for integral evaluation where we use a linear expansion of n GTOs.

Many molecular properties depend on the valence electrons rather than the shape of the wavefunction at a nucleus, two exceptions being properties such as electron spin resonance and nuclear magnetic resonance parameters.

Tables of exponents and expansion coefficients are given in the original reference, and these all refer to an STO exponent of 1. These original GTO basis sets were 'universal' in that they applied to every atom irrespective of the atomic configuration; to convert from the STO exponent 1 to an exponent ζ you simply multiply the primitive exponents by $\zeta^{3/2}$. For reasons of computational efficiency, all basis functions in a given valence shell are taken to have the same primitive GTOs (but with different contraction coefficients).

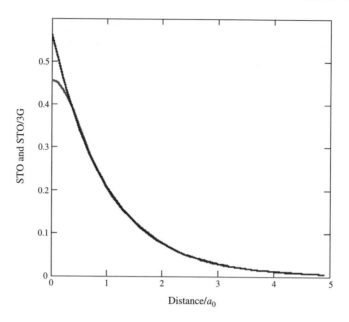

Figure 16.7   The STO/3G expansion

16.9.2 STO/4–31G

Such STO/nG calculations were common in the literature of the 1970s. It soon became apparent that they give poor results in a number of circumstances. There are particular problems for molecules containing atoms toward the end of the first period, such as oxygen and fluorine, where they give poor bond lengths and just about every other property you can think of. Eventually it was realized that whilst most of the energy comes from the inner shell regions, some flexibility ought to be given to the valence regions. The valence orbitals are therefore split into $(n-1)$ primitives and one primitive so we represent a hydrogen atom as two basis functions, as shown in Table 16.3. We think of an inner (3 GTOs) and an outer (1 GTO) basis function. For other atoms, the inner shell basis functions are left in an STO/nG contraction. Again, you might like to read the Synopsis of the keynote paper by R. Ditchfield *et al.* [72]:

Table 16.3   STO/4–31G hydrogen atom basis functions

Orbital exponent	Contraction coefficient
13.00773	0.0334960
1.962079	0.22472720
0.4445290	0.8137573
0.1219492	1

An extended basis set of atomic functions expressed as fixed linear combinations of Gaussian functions is presented for hydrogen and the first row atoms carbon to fluorine. In this set, described as 4–31G, each inner shell is represented by a single basis function taken as a sum over four Gaussians and each valence orbital is split into inner and outer parts described by three and one Gaussian function respectively. The expansion coefficients and Gaussian exponents are determined by minimizing the total calculated energy of the electronic ground state. This basis set is then used in single-determinant molecular-orbital studies of a group of small polyatomic molecules. Optimization of valence-shell scaling factors shows that considerable rescaling of atomic functions occurs in molecules, the largest effects being observed for hydrogen and carbon. However, the range of optimum scale factors for each atom is small enough to allow the selection of a standard molecular set. The use of this standard basis gives theoretical equilibrium geometries in reasonable agreement with experiment.

16.9.3 Gaussian polarization and diffuse functions

I mentioned polarization functions briefly in Section 16.8. The best thing is for me to quote the Synopsis of a keynote paper by J. B. Collins *et al.* [73] at this point:

Three basis sets (minimal s–p, extended s–p and minimal s–p with d functions on the second row atoms) are used to calculate geometries and binding energies of 24 molecules containing second row atoms. d functions are found to be essential in the description of both properties for hypervalent molecules and to be important in the calculations of two-heavy-atom bond lengths even for molecules of normal valence.

The addition of a single set of polarization functions to a heavy atom STO/4-31G basis set gives the so-called STO/4-31G*, and further addition of (p-type) polarization functions to hydrogen gives STO/4-31G** in an obvious notation. There are more explicit conventions when using more than one set of polarization functions per atom. Polarization functions essentially allow spherical atomic charge distributions to distort on molecule formation or in the presence of external electric fields. In order to treat species that carry formal negative charges or deal with molecular properties that depend on regions of the valence shells that are far from the nucleus, it is necessary to include *diffuse* basis functions (primitives with very small exponents), and they are denoted + and ++.

16.9.4 Extended basis sets

Huzinaga's set of uncontracted GTOs provides the classic example for our next topic. These large sets of primitive (uncontracted) GTOs comprise 10 primitive s-type and six primitive p-type, for a first row atom. The orbital exponents were carefully optimized for every first-row atom and for hydrogen. There is no reason in principle why we should not use them as they stand for molecular calculations, but the contraction process is found to give great computational efficiency with little cost in energy. What is needed is a way of effectively grouping them together for a molecular calculation, i.e. a contraction scheme.

Many authors performed HF–LCAO calculations on atoms and small molecules, and looked for groupings of the primitive GTOs that did not change their relative weightings from orbital to orbital and from molecule to molecule. I can illustrate these ideas by mentioning Dunning's 1975 work [74], with the by-now inevitable Synopsis:

> *Contracted [5s3p] and [5s4p] Gaussian basis sets for the first-row atoms are derived from the (10s6p) primitive basis sets of Huzinaga. Contracted [2s] and [3s] sets for the hydrogen atom obtained from primitive sets ranging from (4s) to (6s) are also examined. Calculations on the water and nitrogen molecules indicate that such basis sets when augmented with suitable polarization functions should yield wavefunctions near the Hartree–Fock limit.*

Dunning's notation and ideas can be explained with the example in Table 16.4, an oxygen atom. The first six primitive GTOs with exponents 1805.0 through 0.2558 contribute mostly to what we call 'the' atomic 1s orbital. The two most diffuse s functions (those with exponents 0.7736 and 0.2558) are the main components of what we call the 2s STOs, and they are allowed to vary freely in molecular calculations.

Table 16.4 Dunning's [5s3p] contraction scheme for Huzinaga's (10s6p) GTO set

GTO type	Exponent	Contraction coefficient
s	1805.0	0.000757
	2660.0	0.006066
	585.7	0.032782
	160.9	0.132609
	51.16	0.396839
	17.90	0.542572
s	17.90	0.262490
	6.639	0.769828
s	2.077	1
s	0.7736	1
s	0.2558	1
p	49.83	0.016358
	11.49	0.106453
	3.609	0.349302
	1.321	0.657183
p	0.4821	1
p	0.1651	1

The 1s primitive with exponent 2.077 turns out to make substantial contributions to both the atomic 1s and 2s orbitals, so that one is left free as a separate basis function.

A typical package such as GAUSSIAN98 will have very many basis sets as part of its database; you don't have to rediscover the wheel. On the other hand, some basis sets are good for one particular application and some are poor, and there are implications of cost. The larger the basis set the higher the cost of the calculation, and the proportionality is far from linear. Choice of basis set is a specialist subject, just like many others in our study so far; you simply have to take advice and look up recent literature citations.

17 HF–LCAO Examples

I am going to start with L-phenylanine, as in Chapters 5–7, to exemplify the principles discussed so far for the HF–LCAO model. There are a number of commercial packages that perform (amongst other things) HF–LCAO calculations, and everyone has their favourites. I use GAUSSIAN (http://www.gaussian.com) and HyperChem (http://www.hyper.com), but it's a matter of personal choice. Both can start from a Protein Data Bank .pdb file and in the case of GAUSSIAN we essentially add a few control records to the Cartesian coordinates as follows:

```
%chk = c:\g98w\scratch\phenylanine.chk
# HF/6-31G* Pop = Full SCF = Direct
L-phenylanine
       0    1
N         0.000     0.000     0.000
H         0.000     0.000     1.010
C         1.366     0.000    −0.483
H        −0.476    −0.825    −0.336
H         1.314     0.000    −1.572
C         2.120    −1.229     0.004
C         2.121     1.233     0.005
O         1.560    −2.063     0.714
H         1.586     2.131    −0.303
H         2.160     1.142     1.090
C         3.558     1.321    −0.449
O         2.904    −1.610    −0.972
C         4.348     2.410    −0.061
C         4.101     0.315    −1.257
H         3.398    −2.387    −0.700
H         3.926     3.194     0.568
H         3.486    −0.533    −1.557
C         5.681     2.493    −0.482
C         5.434     0.397    −1.677
H         6.296     3.341    −0.181
H         5.856    −0.386    −2.305
C         6.224     1.487    −1.290
H         7.262     1.551    −1.617
```

HF–LCAO calculations are iterative and resource intensive. The first statement sets up a *checkpoint file* that is used to store information about the calculation in hand, together with the results of the calculation. Calculations can be restarted from checkpoint files.

Next comes the 'route' through the package. I have chosen a closed-shell HF–LCAO calculation using the STO/6-31G* basis set. The major part of resource consumption is concerned with calculating and manipulating the two-electron integrals of which there a maximum of (about) $n^4/8$, where n is the number of basis functions. A number of strategies have been developed to deal with them in an efficient manner. In the early days it was usual to calculate all the integrals once, before the HF iterations began, and store them on magnetic tape or exchangeable disk storage (EDS). At each iteration the integrals are read back into computer memory, the HF matrix constructed and diagonalized. Matrix diagonalization is 'only' an n^2 process and even the old-fashioned Jacobi or Householder algorithms are still found in modern modelling packages.

It usually happens that many of the two-electron integrals are negligibly small, and it is possible to estimate whether an integral will fall into this category before actually calculating it. Only the significant ones have to be calculated and stored. It is usual to store the basis function *label i, j, k, l* packed as bytes into a 32-bit word, together with the value of the integral (often multiplied by an appropriate factor of 10 and stored as an integer).

The maximum number of basis functions determines the amount of storage needed for the label; using one byte for each index limits us to $2^8 - 1 = 255$ basis functions (Figure 17.1). R. C. Raffenetti [75] proposed an alternative way of storing the indices and the integral to make the generation of the HF–LCAO matrix easier.

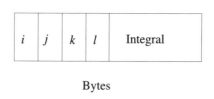

Bytes

Figure 17.1 Integral storage

For molecules with any substantial amount of symmetry, many of the integrals turn out to be plus or minus each other and many can be shown to be identically zero by symmetry arguments, without even having to calculate them. Early packages such as POLYATOM made extensive use of this symmetry feature.

HF–LCAO calculations are usually *input–output bound* because of the finite time taken for disk or magnetic tape transfers compared with the time needed to set up the HF matrix; ideally one would like to calculate the integrals just once and hold them in RAM rather than store them externally, but early computers had very limited RAM.

Technology has moved on, and I feel apologetic when I say that my humble office PC has 'only' 512 Mbytes of RAM. The motherboard will support a mere 3 Gbytes.

Many packages have a so-called *incore* option, where the integrals are calculated once at the start of the HF optimizations, and held in memory. In GAUSSIAN we would put SCF = Incore in the route (usage of the word 'core' is historical, and dates from the time when computer memory was made of magnetized ferrite rings and people spoke about computer cores). Again, processor speed has also increased beyond belief, and it was eventually realized that a cost-effective alternative to calculation of the integrals once and repeated read/write operations involving slow devices such as disks and magnetic tape was to calculate afresh the integrals as and when needed on each cycle. That is the meaning of 'SCF = Direct' in the route. The break-even point between 'Traditional' and 'Direct' calculations depends on many factors, but roughly speaking calculations involving more than 100 basis functions are perhaps best done with the 'Direct' option. It is a trade-off between processor power and disk transfer rate. For the record, my calculation on phenylanine with 202 basis functions gave almost exactly the same execution times when run with the two options.

The 'Pop = Full' is simply an option to print out all the LCAO coefficients and everything else one might possibly need at the end of the calculation. The next two records are the Title then the charge on the molecule (0 in this case) and the spin multiplicity (it is a singlet spin state). Then come the Cartesian coordinates in ångströms, in the usual manner. I could have also have input the molecular geometry as a Z matrix.

17.1 Output

The output is straightforward; I will explain it one piece at a time:

Entering Link 1 = C:\G98W\l1.exe PID = −1822917.
Copyright (c) 1988, 1990, 1992, 1993, 1995, 1998 Gaussian, Inc.
                    All Rights Reserved.
This is part of the Gaussian(R) 98 program. It is based on
the Gaussian 94(TM) system (copyright 1995 Gaussian, Inc.),
the Gaussian 92(TM) system (copyright 1992 Gaussian, Inc.),
the Gaussian 90(TM) system (copyright 1990 Gaussian, Inc.),
the Gaussian 88(TM) system (copyright 1988 Gaussian, Inc.),
the Gaussian 86(TM) system (copyright 1986 Carnegie Mellon
University), and the Gaussian 82(TM) system (copyright 1983
Carnegie Mellon University). Gaussian is a federally registered
trademark of Gaussian, Inc.

Cite this work as:
Gaussian 98, Revision A.7,
M. J. Frisch, G. W. Trucks, H. B. Schlegel, G. E. Scuseria,

M. A. Robb, J. R. Cheeseman, V. G. Zakrzewski, J. A. Montgomery, Jr.,
R. E. Stratmann, J. C. Burant, S. Dapprich, J. M. Millam,
A. D. Daniels, K. N. Kudin, M. C. Strain, O. Farkas, J. Tomasi,
V. Barone, M. Cossi, R. Cammi, B. Mennucci, C. Pomelli, C. Adamo,
S. Clifford, J. Ochterski, G. A. Petersson, P. Y. Ayala, Q. Cui,
K. Morokuma, D. K. Malick, A. D. Rabuck, K. Raghavachari,
J. B. Foresman, J. Cioslowski, J. V. Ortiz, A. G. Baboul,
B. B. Stefanov, G. Liu, A. Liashenko, P. Piskorz, I. Komaromi,
R. Gomperts, R. L. Martin, D. J. Fox, T. Keith, M. A. Al-Laham,
C. Y. Peng, A. Nanayakkara, C. Gonzalez, M. Challacombe,
P. M. W. Gill, B. Johnson, W. Chen, M. W. Wong, J. L. Andres,
C. Gonzalez, M. Head-Gordon, E. S. Replogle, and J. A. Pople,
Gaussian, Inc., Pittsburgh PA, 1998.

****
Gaussian 98: x86-Win32-G98RevA.7 11-Apr-1999
                   19-Jun-2002
****

The first part reminds us that packages such as GAUSSIAN have been developed
over a number of years by a large team of people. There was a time in the 1970s when
scientists freely exchanged their software, but large packages now tend to be commercial in nature and generally subject to the laws of copyright. The text above also
serves the purpose of giving the appropriate citation for my calculations in this book.
   The next piece of relevant output summarizes the iterative calculations:

Standard basis: 6–31G(d) (6D, 7F)
   There are 202 symmetry adapted basis functions of A symmetry.
   Crude estimate of integral set expansion from redundant integrals = 1.000.
   Integral buffers will be 262144 words long.
   Raffenetti 1 integral format.
   Two-electron integral symmetry is turned on.
      202 basis functions   380 primitive gaussians
      44 alpha electrons   44 beta electrons
         nuclear repulsion energy   695.8409407052 Hartrees.
   One-electron integrals computed using PRISM.
   NBasis = 202 RedAO = T NBF = 202
   NBsUse = 202 1.00D – 04 NBFU = 202
   Projected INDO Guess.
   Warning! Cutoffs for single-point calculations used.
   Requested convergence on RMS density matrix = 1.00D – 04 within 64 cycles.
   Requested convergence on MAX density matrix = 1.00D – 02.
   Requested convergence on energy = 5.00D – 05.
   SCF Done: E(RHF) = −551.290681260   A.U. after 7 cycles
   Convg = 0.2411D−04   −V/T = 2.0010
   S**2 = 0.0000

You should by now know about most things in the text above apart from INDO. I will explain INDO in Chapter 18; just accept for the minute that it is a cheap and cheerful way of setting off the HF–LCAO iterative calculations. The $S^{**}2$ item is the expectation value of the spin operator and it shows that we are indeed dealing with a pure singlet spin state. This shouldn't come as a surprise, since we required pairs of electrons to occupy the same spatial orbital. The virial ratio $-V/T$ is the ratio of the expectation values of the potential and kinetic energies, and for an exact wavefunction it would equal 2. It's a quantity of some historical interest.

A sensible choice of the initial electron density can make all the difference between success and failure; one strategy is to allow a few of the highest occupied and the lowest unoccupied orbitals to be partially occupied, in order to try to sort out the ground state from the excited states. Most packages make such choices as a matter of course.

The final energy shown is the total, that is to say electronic plus nuclear repulsion; the quantity quoted is technically the reduced energy, ε/E_h. Next come the orbital energies (called not surprisingly, 'EIGENVALUES' since they are eigenvalues of the generalized HF–LCAO matrix eigenvalue equation) and LCAO coefficients; a small subset is shown in the next excerpt.

EIGENVALUES --			−20.63814	−20.57266	−15.54941	−11.39810	−11.28063
1 1 N	1S	0.00000	0.00000	0.99502	−0.00001	−0.00024
2	2S	0.00000	0.00003	0.02505	−0.00010	−0.00005
3	2PX	0.00000	0.00000	0.00002	0.00002	−0.00004
4	2PY	−0.00001	0.00000	0.00115	−0.00002	0.00004
5	2PZ	0.00000	0.00001	0.00085	−0.00003	0.00002
6	3S	−0.00011	−0.00011	−0.00304	0.00016	0.00431
7	3PX	0.00007	0.00008	0.00046	−0.00019	−0.00156
8	3PY	0.00002	−0.00005	−0.00116	−0.00022	0.00075
9	3PZ	−0.00008	−0.00005	−0.00065	0.00023	0.00033
10	4XX	−0.00001	0.00000	−0.00343	0.00000	−0.00046
11	4YY	0.00001	0.00000	−0.00328	−0.00006	−0.00031
12	4ZZ	−0.00002	0.00000	−0.00329	0.00000	−0.00007
13	4XY	0.00000	0.00000	0.00009	0.00000	0.00024

. . . etc.

Koopmans' theorem applies to all HF wavefunctions, no matter whether they are at the Hartree–Fock limit or just minimal basis set ones. The orbital energies can therefore be used to predict and rationalize ionization energies.

Next come Mulliken population analysis indices. As explained in earlier chapters, these give a rough-and-ready decomposition of the molecular charge density, but the Mulliken partitioning scheme is at best subjective and the numbers have to be treated with caution. Even at best, the numbers should only be used when comparing similar molecules calculated with the same basis set.

Total atomic charges:

		1
1	N	−0.813293
2	H	0.349031
3	C	−0.104570
4	H	0.353108
5	H	0.230505
6	C	0.738642
7	C	−0.330255
8	O	−0.563670
9	H	0.208910
10	H	0.174735
11	C	−0.003026
12	O	−0.695939
13	C	−0.215109
14	C	−0.296860
15	H	0.476022
16	H	0.195021
17	H	0.352644
18	C	−0.204224
19	C	−0.223956
20	H	0.196201
21	H	0.185426
22	C	−0.201527
23	H	0.192182

Sum of Mulliken charges = 0.00000

Finally, we have a number of molecular properties, thought for the day and most important of all, timing for the calculation.

Electronic spatial extent (au): $< R**2 > = 2207.7382$
  Charge = 0.0000 electrons
  Dipole moment (Debye):

| X = −0.5782 | Y = 1.3722 | Z = 0.1690 | Tot = 1.4986 |

Quadrupole moment   (Debye-Ang):

| XX = −74.8112 | YY = −64.5102 | ZZ = −74.5878 |
| XY = 0.9589 | XZ = −3.7721 | YZ = −3.7827 |

Octapole moment (Debye-Ang**2):

XXX = −2.9972	YYY = 33.1091	ZZZ = −0.6338	XYY = −1.5751
XXY = −0.1467	XXZ = −5.5624	XZZ = 11.6054	YZZ = 2.4320
YYZ = 10.4086	XYZ = −7.2147		

Hexadecapole moment (Debye-Ang**3):

XXXX = −1997.2398	YYYY = −551.4398	ZZZZ = −183.6336	XXXY = −16.5841
XXXZ = −61.2789	YYYX = 48.2110	YYYZ = −4.8287	ZZZX = −1.3817
ZZZY = −0.3652	XXYY = −463.2866	XXZZ = −417.5712	YYZZ = −136.5187
XXYZ = −52.3198	YYXZ = 20.4041	ZZXY = 3.5225	

SILVERMAN'S PARADOX — IF MURPHY'S LAW CAN GO WRONG, IT WILL.
Job cpu time: 0 days   0 hours   5 minutes   36.0 seconds.
File lengths (MBytes):   RWF = 27   Int = 0   D2E = 0   Chk = 7

17.2 Visualization

Visualization follows the path discussed in Chapter 15, but naturally there are very many more HF–LCAOs than in the case of dihydrogen. In any case, the HF–LCAOs are not uniquely determined by the HF procedure. This is because the electronic wavefunction has to be written as a Slater determinant; one of the properties of determinants is that we can add multiples of rows and/or columns without changing their value. This is equivalent to mixing the HF–LCAO orbitals, and at one time it was fashionable to use this idea in order to produce *localized* orbitals. Organic chemists place great emphasis on the highest occupied molecular orbital (HOMO) and the lowest unoccupied molecular orbital (LUMO). I have shown the HOMO and the LUMO for L-phenylanine in Figures 17.2 and 17.3 as the three-dimensional objects that they are. The rendering chosen for the squares of the square of the wavefunctions is called (in HyperChem) a 3D-isosurface (shaded surface). The plots look very attractive in colour, unfortunately wasted here in the greyscale illustrations. GAUSSIAN Inc. market a package called GAUSSVIEW that also produces brilliant images.

The HOMO comprises large contributions from the benzene ring. By contrast, the LUMO has a nodal plane through the benzene ring as shown, and a much larger contribution from the remainder of the molecule.

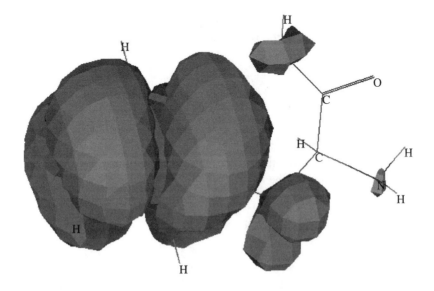

Figure 17.2   The HOMO of L-phenylanine

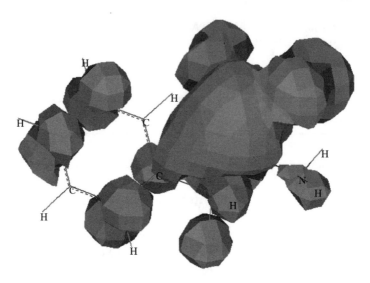

Figure 17.3   The LUMO of L-phenylanine

17.3 Properties

Once the wavefunction has been determined, then various molecular properties can be calculated. My final section of standard output shows a selection of molecular electric moments. The so-called 'quadrupole moment' diagonal elements don't sum to zero, which gives the game away; they are actually second moments.

Such properties are often called *primary* properties because their electronic part can be obtained directly from the wavefunction Ψ_{el}

$$\langle \hat{X} \rangle = \int \Psi_{el}^* \left(\sum_{i=1}^{n} \hat{X}_i \right) \Psi_{el} \, d\tau \tag{17.1}$$

where the operators refer to each of the n electrons. Usually it is necessary to add a corresponding nuclear contribution, since we work within the Born–Oppenheimer approximation. For example, the electric dipole moment operator is

$$\hat{\mathbf{p}}_e = e \sum_{\alpha=1}^{N} Z_\alpha \mathbf{R}_\alpha - e \sum_{i=1}^{n} \mathbf{r}_i$$

where the first sum runs over the N nuclei and the second sum over the n electrons. All electrons enter the sum on an equal footing, as they should, and the expectation value can be written in terms of the charge density $P(\mathbf{r})$

$$\langle \hat{\mathbf{p}}_e \rangle = e \sum_{\alpha=1}^{N} Z_\alpha \mathbf{R}_\alpha - e \int \mathbf{r} P(\mathbf{r}) \, d\tau \tag{17.2}$$

Such electric moments are often reported in non-SI units; the old-fashioned unit of length is the ångström and the debye (itself a relic from the days of electrostatic units), is the dipole moment corresponding to a pair of equal and opposite charges of magnitude 10^{-10} electrostatic units (esu $= \mathrm{g}^{-1/2} \, \mathrm{cm}^{3/2} \, \mathrm{s}^{-1}$) separated by $1 \, \text{Å} \, (= 10^{-10} \, \mathrm{m})$. There are 2.9989×10^9 esu per Coulomb, and so $1 \, \mathrm{D} = 10^{-10} \, \mathrm{esu} \times 10^{-10} \, \mathrm{m}$ or $3.336 \times 10^{-30} \, \mathrm{C} \, \mathrm{m}$. The atomic unit of electric dipole is $e a_0 = 8.4784 \times 10^{-30} \, \mathrm{C} \, \mathrm{m}$, which is $2.5418 \, \mathrm{D}$.

17.3.1 The electrostatic potential

One of the fundamental objectives of chemistry is to predict and rationalize chemical reactivity. In principle, this involves searching a potential energy surface for saddle points (i.e. transition states) and minima (reactants and products), and this kind of detailed investigation has only become possible in the last decade. Most of the traditional theories of chemical reactivity have concerned themselves with organic molecules, and the simplest theories have attempted to extract useful information from the electronic properties of reactants. We can distinguish *static* theories, which in essence make use of the electronic wavefunction and/or electronic properties appropriate to an isolated molecule in the gas phase, and *dynamic* theories. Dynamic theories aim (for example) to predict the likely reaction sites for the approach of a charged reagent, usually modelled as a point charge.

The electrostatic potential gives an index that has been widely used since the 1970s for just this purpose (see, for example, E. Scrocco and J. Tomasi [76]).

Figure 17.4 shows benzene; within the Born–Oppenheimer approximation, molecules are thought of as point positive charges (the nuclei) surrounded by continuous distributions of electron charge (shown as a surrounding sphere). I can therefore calculate the electrostatic potential at points in space around the molecule, using the methods of classical electromagnetism. I have placed a point charge Q at the origin; the electrostatic potential ϕ at this point will contain contributions from the nuclei such as

$$\frac{1}{4\pi\epsilon_0} \frac{e Z_\mathrm{H}}{R_{\mathrm{H}_1}}$$

where $Z_\mathrm{H} = 1$, and a contribution from the electrons, which I can write in terms of the electron density $-eP(\mathbf{r})$ as

$$-\frac{e}{4\pi\epsilon_0} \int \frac{P(\mathbf{r})}{r} \, d\tau$$

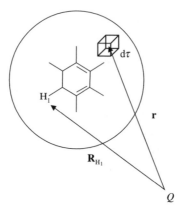

Figure 17.4   Electrostatic potential due to benzene

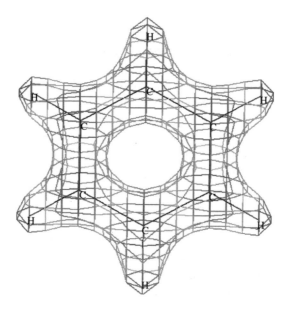

Figure 17.5   Benzene electrostatic potential

It is conventional to record the mutual potential energy of the molecule and a unit positive charge Q, rather than the electrostatic potential, but authors in the field still speak of potential energy maps without making the distinction. Again, it is usual to present the calculations as contour maps, which make beautiful illustrations for textbooks like this one. The electrostatic potential maps are used to give a rough-and-ready survey of a molecule, and the spirit of the calculation is that one does not need a particularly sophisticated basis set. Figure 17.5 shows a three-dimensional isosurface representation of the electrostatic potential for benzene (calculated with a STO/6–31G* basis set and geometry optimized).

Such objects are usually colour-coded. Obviously I can't show you this in a mono-chrome illustration, but the important point is that there are no local areas where the electrostatic function is negative. If we make a similar calculation on pyridine, then we see a very different story; there is an area around the bottom end of the nitrogen atom where electrostatic contours are negative and where therefore a positively charged species would be expected to attack. This isn't a terribly exciting example, any chemist would have been able to give the conclusion without having to perform a calculation.

17.4 Geometry Optimization

The next step might well be a geometry optimization. In contrast to MM calculations, geometry optimization is an expensive business because a HF–LCAO calculation has to be done for every point investigated on the molecular potential energy surface. I want to first tell you about an early and interesting attempt at a short-cut.

17.4.1 The Hellmann–Feynman Theorem

Suppose that an electronic wavefunction Ψ depends on a single parameter c such as a single bond length or an orbital exponent. According to the variation principle, the best value of c is the one that minimizes the energy

$$\varepsilon(c) = \int \Psi^*(c) \hat{H} \Psi(c) \, d\tau \tag{17.3}$$

subject to the requirement that the wavefunction is normalized

$$\int \Psi^*(c) \Psi(c) \, d\tau = 1 \tag{17.4}$$

If I differentiate these two equations with respect to c we have

$$\frac{d\varepsilon(c)}{dc} = \int \frac{\partial \Psi^*(c)}{\partial c} \hat{H} \Psi(c) \, d\tau + \int \Psi^*(c) \frac{\partial \hat{H}}{\partial c} \Psi(c) \, d\tau + \int \Psi^*(c) \hat{H} \frac{\partial \Psi(c)}{\partial c} \, d\tau$$

$$\int \frac{\partial \Psi^*(c)}{\partial c} \Psi(c) \, d\tau + \int \Psi^*(c) \frac{\partial \Psi(c)}{\partial c} \, d\tau = 0 \tag{17.5}$$

Normally $\Psi(c)$ will be an approximate wavefunction, but suppose that $\Psi(c)$ happens to be an eigenfunction of the Hamiltonian. A little analysis shows that

$$\frac{d\varepsilon(c)}{dc} = \int \Psi^*(c) \frac{\partial \hat{H}}{\partial c} \Psi(c) \, d\tau \tag{17.6}$$

This result is known as the Hellmann−Feynman Theorem, and it was hoped in the early days that it could be used to optimize geometries. The first applications were to isoelectronic processes

$$X \rightarrow Y$$

such as the internal rotation in ethane, where it was believed that the only non-zero contributions to the right-hand side of Equation (17.6) would be due to changes in the nuclear positions and hence in the one-electron integrals. It was hoped that such calculations would be useful for geometry optimizations, where the changes also involve two-electron integrals, but enthusiasm vanished once it became clear that approximate wavefunctions also depend on the parameter(s) contained either explicitly or implicitly in the wavefunction and so wavefunction gradient terms such as $\frac{\partial \Psi}{\partial c}$ cannot be ignored.

17.4.2 Energy minimization

All the considerations of Chapter 16 apply here, with the added difficulty that the energy calculation for each point on the molecular potential energy surface is now much more time consuming than for molecular mechanics because a HF−LCAO calculation is involved.

I explained in Chapter 16 how the Hartree−Fock energy could be written in terms of the electron density and various one- and two-electron integrals over the basis functions χ. The HF−LCAO matrix is

$$
h_{ij}^F = \int \chi_i(\mathbf{r}_1)\hat{h}^{(1)}(\mathbf{r}_1)\chi_j(\mathbf{r}_1)\,\mathrm{d}\tau
$$

$$
+ \sum_{k=1}^{n}\sum_{l=1}^{n} P_{kl} \iint \chi_i(\mathbf{r}_1)\chi_j(\mathbf{r}_1)\hat{g}(\mathbf{r}_1,\mathbf{r}_2)\chi_k(\mathbf{r}_2)\chi_l(\mathbf{r}_2)\,\mathrm{d}\tau_1\mathrm{d}\tau_2
$$

$$
- \frac{1}{2}\sum_{k=1}^{n}\sum_{l=1}^{n} P_{kl} \iint \chi_i(\mathbf{r}_1)\chi_k(\mathbf{r}_1)\hat{g}(\mathbf{r}_1,\mathbf{r}_2)\chi_j(\mathbf{r}_2)\chi_l(\mathbf{r}_2)\,\mathrm{d}\tau_1\mathrm{d}\tau_2 \quad (17.7)
$$

and the energy gradient will therefore involve terms like

$$
\frac{\partial P_{kl}}{\partial c}, \quad \int \frac{\partial \chi_i(\mathbf{r}_1)}{\partial c}\hat{h}(\mathbf{r}_1)\chi_j(\mathbf{r}_1)\,\mathrm{d}\tau \quad \text{and}
$$

$$
\iint \frac{\partial \chi_i(\mathbf{r}_1)}{\partial c}\chi_j(\mathbf{r}_2)\hat{g}(\mathbf{r}_1,\mathbf{r}_2)\chi_k(\mathbf{r}_2)\chi_l(\mathbf{r}_2)\,\mathrm{d}\tau_1\mathrm{d}\tau_2
$$

where c is a parameter to be varied. Advanced methods of geometry optimization usually require both the gradient and the hessian, and these can either be calculated numerically or analytically. Over the years, a great deal of effort has gone into the elucidation of analytical expressions for the energy gradient and the hessian and all molecular structure packages now use analytical gradients wherever these are available; in the case of the HF–LCAO model, the gradients do not involve the LCAO coefficients directly, just the electron density matrix **P**. This speeds up the calculation by a large factor.

Let me use L-phenylanine for my example. We simply modify the control cards

```
%chk = c:\g98w\scratch\phenylanine.chk
# HF/6−31G* Opt Guess = Read Geom = Check
```

This picks up the single point HF–LCAO results discussed above from the check-point file, and uses them as the starting points in a geometry optimization using the default algorithm. In the GAUSSIAN package, this uses analytical expressions for the gradient and the hessian. Here are some relevant parts of the output file. First of all, an estimate of the hessian has to be made.

```
!    Initial Parameters    !
! (Angstroms and Degrees) !
```

!	Name	Definition	Value	Derivative Info.	!
!	R1	R(1,2)	1.01	estimate D2E/DX2	!
!	R2	R(1,3)	1.4489	estimate D2E/DX2	!
!	R3	R(1,4)	1.01	estimate D2E/DX2	!
!	R4	R(3,5)	1.0902	estimate D2E/DX2	!
!	R5	R(3,6)	1.5219	estimate D2E/DX2	!
!	R6	R(3,7)	1.5259	estimate D2E/DX2	!
!	R7	R(6,8)	1.2301	estimate D2E/DX2	!
...etc.					

The first estimate is made numerically. Next, the geometry iterations begin.

Variable	Old X	−DE/DX	Delta X (Linear)	Delta X (Quad)	Delta X (Total)	New X
R1	1.90862	−0.00885	0.00000	−0.00588	−0.00588	1.90275
R2	2.73798	−0.00371	0.00000	−0.00281	−0.00281	2.73517
R3	1.90862	−0.00961	0.00000	−0.00639	−0.00639	1.90224
R4	2.06026	−0.00681	0.00000	−0.00555	−0.00555	2.05471
R5	2.87594	0.01442	0.00000	0.01525	0.01406	2.89000
R6	2.88359	0.02351	0.00000	0.02423	0.02220	2.90578
R7	2.32464	−0.06282	0.00000	−0.02391	−0.02391	2.30072
...etc.						

The calculation proceeds (hopefully) downwards on the molecular potential energy surface until eventually a stationary point is reached (that is, a point on the molecular potential energy surface where the gradient is zero).

	Item	Value	Threshold	Converged?
Maximum	Force	0.000017	0.000450	YES
RMS	Force	0.000004	0.000300	YES
Maximum	Displacement	0.000551	0.001800	YES
RMS	Displacement	0.00012	0.001200	YES

Predicted change in Energy=−5.195054D-09
Optimization completed.
  -- Stationary point found.

```
                  -----------------------------------------
                  !    Optimized Parameters     !
                  !   (Angstroms and Degrees)   !
 ---------------------------------------               ----------------------------
 !   Name    Definition    Value            Derivative Info.          !
 ----------------------------------------------------------------------------------
 !    R1      R(1,2)        1.0026           −DE/DX=0.                 !
 !    R2      R(1,3)        1.4505           −DE/DX=0.                 !
 !    R3      R(1,4)        1.002            −DE/DX=0.                 !
 !    R4      R(3,5)        1.0818           −DE/DX=0.                 !
 !    R5      R(3,6)        1.5231           −DE/DX=0.                 !
 !    R6      R(3,7)        1.5393           −DE/DX=0.                 !
 !    R7      R(6,8)        1.1908           −DE/DX=0.                 !
 ...etc.
```

17.5 Vibrational Analysis

It is wise to calculate the force constants at this point on the surface, in order to characterize the stationary point. Just to remind you, harmonic force constants correspond to the eigenvalues of the hessian, calculated at the stationary point. A minimum on the molecular potential energy curve should have $3N-6$ positive eigenvalues ($3N-5$ for a linear molecule). A transition state of the chemical kind will have just one negative eigenvalue of the hessian and so on. In this example there are 23 nuclei and so 69 vibrational coordinates in total. This gives 63 vibrational coordinates, but because I have chosen to do the calculation in redundant internal coordinates rather than starting from a Z matrix, there should be six redundant coordinates corresponding to the three translational coordinates of the centre of mass and three for the rotations. Each of these should have a force constant of zero, but there is a question as to how small a number needs to be before it is taken to be nonzero. This has to do with the optimization cut-off point, the convergence criterion for the HF–LCAO calculation and so on.

Here is what I found

Full mass-weighted force constant matrix:
Low frequencies --- −0.9544 −0.4840 −0.0007 0.0006 0.0008 0.8880
Low frequencies --- 37.1567 46.9121 62.6582

The first six are essentially zero and are therefore taken to represent the redundant coordinates. The next piece of output gives the normal modes, as discussed in Chapters 4 and 5.

		61			62			63		
		?A			?A			?A		
Frequencies --		3727.6900			3813.0599			4045.0678		
Red. masses --		1.0506			1.0941			1.0650		
Frc consts --		8.6014			9.3721			10.2671		
IR Inten --		2.6994			4.8499			102.8922		
Raman Activ --		107.7964			71.9141			74.7041		
Depolar --		0.1384			0.6803			0.3093		
Atom AN		X	Y	Z	X	Y	Z	X	Y	Z
1	7	−0.05	0.02	−0.01	0.02	0.02	−0.08	0.00	0.00	0.00
2	1	0.22	−0.29	0.61	0.24	−0.31	0.59	0.00	0.00	0.00
3	6	0.00	0.00	0.00	0.00	0.00	0.00	0.00	0.00	0.00
4	1	0.49	−0.06	−0.51	−0.50	0.08	0.49	0.00	0.00	0.00
5	1	0.00	0.00	0.01	0.00	0.00	0.02	0.00	0.00	0.00
6	6	0.00	0.00	0.00	0.00	0.00	0.00	0.00	0.00	0.00
7	6	0.00	0.00	0.00	0.00	0.00	0.00	0.00	0.00	0.00
8	8	0.00	0.00	0.00	0.00	0.00	0.00	0.00	0.00	0.00
9	1	0.00	0.00	0.00	0.00	0.00	0.00	0.00	0.00	0.00
10	1	0.00	0.00	0.00	0.00	0.00	0.00	0.00	0.00	0.00
11	6	0.00	0.00	0.00	0.00	0.00	0.00	0.00	0.00	0.00
12	8	0.00	0.00	0.00	0.00	0.00	0.00	0.00	−0.06	0.00
13	6	0.00	0.00	0.00	0.00	0.00	0.00	0.00	0.00	0.00
14	6	0.00	0.00	0.00	0.00	0.00	0.00	0.00	0.00	0.00
15	1	0.00	0.00	0.00	0.00	0.00	0.00	−0.07	1.00	−0.02
16	1	0.00	0.00	0.00	0.00	0.00	0.00	0.00	0.00	0.00
17	1	0.00	0.00	0.00	0.00	0.00	0.00	0.00	0.00	0.00
18	6	0.00	0.00	0.00	0.00	0.00	0.00	0.00	0.00	0.00
19	6	0.00	0.00	0.00	0.00	0.00	0.00	0.00	0.00	0.00
20	1	0.00	0.00	0.00	0.00	0.00	0.00	0.00	0.00	0.00
21	1	0.00	0.00	0.00	0.00	0.00	0.00	0.00	0.00	0.00
22	6	0.00	0.00	0.00	0.00	0.00	0.00	0.00	0.00	0.00
23	1	0.00	0.00	0.00	0.00	0.00	0.00	0.00	0.00	0.00

Output includes the vibration frequencies, reduced masses, force constants, infrared intensities, Raman activities and depolarization ratios, together with the normal modes expressed as linear combinations of the Cartesian coordinates of the atoms.

Normal mode 63, for example, comprises almost entirely the z-component for atom 15, which is a hydrogen atom.

Some packages are more visually friendly than others, and will simulate the infrared spectrum. Often it is possible to animate the normal modes of vibration on-screen. I can illustrate various points by considering a much smaller molecule, carbon dioxide.

It is a linear molecule and so has four vibrational degrees of freedom. To emphasize a point already made, vibrational frequencies are defined as the second derivatives calculated at the appropriate minimum of the molecular potential energy curve, so it is mandatory that the geometry be optimized before evaluation. Figure 17.6 shows the result of a HF/6–31G* calculation using HyperChem.

The top part of the display shows the calculated frequencies, the bottom part their infrared intensities (given by the integrated molar absorption coefficient). Agreement with experiment is not particularly good, as Table 17.1 reveals. The intensities are qualitatively correct; the symmetric stretch is not infrared active because the dipole

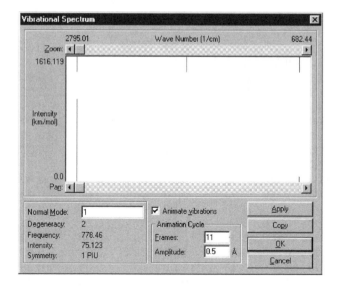

Figure 17.6 Vibrational infrared spectrum for carbon dioxide

Table 17.1 Calculated (HF–LCAO) and experimental vibration wavenumbers (cm^{-1})

Mode	Expt	STO/6–31G*	STO-3G	STO/6–311G*
Symmetric stretch (σ_g)	1388.17	1752.5	1656.3	1751.5
Asymmetric stretch (σ_u)	2349.16	2699.0	2647.4	2683.7
Degenerate bend (π)	667.40	778.5	591.0	778.9

moment does not change during the vibration, and the doubly degenerate bending vibration has a lower intensity than the asymmetric stretch.

One decision that has to be made with a HF–LCAO calculation is the choice of basis set, and this is usually a matter of experience. I repeated the calculation with a smaller basis set (STO-3G) and a larger one (STO/6–311G*) to give the remaining two columns of the table. It is a common experience that HF–LCAO calculations overestimate force constants. Provided the same basis set is used from molecule to molecule, it is possible to scale by a fixed factor the calculated results with remarkable success.

17.6 Thermodynamic Properties

In Chapter 8 I discussed the canonical ensemble and the canonical partition function

$$Q = \sum_i \exp\left(-\frac{E_i^*}{k_B T}\right)$$

I added an asterisk to focus attention on the fact that E^* refers to the collection of particles in each cell of the ensemble. There will be many possible allowed values for E^*, and the N particles in each cell will contribute in some way to make up the total. So far, I have made no assumptions whatever about these particles, neither have I made any assumption about the way in which they might interact with each other.

Suppose now that the N particles are identical molecules, but that they form essentially an ideal gas. One characteristic of an ideal gas is that there is no interaction between the particles, so the total energy E^* of the N particles will be a simple sum of the particle energies. If I label the molecules $1, 2, 3, \ldots, N$, then we can write

$$E^* = \varepsilon^{(1)} + \varepsilon^{(2)} + \cdots + \varepsilon^{(N)} \tag{17.8}$$

Each molecular energy will contain a kinetic and a potential part, but there are no intermolecular interactions because of the ideal gas behaviour. So for each possible value of E^* we have

$$E_i^* = \varepsilon_i^{(1)} + \varepsilon_i^{(2)} + \cdots + \varepsilon_i^{(N)} \tag{17.9}$$

Each allowed value of E^* will correspond to different values of the constituent εs and a simple rearrangement of Q shows that

$$Q = \sum_i \exp\left(-\frac{\varepsilon_i^{(1)}}{k_B T}\right) \sum_j \exp\left(-\frac{\varepsilon_j^{(2)}}{k_B T}\right) \cdots \sum_n \exp\left(-\frac{\varepsilon_n^{(N)}}{k_B T}\right) \tag{17.10}$$

I have used a different dummy index $i, j, \ldots, n$ but the sums are all the same since they refer to identical particles. At first sight we should therefore write

$$Q = \left(\sum_i \exp\left(-\frac{\varepsilon_i^{(1)}}{k_B T} \right) \right)^N$$

This assumes that the N particles can somehow be distinguished one from another (we say that such particles are *distinguishable*). The laws of quantum mechanics tell us that seemingly identical particles are truly identical, they are *indistinguishable* and to allow for this we have to divide Q by the number of ways in which N things can be permuted

$$Q = \frac{1}{N!} \left(\sum_i \exp\left(-\frac{\varepsilon_i}{k_B T} \right) \right)^N \tag{17.11}$$

I have dropped all reference to particle '1', since all N are identical. The summation

$$q = \sum_i \exp\left(-\frac{\varepsilon_i}{k_B T} \right) \tag{17.12}$$

refers only to the energy states of a representative particle, and it is a quantity of great interest in our theories. It is called the *molecular partition function* and given the symbol q (sometimes z).

17.6.1 The ideal monatomic gas

Consider now an even more restricted case, namely a monatomic ideal gas. Each atom has mass m and the gas is constrained to a cubic container whose sides are a, b and c (so that the volume V is abc). We will ignore the fact that each atom could well have different electronic energy, and concentrate on the translational energy. Elementary quantum mechanics texts (and Chapter 11) show that the available (translational) energy levels are characterized by three quantum numbers, n_x, n_y and n_z, which can each take integral values ranging from 1 to infinity. The translational energy is given by

$$\varepsilon_{n_x, n_y, n_z} = \frac{h^2}{8m} \left(\frac{n_x^2}{a^2} + \frac{n_y^2}{b^2} + \frac{n_z^2}{c^2} \right)$$

and so each particle will make a contribution

$$\sum_{n_x}\sum_{n_y}\sum_{n_z} \exp\left(-\frac{\varepsilon_{n_x,n_y,n_z}}{k_BT}\right)$$

$$= \sum_{n_x}\sum_{n_y}\sum_{n_z} \exp\left(-\frac{1}{k_BT}\frac{h^2}{8m}\left(\frac{n_x^2}{a^2}+\frac{n_y^2}{b^2}+\frac{n_z^2}{c^2}\right)\right)$$

$$= \sum_{n_x} \exp\left(-\frac{1}{k_BT}\frac{h^2}{8m}\frac{n_x^2}{a^2}\right)\sum_{n_y} \exp\left(-\frac{1}{k_BT}\frac{h^2}{8m}\frac{n_y^2}{b^2}\right)\sum_{n_z} \exp\left(-\frac{1}{k_BT}\frac{h^2}{8m}\frac{n_z^2}{c^2}\right)$$

$$(17.13)$$

to the translational molecular partition function. Each of the three summations can be treated as follows. A simple calculation shows that, for an ordinary atom such as argon at room temperature constrained to such a macroscopic container, typical quantum numbers are of the order of 10^9, by which time the allowed energy states essentially form a continuum. Under these circumstances we replace the summation by an integral and treat n_x as a continuous variable. So, for example,

$$\sum_{n_x} \exp\left(-\frac{1}{k_BT}\frac{h^2}{8m}\frac{n_x^2}{a^2}\right) \approx \int_0^\infty \exp\left(-\frac{1}{k_BT}\frac{h^2}{8m}\frac{n_x^2}{a^2}\right)dn_x$$

$$= \sqrt{2\pi m k_B T}\frac{a}{h}$$

The translational partition functions are then given by

$$q_{trans} = \left(\frac{2\pi m k_B T}{h^2}\right)^{3/2} V$$

$$Q_{trans} = \frac{1}{N!}\left(\left(\frac{2\pi m k_B T}{h^2}\right)^{3/2} V\right)^N$$

$$(17.14)$$

The thermodynamic functions can be calculated using the equations given in Chapter 8. We have, for example,

$$U = k_B T^2 \left(\frac{\partial \ln Q}{\partial T}\right)_V$$

and so

$$U = \tfrac{3}{2} N k_B T$$

For a monatomic ideal gas, the translational kinetic energy is the only energy that an atom will possess. The internal energy is therefore just what we would expect from the equipartition of energy principle. In a similar way we find that the entropy is

$$S = Nk_B \left(\frac{5}{2} + \ln \left(\left(\frac{2\pi m k_B T}{h^2} \right)^{3/2} \frac{V}{N} \right) \right) \tag{17.15}$$

and this is called the *Sackur–Tetrode equation.*

17.6.2 The ideal diatomic gas

Monatomic gas particles can only store translational kinetic energy, whilst we have to consider

- rotational energy and
- vibrational energy

for a polyatomic molecule. We also need to consider electronic energies. In order to progress, we make the assumption that the different energy modes are independent, and that each set has its own Boltzmann distribution. We therefore write for a typical molecule

$$\varepsilon = \varepsilon_{trans} + \varepsilon_{rot} + \varepsilon_{vib} + \varepsilon_{elec} \tag{17.16}$$

and it is easily seen that the molecular partition function is the product of a translational, rotational, vibrational and electronic molecular partition function. These are related to the canonical partition function by

$$Q = \frac{1}{N!} \left(q_{trans} q_{rot} q_{vib} q_{elec} \right)^N \tag{17.17}$$

17.6.3 q_{rot}

The rotational energy levels for a rigid diatomic rotator are given by

$$\varepsilon_J = J(J+1) \frac{h^2}{8\pi^2 I} \tag{17.18}$$
$$= B h c_0 J(J+1)$$

where the rotational quantum number J takes integral values $0, 1, 2, \ldots, I$ is the moment of inertia about an axis through the centre of mass of the molecule and c_0 the speed of light *in vacuo*. B is called the rotation constant. Each energy level is $2J + 1$-fold degenerate, and so all $2J + 1$ individual quantum states must be counted in the Boltzmann formula.

$$q_{rot} = \sum_{J=0}^{\infty} (2J + 1) \exp\left(-\frac{Bhc_0 J(J + 1)}{k_B T}\right)$$

The sum cannot be found in a closed form, but we note that for molecules with small rotational constants, the rotational energy levels crowd together and the summation can be replaced by an integral, treating J as a continuous variable

$$\begin{aligned} q_{rot} &= \int_0^{\infty} (2J + 1) \exp\left(-\frac{Bhc_0 J(J + 1)}{k_B T}\right) dJ \\ &= \frac{k_B T}{hc_0 B} \end{aligned} \quad (17.19)$$

Equation (17.19) works with complete reliability for all heteronuclear diatomic molecules, subject to the accuracy of the rotational energy level formula and the applicability of the continuum approximation. For homonuclear diatomics we have to think more carefully about indistinguishability of the two nuclei which results in occupancy of either the odd J or the even J levels; this depends on what the nuclei are, and need not concern us in detail here. It is dealt with by introducing a *symmetry factor* σ that is 1 for a heteronuclear diatomic and 2 for a homonuclear diatomic. q_{rot} is written

$$q_{rot} = \frac{1}{\sigma} \frac{k_B T}{hc_0 B} \quad (17.20)$$

The so-called *rotational temperature* θ_{rot} is often used in discussions of statistical mechanics; we re-write Equation (17.20) as

$$q_{rot} = \frac{1}{\sigma} \frac{T}{\theta_{rot}} \quad (17.21)$$

17.6.4 q_{vib}

Vibrational energy levels have separations that are at least an order of magnitude greater than the rotational modes, which are in turn some 20 orders of magnitude greater than the translational modes. As a consequence, the spacing is comparable with $k_B T$ for everyday molecules and temperatures. If we consider a single harmonic vibrational mode for which

$$\varepsilon_v = hc_0 \omega_e \left(v + \tfrac{1}{2}\right)$$

then

$$\omega_e = \frac{1}{2\pi c_0} \sqrt{\frac{k_s}{\mu}}$$

If we take the energy zero as that of the energy level $v = 0$, then we have

$$q_{vib} = \sum_{v=0}^{\infty} \exp\left(-\frac{hc_0 v \omega_e}{k_B T}\right)$$

This simple geometric series can be summed exactly to give

$$q_{vib} = \frac{1}{1 - \exp\left(-\frac{hc_0 \omega_e}{k_B T}\right)} \tag{17.22}$$

and once again it is usually written in terms of the vibrational temperature θ_{vib}

$$q_{vib} = \frac{1}{1 - \exp\left(-\frac{\theta_{vib}}{T}\right)} \tag{17.23}$$

Finally, we need to know the electronic partition function. This is almost always 1, unless the electronic ground state is degenerate (in which case it is equal to the degeneracy) or unless there is a very low-lying excited state.

17.7 Back to L-phenylanine

Let me now return to L-phenylanine. The GAUSSIAN 'frequency' calculation also gives thermodynamic properties, calculated along the lines discussed above. Here is an abbreviated output. First we have moments of inertia (not shown) and then a calculation of the rotational and vibrational temperatures.

ROTATIONAL SYMMETRY NUMBER 1.

ROTATIONAL TEMPERATURES (KELVIN)   0.08431   0.02883   0.02602
ROTATIONAL CONSTANTS (GHZ)   1.75683   0.60063   0.54210
Zero-point vibrational energy   537317.4 (Joules/Mol)
                                128.42193 (Kcal/Mol)

VIBRATIONAL TEMPERATURES: 53.46   67.50   90.15   142.09   281.61
     (KELVIN)   410.64   428.42   484.99   496.50   548.48
                         656.89   715.21   814.98   933.21   964.82
                         980.06   1013.28   1122.16 1196.52 1225.56
                        1276.27   1299.48   1380.29 1429.07 1493.75
. . . etc.

Next come the various contributions to the internal energy U (here called E), the heat capacity C and the entropy S. The partition functions are also printed.

```
Zero-point correction = 0.204653 (Hartree/Particle)
Thermal correction to Energy = 0.215129
Thermal correction to Enthalpy = 0.216073
Thermal correction to Gibbs Free Energy = 0.166730
Sum of electronic and zero-point Energies = −551.206897
Sum of electronic and thermal Energies = −551.196422
Sum of electronic and thermal Enthalpies = −551.195478
Sum of electronic and thermal Free Energies = −551.244821
```

	E (Thermal) KCAL/MOL	CV CAL/MOL-KELVIN	S CAL/MOL-KELVIN
TOTAL	134.995	38.687	103.852
ELECTRONIC	0.000	0.000	0.000
TRANSLATIONAL	0.889	2.981	41.212
ROTATIONAL	0.889	2.981	30.709
VIBRATIONAL	133.218	32.725	31.930
VIBRATION 1	0.594	1.982	5.405
VIBRATION 2	0.595	1.979	4.944
VIBRATION 3	0.597	1.972	4.372
... etc.			

	Q	LOG10(Q)	LN(Q)
TOTAL BOT	0.204249D−76	−76.689839	−176.584881
TOTAL V=0	0.277678D+18	17.443541	40.165238
VIB (BOT)	0.213500D−90	−90.670602	−208.776777
VIB (BOT) 1	0.556962D+01	0.745825	1.717326
VIB (BOT) 2	0.440790D+01	0.644232	1.483398
VIB (BOT) 3	0.329468D+01	0.517813	1.192309
... etc.			
VIB (V=0)	0.290254D+04	3.462779	7.973343
VIB (V=0) 1	0.609201D+01	0.784761	1.806979
VIB (V=0) 2	0.493617D+01	0.693390	1.596589
VIB (V=0) 3	0.383240D+01	0.583471	1.343492
... etc.			

17.8 Excited States

Electronic ground states are all well and good, but there is a wealth of experimental spectroscopic data that relate to the excited states of molecules. One of the simplest and most widely studied sequence of organic molecules studied by spectrophotometry is afforded by the set of monocyclic azines, of which pyridine forms the first member. It is instructive to compare this series with benzene.

All the molecules have 42 electrons, 21 electron pairs, and I can write their ground state electronic configurations

$$\Psi_0 : (\psi_1)^2 (\psi_2)^2 \cdots (\psi_{21})^2$$

We refer to Ψ_0 as the *reference state*. Single excitations can be produced by promoting an electron from one of the occupied orbitals to an unoccupied (virtual) one, just as we did in Chapter 15 for dihydrogen, so the first few single excitations have configurations

$$\Psi_1 : \psi_1^2 \psi_2^2 \cdots \psi_{21}^1 \psi_{22}^1$$
$$\Psi_2 : \psi_1^2 \psi_2^2 \cdots \psi_{21}^1 \psi_{23}^1$$
$$\Psi_3 : \psi_1^2 \psi_2^2 \cdots \psi_{20}^1 \psi_{21}^2 \psi_{22}^1$$
$$\Psi_4 : \psi_1^2 \psi_2^2 \cdots \psi_{20}^1 \psi_{21}^2 \psi_{23}^1$$

Electron spin has to be taken into account, and proper Slater determinants constructed. The ground state wavefunction will be a single Slater determinant but there are four possibilities (say, (a), (b), (c) and (d)) for each of the four excited states. In our dihydrogen calculation, I grouped the determinants together so that they were spin eigenfunctions. In modern configuration interaction (CI) calculations, we work directly with the individual Slater determinants and classify the resulting wavefunctions at the end of the calculation.

We write the CI wavefunction as

$$\Psi_{CI} = c_0 \Psi_0 + c_{1a} \Psi_{1a} + c_{1b} \Psi_{1b} + \cdots$$

where the CI expansion coefficients have to be determined from a variational calculation. This involves finding the eigenvalues and eigenvectors of a matrix whose elements are typically

$$\int \Psi_i \Psi_j \, d\tau$$

The matrix elements can be determined from the Slater–Condon–Shortley rules, giving typically HF–LCAO orbital energies and various two-electron integrals such as

$$\frac{e^2}{4\pi\epsilon_0} \int \psi_{21}(\mathbf{r}_1)\psi_{22}(\mathbf{r}_2) \frac{1}{r_{12}} \psi_{21}(\mathbf{r}_1)\psi_{22}(\mathbf{r}_2) \, d\tau_1 d\tau_2$$

These have to be calculated from the two-electron integrals over the HF–LCAO basis functions, at first sight a four-dimensional sum known in the trade as the *four-index transformation*.

Two of each of the (a), (b), (c) and (d) Slater determinants correspond to spin eigenfunctions having spin quantum number $M_s = 0$, one to $M_s = +1$ and one to

$M_s = -1$ and so need not be considered because states of different spin symmetry do not mix. In addition, Brillouin's Theorem [77] tells us that singly excited states constructed using HF wavefunctions don't mix with the ground state for a closed shell system, so don't need to include the ground state Ψ_0 in the variational calculation.

If we take all possible excited Slater determinants and solve the variational problem we reach a level of theory known as *CI singles (CIS)*. A keynote paper is that by J. B. Foresman *et al.* [78], and their Synopsis puts everything into perspective.

This work reviews the methodological and computational considerations necessary for the determination of the ab initio energy, wavefunction and gradient of a molecule in an electronically excited state using molecular orbital theory. In particular, this paper re-examines a fundamental level of theory which was employed several years ago for the interpretation of the electronic spectra of simple organic molecules: configuration interaction (CI) among all singly substituted determinants using a Hartree Fock reference state. This investigation presents several new enhancements to this general theory. First, it is shown how the CI singles wavefunction can be used to compute efficiently the analytic first derivative of the energy. ... Second, a computer program is described which allows these computations to be done in a 'direct' fashion.

You should have picked up the words 'direct' and 'gradient'.

I now return to my example of the azines. In the earliest and necessarily qualitative treatments, the first four transitions were identified with the single excitations shown in Figure 17.7. Benzene can be regarded as a special case of the series and the highest occupied orbitals are doubly degenerate, as are the lowest unoccupied ones.

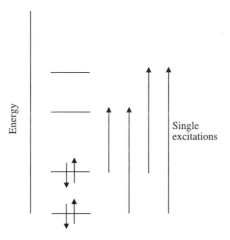

Figure 17.7 Simple CI treatment of azines

The assignment of bands in the benzene spectrum led to considerable discussion in the primary literature.

Excitation energies and oscillator strengths:

Excited State	1:	Singlet-A"	6.0836 eV	203.80 nm	f=0.0063
19 –> 22		0.64639			
19 –> 29		0.18254			

Excited State	2:	Singlet-A'	6.1539 eV	201.47 nm	f=0.0638
20 –> 23		−0.39935			
21 –> 22		0.65904			

Excited State	3:	Singlet-A'	6.3646 eV	194.80 nm	f=0.0159
20 –> 22		0.47433			
21 –> 23		0.57521			

Excited State	4:	Singlet-A"	7.3369 eV	168.98 nm	f=0.0000
19 –> 23		0.69159			

Excited State	5:	Singlet-A'	8.1128 eV	152.82 nm	f=0.6248
20 –> 22		0.49328			
21 –> 23		−0.36631			

Excited State	6:	Singlet-A'	8.2142 eV	150.94 nm	f=0.5872
20 –> 23		0.58185			
21 –> 22		0.24477			

Excited State	7:	Singlet-A"	8.4510 eV	146.71 nm	f=0.0000
17 –> 25		−0.10404			
20 –> 25		−0.10957			
21 –> 24		0.66688			
21 –> 26		−0.11328			

Excited State	8:	Singlet-A"	8.9454 eV	138.60 nm	f=0.0383
17 –> 24		−0.14068			
20 –> 24		−0.45970			
21 –> 25		0.49000			

The text above shows sample output from a CIS singlet state calculation on pyridine, and I have only included the first eight excited states. The MOs ψ_{20} and ψ_{21} are both π type, whilst ψ_{22} and ψ_{23} are antibonding π^*. The 19th occupied orbital corresponds to the N lone pair. The lowest singlet excited state is $n \rightarrow \pi^*$ as is the fourth. The remainder in my small sample are all $\pi \rightarrow \pi^*$.

I have collected a small number of calculated data into Table 17.2. In each case I have recorded the first one or two $n \rightarrow \pi^*$ and first four $\pi \rightarrow \pi^*$ transitions, together with the oscillator strengths.

A satisfactory theory of the spectra of aromatic hydrocarbons was not developed until there had been a detailed classification of their absorption bands. This important

Table 17.2 Spectroscopic data for various cyclic molecules, λ (nm) and oscillator strength f

	benzene	pyridine	1,2-diazine	1,3-diazine	1,4-diazine
$n \rightarrow \pi^*$	–	203.80	259.80	213.41	247.08
		0.0063	0.0088	0.0099	0.0081
$n \rightarrow \pi^*$	–	–	210.86	190.58	184.52
			0.0000	0.0000	0.0000
$\pi \rightarrow \pi^*$	204.79	201.47	194.08	190.93	208.67
	0.0000	0.0638	0.0423	0.0696	0.1525
$\pi \rightarrow \pi^*$	203.59	194.80	190.68	182.87	190.92
	0.0000	0.0159	0.0000	0.0294	0.0584
$\pi \rightarrow \pi^*$	159.08	152.82	152.26	147.01	145.63
	0.7142	0.6248	0.5952	0.5473	0.5010
$\pi \rightarrow \pi^*$	159.08	150.94	146.87	143.18	138.47
	0.7142	0.5872	0.5133	0.5139	0.4620

work was first undertaken by E. Clar [79]. There are three types of absorption bands that are classified mainly by their intensity but also by their vibrational fine structure and the frequency shifts within a family of compounds. The ground state of benzene (with D_{6h} symmetry) has closed shell configuration $a_{1u}^2 e_{1g}^4$ which gives symmetry A_{1g}. The first excited configuration $a_{1u}^2 e_{1g}^3 e_{2u}^1$ gives rise to states of symmetry B_{1u}, B_{2u} and E_{1u} which can be either singlet or triplet. The experimental absorption spectrum shows a weak band near 260 nm, and a strong band at 185.0 nm. Our CIS calculation therefore accounts only qualitatively for the features of the benzene spectrum.

The lowest singlet excitation for the azines is $n \rightarrow \pi^*$ in character, and the experimental values are 270, 340, 298 and 328 nm (S. F. Mason [80]). The CIS calculation gives the trend correctly but the absolute agreement with experiment is poor. Similar conclusions apply to the $\pi \rightarrow \pi^*$ transitions. The general conclusion is that the CIS method gives a useful first approximation to the excited states.

17.9 Consequences of the Brillouin Theorem

I mentioned primary properties in Section 17.3; these are properties such as the electric dipole moment, which is given as the expectation value of a sum of one-electron operators (together with the corresponding nuclear contribution)

$$\langle X \rangle = \int \Psi_{\text{el}}^* \left(\sum_{i=1}^{n} \hat{X}_i \right) \Psi_{\text{el}} \, d\tau$$

Suppose that we have a singlet ground state closed shell HF wavefunction Ψ_0 and we try to improve it by adding singly excited states; if I denote occupied orbitals

A, B, ... and virtual ones X, Y, ... then

$$\Psi = c_0 \Psi_0 + \sum_{A,X} c_A^X \Psi_A^X \qquad (17.24)$$

where c_A^X is the CI expansion coefficient and Ψ_A^X the appropriate combination of Slater determinants describing a singlet singly excited state. We find

$$\langle X \rangle = c_0^2 \int \Psi_0 \sum_{i=1}^{n} \hat{X}_i \Psi_0 d\tau + 2 \sum_{A,X} c_A^X c_0 \int \Psi_0 \sum_{i=1}^{n} \hat{X}_i \Psi_A^X d\tau$$

$$+ \sum_{A,X} \sum_{B,Y} c_A^X c_B^Y \int \Psi_B^Y \sum_{i=1}^{n} \hat{X}_i \Psi_A^X d\tau \qquad (17.25)$$

and all terms on the right-hand side apart from the first are zero, since the cs are zero by Brillouin's theorem. In addition, the only non-zero integrals in the third sum are those where the excited states differ by no more than one spin orbital. This can be easily proved by application of the Slater–Condon–Shortley rules.

If we now repeat the argument but include doubly excited rather than singly excited states

$$\Psi = c_0 \Psi_0 + \sum_{A,B,X,Y} c_{AB}^{XY} \Psi_{AB}^{XY}$$

$$\langle X \rangle = c_0^2 \int \Psi_0 \sum_{i=1}^{n} \hat{X}_i \Psi_0 d\tau + 2 \sum_{A,B,X,Y} c_{AB}^{XY} c_0 \int \Psi_0 \sum_{i=1}^{n} \hat{X}_i \Psi_{AB}^{XY} d\tau$$

$$+ \sum_{A,B,X,Y} \sum_{C,D,U,V} c_{AB}^{XY} c_{CD}^{UV} \int \Psi_{CD}^{UV} \sum_{i=1}^{n} \hat{X}_i \Psi_{AB}^{XY} d\tau \qquad (17.26)$$

then the CI expansion coefficients are no longer zero. However, the integrals in the second sum are all zero because the doubly excited states differ from Ψ_0 by two spin orbitals. All the integrals in the third sum except those where $\Psi_{AB}^{XY} = \Psi_{CD}^{UV}$ are zero for the same reason, and the final expression is

$$\langle X \rangle = c_0^2 \int \Psi_0 \sum_{i=1}^{n} \hat{X}_i \Psi_0 d\tau + \sum_{A,B,X,Y} \int \Psi_{AB}^{XY} \sum_{i=1}^{n} \hat{X}_i \Psi_{AB}^{XY} d\tau \qquad (17.27)$$

At first sight, one-electron properties calculated from HF wavefunctions ought to be very accurate because the lowest 'correction' terms are certain doubly excited configurations. The argument is correct as far as it goes, but if we include *both* the singly and doubly excited states together

$$\Psi = c_0 \Psi_0 + \sum_{A,X} c_A^X \Psi_A^X + \sum_{A,B,X,Y} c_{AB}^{XY} \Psi_{AB}^{XY} \qquad (17.28)$$

Table 17.3 Electric dipole moment for pyridine. HF–LCAO model

Level of theory	p_e $(10^{-30}\,\mathrm{C\,m})$
HF/STO–3G	6.8695
HF/6–31G	8.8114
HF/6–31G*	7.7195
HF/6–311G**	7.5530
Experiment	$7.31 \pm 2\%$

then the singly excited states couple with the doubly excited ones, which couple with the ground state and so the singly excited states do make a contribution to the ground state.

In any case, the key choice when performing HF–LCAO calculations is the choice of atomic orbital basis set, as can be seen from my representative HF–LCAO calculations on pyridine in Table 17.3. In each case I optimized the geometry.

17.10 Electric Field Gradients

I have made mention at several points in the text to electron spin, the internal angular momentum possessed by electrons just because they are electrons. The total angular momentum of an atomic electron is found by summing the electron spin and the electron orbital angular momentum.

Many nuclei have a corresponding internal angular momentum, which is given the symbol $\mathbf{I}$ and the vector is characterized by two quantum numbers, I and m_I. The vector obeys the laws of quantum mechanics appropriate to angular momenta; for each value of I, $m_I = -I, -I+1, \ldots, +I$, the size of the vector is

$$\sqrt{I(I+1)}\,\frac{h}{2\pi}$$

and we can only measure simultaneously the size of the vector and one component that we conventionally take as the z-component. The z-component has size

$$m_I\,\frac{h}{2\pi}$$

The nuclear spin quantum number is characteristic of a given nucleus and can have different values for different isotopic species. A sample is shown in Table 17.4.

Many nuclei with $I \geq 1$ also possess a nuclear quadrupole $\mathbf{Q}_n$, and it is usually defined in terms of the nuclear charge distribution $\rho_n(\mathbf{r})$ as

$$\mathbf{Q}_n = \frac{1}{e} \begin{pmatrix} \int \rho_n(3x^2 - r^2)\,d\tau & 3\int \rho_n xy\,d\tau & 3\int \rho_n xz\,d\tau \\ 3\int \rho_n yx\,d\tau & \int \rho_n(3y^2 - r^2)\,d\tau & 3\int \rho_n yz\,d\tau \\ 3\int \rho_n zx\,d\tau & 3\int \rho_n zy\,d\tau & \int \rho_n(3z^2 - r^2)\,d\tau \end{pmatrix} \tag{17.29}$$

Table 17.4   Representative nuclear spin quantum numbers

Isotopic species	I
^{1}H	1/2
^{2}D	1
^{12}C	0
^{13}C	1/2
^{17}O	5/2

This definition gives a traceless tensor. Here the integration is over the nuclear charge distribution. Nuclear wavefunctions are pretty hard to come by, and we normally have to determine the components of $\mathbf{Q}_n$ by experiment.

In a molecule, a given nucleus will generally experience an electric field gradient due to the surrounding electrons. Electric fields are vector quantities and so an electric field gradient is a tensor quantity. The electric field gradient at nucleus n is usually written $\mathbf{q}_n$. The energy of interaction U between the nuclear quadrupole and the electric field gradient is

$$U = -\frac{e}{6}\sum(Q_n)_{ij}(q_n)_{ij} \tag{17.30}$$

In principal axes, the interaction is determined by $Q_n = (Q_n)_{zz}$ and $q_n = $ the largest of the diagonal components of $\mathbf{q}_n$. The quantity $eQ_n q_n/h$ is referred to as the *quadrupole coupling constant*. According to C. H. Townes and B. P. Dailey [81], since filled shells and s orbitals have spherical symmetry, and since d and f orbitals do not penetrate near the nucleus, the quadrupole coupling constant should be largely due to any p electrons present in the valence shell.

Molecular quadrupole coupling constants are usually determined from the hyperfine structure of pure rotational spectra or from electric beam and magnetic beam resonance spectroscopies. Nuclear magnetic resonance, electron spin resonance and Mossbäuer spectroscopies are also routes to this property.

I can use the well-studied series HCN, FCN and ClCN to illustrate a couple of points. These molecules, including many isotopic species, have been exhaustively studied by spectroscopic techniques. An interesting feature of the experimental geometries is that the CN bond lengths are almost identical in length yet the ^{14}N quadrupole coupling constant is quite different (Table 17.5).

Table 17.5   Experimental results for XCN

Molecule	R(C≡N) (pm)	^{14}N QCC (MHz)
HCN	115.5	− 4.58
FCN	115.9	− 2.67
ClCN	115.9	− 3.63

Table 17.6 HF/6–311G** results for XCN

Molecule	R(C≡N) (pm)	^{14}N QCC (MHz)
HCN	112.7	− 4.53
FCN	112.5	− 2.95
ClCN	112.8	− 3.85

Table 17.6 shows results calculated at the HF/6–311G** level of theory, including geometry optimization. The agreement with experiment is good to within a few percent. What chemical insight these calculations give into the reason why the three coupling constants are so different is not clear!

18 Semi-empirical Models

I have spent a good deal of time discussing the current state of the art in quantum chemical calculations, and I have deliberately focused on techniques where we perform all the calculations as exactly as possible. Over the years, it has become fashionable to refer to such models as *ab initio*. This Latin phrase translates into 'from the beginning'; it certainly doesn't necessarily mean 'correct'. Once we have stated the Hamiltonian and, if necessary, the atomic orbital basis set, then all the integrals are evaluated exactly without any attempt to calibrate them against experiment. Thus, the atomic Hartree–Fock model is an *ab initio* model, and the calculations I described for the hydrogen molecule ion and dihydrogen are *ab initio* calculations.

These early calculations were done in the 1920s and the 1930s, and you probably noticed that I then jumped to the 1970s to discuss modern HF–LCAO calculations with GTOs. Scientists first got their hands on computers in a meaningful way in the 1970s, and one unfortunate characteristic of *ab initio* calculations is that they are compute-intensive. From the early days, by necessity rather than choice, workers tried to develop simple HF–LCAO based models that applied to certain classes of molecule, and a characteristic of these models is that they avoid all the difficult integrals. There usually is a price to pay, and the models have to be calibrated against experiment; some workers regard this as a major advantage in order to achieve meaningful chemical predictions from cheap calculations. We often refer to such calibrated models as *semi-empirical*. This isn't a Latin phrase but it translates into 'based on a correct model but calibrated and/or parameterized'. We can of course argue for ever about what constitutes a 'correct' model; Hartree–Fock theory is certainly not absolutely 'correct' because it averages over the electron interactions and so it usually runs into trouble when trying to describe bond-breaking.

18.1 Hückel π-Electron Theory

Chemists describe the structure of planar organic molecules such as benzene in terms of two types of bond: σ and π. The σ bonds are thought of as localized in bonding regions between atoms, whilst the π bonds are delocalized over large parts of the molecule. Much of the interesting chemistry of these compounds appears to derive

Figure 18.1 Pyridine and pyrrole

from the π electrons and chemists imagine that these π electrons see a fairly constant potential due to the σ electrons and the nuclei.

One of the earliest models for π-electron compounds is afforded by Hückel π-electron theory that dates from the 1930s. The ideas are chemically simple and appealing, and the model enjoyed many years of successful application. Imagine two simple conjugated molecules such as pyridine and pyrrole, shown in Figure 18.1. I haven't indicated the hydrogen atoms because they are excluded from such calculations, but I have numbered the ring heavy atoms.

Hückel's method was developed long before modern HF–LCAO theory, but it turns out that the two can be related. You will recall from Chapter 16 that the HF–LCAO Hamiltonian matrix for a closed shell molecule is written

$$\hat{h}_{ij}^F = \int \chi_i(\mathbf{r}_1)\hat{h}^{(1)}(\mathbf{r}_1)\chi_j(\mathbf{r}_1)\,d\tau$$

$$+ \sum_{k=1}^{n}\sum_{l=1}^{n} P_{kl} \iint \chi_i(\mathbf{r}_1)\chi_j(\mathbf{r}_1)\hat{g}(\mathbf{r}_1,\mathbf{r}_2)\chi_k(\mathbf{r}_2)\chi_l(\mathbf{r}_2)\,d\tau_1 d\tau_2$$

$$- \frac{1}{2}\sum_{k=1}^{n}\sum_{l=1}^{n} P_{kl} \iint \chi_i(\mathbf{r}_1)\chi_k(\mathbf{r}_1)\hat{g}(\mathbf{r}_1,\mathbf{r}_2)\chi_j(\mathbf{r}_2)\chi_l(\mathbf{r}_2)\,d\tau_1 d\tau_2 \quad (18.1)$$

and the HF–LCAO orbital coefficients $\mathbf{c}_i$ and orbital energies ε_i are found as the n solutions of the generalized matrix eigenvalue problem

$$\mathbf{h}^F\mathbf{c}_i = \varepsilon_i \mathbf{S}\mathbf{c}_i \qquad (18.2)$$

There are n basis functions, and $1 \leq i \leq n$. The lowest energy orbitals each accommodate a pair of electrons with antiparallel spin, in the simplest version of HF–LCAO theory.

In Hückel π-electron theory, we treat only the π electrons explicitly. At first sight, this is a big approximation but it can be justified by including the effects of the remaining electrons in the one-electron terms. Naturally, we allocate pairs of π electrons to the lowest energy π orbitals. Each carbon in the molecules above contributes one π electron as does the nitrogen in pyridine. The nitrogen in pyrrole contributes two π electrons. The basis functions are not rigorously defined, but they are usually visualized as ordinary STO $2p_\pi$ atomic orbitals. For first-row atoms, we therefore use a single atomic orbital per centre. If considering sulphur compounds, it has long been argued that we should use two atomic orbitals, a $3p_\pi$ and a suitable $3d_\pi$. The atomic orbitals are nevertheless taken to be normalized and orthogonal so that the overlap matrix $\mathbf{S}$ is the unit matrix.

The Hamiltonian is not rigorously defined, although we can think of it as Hartree–Fock like. No attempt is made to evaluate any integrals exactly, and no HF–LCAO iterations are performed. We simply assume that

1. The diagonal elements of the Hamiltonian matrix depend only on the atom type (X), and are written α_X. Of the two molecules shown above, there will be three different αs; one for carbon, one for the pyridine nitrogen and one for the pyrrole nitrogen. Hydrogen atoms are excluded from the calculation because they don't formally have occupied $2p_\pi$ orbitals in their electronic ground state.

2. Off-diagonal elements of the matrix are zero, unless a pair of atoms X and Y is directly bonded in which case we call the matrix element β_{XY}.

The idea is to avoid giving numerical values to the parameters until absolutely necessary, and when it becomes necessary the values are assigned by comparing calculated properties with experimental results. The first calculations were done on conjugated hydrocarbons, for which we only have to consider two parameters α_C and β_{CC}. For heteroatomic systems we write

$$\alpha_X = \alpha_C + h_X \beta_{CC}$$
$$\beta_{XY} = k_{XY} \beta_{CC} \tag{18.3}$$

Table 18.1 gives a small selection of h_X and k_{XY}, taken from A. J. Streitwieser's classic book [82]. With these values, the Hückel Hamiltonian matrices for pyridine and pyrrole are

$$\begin{pmatrix} \alpha_C+0.5\beta_{CC} & 0.8\beta_{CC} & 0 & 0 & 0 & 0.8\beta_{CC} \\ 0.8\beta_{CC} & \alpha_{CC} & \beta_{CC} & 0 & 0 & 0 \\ 0 & \beta_{CC} & \alpha_{CC} & \beta_{CC} & 0 & 0 \\ 0 & 0 & \beta_{CC} & \alpha_{CC} & \beta_{CC} & 0 \\ 0 & 0 & 0 & \beta_{CC} & \alpha_{CC} & \beta_{CC} \\ 0.8\beta_{CC} & 0 & 0 & 0 & \beta_{CC} & \alpha_{CC} \end{pmatrix} \text{ and } \begin{pmatrix} \alpha_C+1.5\beta_{CC} & \beta_{CC} & 0 & 0 & \beta_{CC} \\ \beta_{CC} & \alpha_C & \beta_{CC} & 0 & 0 \\ 0 & \beta_{CC} & \alpha_C & \beta_{CC} & 0 \\ 0 & 0 & \beta_{CC} & \alpha_C & \beta_{CC} \\ \beta_{CC} & 0 & 0 & \beta_{CC} & \alpha_C \end{pmatrix}$$

Table 18.1 Hückel parameters

Atom X	h_X	k_{CX}
C	1	1
N (1 π electron)	0.5	0.8
N (2 π electrons)	1.5	1.0
B	−1.0	0.7
O (1 π electron)	1.0	0.8
CH$_3$	2.0	0.7

Table 18.2 Symmetry-adapted orbitals for pyridine by irreducible representation (IR)

a_2 IR	b_1 IR
χ_1	
χ_4	
$\sqrt{\frac{1}{2}}(\chi_2 + \chi_6)$	$\sqrt{\frac{1}{2}}(\chi_2 - \chi_6)$
$\sqrt{\frac{1}{2}}(\chi_3 + \chi_5)$	$\sqrt{\frac{1}{2}}(\chi_3 - \chi_5)$

Matrix eigenvalue problems, even of such low order, are not easy to solve with pencil and paper, and so we find that Streitwieser's book (written in 1961) places a great deal of emphasis on the use of molecular symmetry. If we take pyridine and label the $2p_\pi$ atomic orbitals χ_1 through χ_6, then the six normalized combinations shown in Table 18.2 are called *symmetry-adapted orbitals*.

Molecular symmetry operators commute with molecular Hamiltonians and as a consequence the off-diagonal Hamiltonian matrix elements between symmetry-adapted orbitals of different symmetry are zero. Thus, the Hamiltonian matrix takes a block diagonal form, giving essentially two eigenvalue problems one of order 4 (for the a_2 IR) and one of order 2 (for the b_1 IR)

$$
\begin{pmatrix}
\alpha_{CC} + 0.5\beta_{CC} & 0 & 0.8\sqrt{2}\beta_{CC} & 0 & 0 & 0 \\
0 & \alpha_{CC} & 0 & \sqrt{2}\beta_{CC} & 0 & 0 \\
0.8\sqrt{2}\beta_{CC} & 0 & \alpha_{CC} & \beta_{CC} & 0 & 0 \\
0 & \sqrt{2}\beta_{CC} & \beta_{CC} & \alpha_{CC} & 0 & 0 \\
0 & 0 & 0 & 0 & \alpha_{CC} & \beta_{CC} \\
0 & 0 & 0 & 0 & \beta_{CC} & \alpha_{CC}
\end{pmatrix}
$$

Despite its advanced age, Hückel π-electron theory appears in the literature from time to time, even today.

18.2 Extended Hückel Theory

The 1952 pioneering calculations of M. Wolfsberg and L. Helmholtz [83] are usually cited as the first real attempt to extend Hückel theory to inorganic molecules. It is worth reading the Synopsis to their paper, in order to catch the spirit of the times:

> We have made use of a semiempirical treatment to calculate the energies of the molecular orbitals for the ground state and the first few excited states of permanganate, chromate and perchlorate ions. The calculations of the excitation energies is in agreement with the qualitative features of the observed spectra, i.e. absorption in the far ultraviolet for ClO_4^- with two strong maxima in the visible or near ultraviolet for MnO_4^- and CrO_4^{2-} with the chromate

spectrum displaced towards higher energies. An approximate calculation of the relative f-values for the first two transitions in CrO_4^{2-} and MnO_4^{-} is also in agreement with experiment.

The data on the absorption spectra of permanganate ion in different crystalline fields is interpreted in terms of the symmetries of the excited states predicted by our calculations.

The extended Hückel model treats all valence electrons within the spirit of the original π-electron version. Once again it is helpful to think of the HF–LCAO model; the basis functions are taken to be valence shell STOs. For Li through Ne we include 2s and 2p STOs. For Na through Al, we just use 3s and 3p orbitals and then we have to consider whether to include 3d orbitals for Si through Ar. This seems to be a matter of personal preference.

The diagonal elements of the Hamiltonian are taken as the negatives of the valence shell ionization energies. Such quantities can be deduced, with certain subjectivity, from atomic spectral data. The off-diagonal elements are usually related to these very simply as

$$h_{ij} = kS_{ij}\frac{h_{ii} + h_{jj}}{2} \tag{18.4}$$

where S_{ij} is the overlap integral between the STOs (which is calculated exactly) and k is a constant that has to be adjusted to give best agreement with experiment. Despite using overlap integrals in the formula above, the basis functions are treated as orthogonal for the purposes of solving the eigenvalue problem.

Wolfsberg's and Helmholtz's classic paper was a serious attempt to explain why certain tetrahedral anions are coloured and others aren't, but this actually highlights a dramatic failing of Hückel theory: the formal neglect of electron repulsion. In simple terms, excitation of an electron in a closed shell molecule from a bonding orbital to an empty orbital leads to singlet and triplet excited states. As a rule of thumb (Hund's rule A in fact) the triplet will have the lower energy, yet in Hückel theory transitions to both singlet and triplet have exactly the same energy given by a difference of the same pair of orbital energies.

18.2.1 Roald Hoffman

The next milestone was the work of R. Hoffmann [84], who made a systematic study of organic compounds along the lines of Wofsberg and Helmholtz. Once again I will let the author put his work in his own words:

The Hückel theory, with an extended basis set consisting of 2s and 2p carbon and 1s hydrogen orbitals, with inclusion of overlap and all interactions, yields a good qualitative solution of most hydrocarbon conformational problems. Calculations have been performed within the same parameterisation for nearly all simple saturated and unsaturated compounds, testing a variety of geometries for each. Barriers to internal rotation, ring conformations, and geometrical isomerism are among the topics treated. Consistent σ and π charge distributions and overlap populations are obtained for aromatics and their relative roles discussed. For alkanes and alkenes charge distributions are also presented. Failures include overemphasis on steric factors, which leads to some incorrect isomerization energies; also the failure to predict strain energies. It is stressed that the geometry of a molecule appears to be its most predictable property.

18.3 Pariser, Parr and Pople

The next advances came in the 1950s, with a more systematic treatment of electron repulsion in π-electron molecules. These two keynote papers due to R. Pariser and G. R. G Parr, and to J. A. Pople are so important that I will give the Abstracts (almost) in full. Pople wrote [85]

An approximate form of the molecular orbital theory of unsaturated hydrocarbon molecules in their ground states is developed. The molecular orbital equations rigorously derived from the correct many-electron Hamiltonian are simplified by a series of systematic approximations and reduce to equations comparable with those used in the semi-empirical method based on an incompletely defined one-electron Hamiltonian. The two sets of equations differ, however, in that those of this paper include certain important terms representing electronic interactions. The theory is used to discuss the resonance energies, ionization potentials, charge densities, bond orders and bond lengths of some simple hydrocarbons. The electron interaction terms introduced in the theory are shown to play an important part in determining the ionization potentials, etc.

You should have picked up many of the key phrases. He started from the HF–LCAO equations and made what is now known as the σ–π separation approximation; the π electrons are treated separately and the effect of the remaining σ electrons is absorbed into the HF–LCAO Hamiltonian. The HF–LCAO equations have to be solved iteratively in order to get the HF–LCAO π-electron molecular orbitals, and many of the two-electron integrals (the 'electronic interaction terms') are retained. In order to take account of the effect of σ–π separation, most integrals are calibrated by appeal to experiment. The 'charges and bond orders' are simply the Mulliken populations calculated with an overlap matrix equal to the unit matrix, and ionization energies are calculated according to Koopmans' theorem.

The second keynote paper, by Pariser and Parr [86], also gives a snapshot of the times, when there was a great deal of interest in the electronic spectra of conjugated molecules. They wrote

A semi-empirical theory is outlined which is designed for the correlation and prediction of the wavelengths and intensities of the first main visible or

ultraviolet bands and other properties of complex unsaturated molecules, and preliminary application of the theory is made to ethylene and benzene.

The theory is formulated in the language of the purely theoretical method of the antisymmetrized products of molecular orbitals (in LCAO approximation) including configuration interaction, but departs from this theory in several essential respects. First, atomic orbital integrals involving the core Hamiltonian are expressed in terms of quantities which may be regarded as semi-empirical. Second, an approximation of zero differential overlap is employed and an optionally uniformly charged sphere representation of atomic π-orbitals is introduced, which greatly simplify the evaluation of electronic repulsion integrals and make applications to complex molecules containing heteroatoms relatively simple. Finally, although the theory starts from the π-electron approximation, in which the unsaturated electrons are treated apart from the rest, provision is included for the adjustment of the σ-electrons to the π-electron distribution in a way which does not complicate the mathematics.

Once again you should have picked up many of the key phrases. We often speak of the *PPP method* in honour of its three originators.

18.4 Zero Differential Overlap

A key phrase that I have yet to explain is *zero differential overlap (ZDO)*; the atomic orbital basis set is not rigorously defined but we can imagine it to comprise the relevant STO with one atomic orbital per heavy atom. Hydrogen atoms don't enter into π-electron models. The basis functions are taken to be orthonormal, so we have

$$\int \chi_i(\mathbf{r})\chi_j(\mathbf{r})\, d\tau = \begin{cases} 1 & \text{if } i = j \\ 0 & \text{otherwise} \end{cases}$$

ZDO extends this idea to the two-electron integrals; if we have a two-electron integral of the type

$$\frac{e^2}{4\pi\epsilon_0} \int \dots \frac{1}{r_{12}} \chi_i(\mathbf{r}_1)\chi_j(\mathbf{r}_1)\, d\tau_1$$

then the integral is set to zero unless $i = j$. So all two-electron integrals

$$\frac{e^2}{4\pi\epsilon_0} \int \chi_i(\mathbf{r}_1)\chi_j(\mathbf{r}_1)\frac{1}{r_{12}}\chi_k(\mathbf{r}_2)\chi_l(\mathbf{r}_2)\, d\tau_1 d\tau_2 \qquad (18.5)$$

are zero unless $i = j$ and $k = l$.

At first sight the basis functions are STO $2p_\pi$, and the remaining integrals can actually be calculated exactly. When Pariser and Parr first tried to calculate the ionization energies and spectra of simple conjugated hydrocarbons such as benzene with exact two-electron integrals, they got very poor agreement with experiment. They therefore proposed that the two-electron integrals should be treated as parameters to be calibrated against spectroscopic data. Once again, you have to remember that computers were still in their infancy and that any simple formula was good news. We had electromechanical calculators and log tables in those days, but nothing much else. One popular such formula was

$$\frac{e^2}{4\pi\epsilon_0} \int \chi_i(\mathbf{r}_1)\chi_i(\mathbf{r}_1) \frac{1}{r_{12}} \chi_k(\mathbf{r}_2)\chi_k(\mathbf{r}_2)\, d\tau_1 d\tau_2 = \frac{e^2}{4\pi\epsilon_0} \frac{1}{R_{ik} + \alpha_{ik}} \tag{18.6}$$

where R_{ik} is the separation between the orbital centres. All other two-electron integrals were taken as zero and the α_{ik} term has to be fixed by appeal to spectroscopic experiment. The expression is what you would get for the mutual potential energy of a pair of electrons separated by distance $R_{ik} + \alpha_{ik}$, and these repulsion integrals are traditionally given a symbol γ_{ik}.

Next we have to address the one-electron terms, and the treatment depends on whether the term is diagonal ($i = j$) or off-diagonal ($i \neq j$). In standard HF–LCAO theory the one-electron terms are given by

$$\int \chi_i \left(-\frac{h^2}{8\pi^2 m_e} \nabla^2 - \frac{e^2}{4\pi\epsilon_0} \sum_{I=1}^{N} \frac{Z_I}{R_I} \right) \chi_j\, d\tau \tag{18.7}$$

where the first term represents the kinetic energy of the electron and the second term gives the mutual potential energy between the electron and each nucleus.

In PPP theory, the off-diagonal elements are taken as zero unless the atom pairs are directly bonded. If the atom pairs are directly bonded, then the matrix element is given a constant value β for each particular atom pair, which is certainly not the same value as the β value in ordinary Hückel theory. The value of either β is found by calibration against experiment. The one-electron diagonal terms are written so as to separate out the contribution from the nuclear centre I on which the atomic orbital is based (usually taken as minus the valence state ionization energy ω_I) and the other nuclear centres J (usually taken as $-Z_J\gamma_{IJ}$, where Z is the formal number of electrons that an atom contributes to the π system). There is no problem with hydrocarbons in that all carbon atoms are assigned the same valence state ionization energies, irrespective of their chemical environment, but many papers were written to justify one particular choice of ω_I against another for nitrogen.

For the record, the PPP HF–LCAO Hamiltonian $\mathbf{h}^F$ for hydrocarbons and other first-row systems has dimension equal to the number of conjugated atoms and can be written

$$h_{ii}^F = \omega_i + \tfrac{1}{2} P_{ii}\gamma_{ii} + \sum_{j \neq i} \left(P_{jj} - Z_j \right)\gamma_{ij}$$

$$h_{ij}^F = \beta_{ij} - \tfrac{1}{2} P_{ij}\gamma_{ij} \tag{18.8}$$

Here **P** is the usual matrix of charges and bond orders. The HF–LCAO equations are solved by the usual techniques.

In those early days, no one dreamed of geometry optimization; benzene rings were assumed to be regular hexagons with equal C—C bond length of 140 pm. PPP calculations were state of the art in the early 1970s and consumed many hours of Ferranti Mercury, IBM 7040 and ICT Atlas time. Once again there was an argument as to the role of d orbitals on sulphur, and a number of long forgotten schemes and methodologies were advanced.

18.5 Which Basis Functions Are They?

I seem to have made two contradictory statements about the basis functions used in semi-empirical work. On the one hand, I have said that they are orthonormal and so their overlap matrix is the unit matrix, and on the other hand I have used overlap integrals to calculate certain integrals.

Think of a Hückel π-electron treatment of ethene, and call the carbon $2p_\pi$ orbitals χ_1 and χ_2. The matrix of overlap integrals is

$$\mathbf{S} = \begin{pmatrix} 1 & p \\ p & 1 \end{pmatrix}$$

where p is the overlap integral of the two atomic (STO) orbitals in question. The eigenvalues of this matrix are $1 \pm p$ and the normalized eigenvectors are

$$\mathbf{v}_1 = \sqrt{\frac{1}{2}}\begin{pmatrix} 1 \\ 1 \end{pmatrix} \quad \text{and} \quad \mathbf{v}_2 = \sqrt{\frac{1}{2}}\begin{pmatrix} 1 \\ -1 \end{pmatrix}$$

A little matrix algebra will show that

$$\mathbf{S} = (1+p)\mathbf{v}_1\mathbf{v}_1^T + (1-p)\mathbf{v}_2\mathbf{v}_2^T$$

Mathematicians have a rather grand expression for this; they talk about the *spectral decomposition* of a matrix. We can make use of the expression to calculate powers of matrices, such as the negative square root

$$\mathbf{S}^{-1/2} = (1+p)^{-1/2}\mathbf{v}_1\mathbf{v}_1^T + (1-p)^{-1/2}\mathbf{v}_2\mathbf{v}_2^T$$

To cut a long story short, we regard the basis functions used in semi-empirical calculations as related to ordinary STO $\chi_1, \chi_2, \ldots, \chi_n$ by the matrix transformation

$$(\chi_1 \quad \chi_2 \cdots \chi_n)\mathbf{S}^{-1/2} \tag{18.9}$$

These have the property that they are orthonormal, yet resemble the ordinary STO as closely as possible, in a least-squares sense.

18.6 All Valence Electron ZDO Models

The early π-electron semi-empirical models proved a great success, and they are still sometimes encountered in current publications. Attempts to extend them to the σ systems or to inorganic molecules met with mixed fortune for a variety of reasons, especially the following three.

1. If we draw a molecule and then arbitrarily choose an axis system, then physical properties such as the energy should not depend on the choice of axis system. We speak of *rotational invariance*; the answers should be the same if we rotate a local molecular axis system.

2. Also, despite what one reads in elementary organic texts, calculated physical properties ought to be the same whether one works with ordinary atomic orbitals or the mixtures we call hybrids.

3. Finally, we should get the same answers if we work with symmetry-adapted combinations of atomic orbitals rather than the 'raw' AOs. Whatever the outcome, we won't get a different energy.

18.7 Complete Neglect of Differential Overlap

Pople and co-workers seem to be the first authors to have addressed these problems. The most elementary theory that retains all the main features is the complete neglect of differential overlap (CNDO) model. The first paper dealt mainly with hydrocarbons, and only the valence electrons were treated. The inner shells contribute to the core that modifies the one-electron terms in the HF–LCAO Hamiltonian. The ZDO approximation is applied to all two-electron integrals so that

$$\iint \chi_i(\mathbf{r}_1)\chi_j(\mathbf{r}_1)\frac{1}{r_{12}}\chi_k(\mathbf{r}_2)\chi_l(\mathbf{r}_2)\,\mathrm{d}\tau_1\mathrm{d}\tau_2 \tag{18.10}$$

is zero unless $i=j$ and $k=l$. Suppose now that atoms A and B are both carbon, and so we take 2s, $2p_x$, $2p_y$ and $2p_z$ basis functions for either atom. In addition to the ZDO approximation, the CNDO model requires that all remaining two-electron integrals of type (18.10) involving basis functions on A and B are equal to a common value, denoted γ_{AB}. γ_{AB} depends on the nature of the atoms A and B but not on the details of the atomic orbitals. This simplification guarantees rotational invariance.

Similar considerations apply to the one-electron terms

$$\int \chi_i \left(-\frac{h^2}{8\pi^2 m_e} \nabla^2 - \frac{1}{4\pi\epsilon_0} \sum_{\alpha=1}^{N} \frac{Z_\alpha}{R_\alpha} \right) \chi_j \, d\tau$$

When $i = j$ and the basis function χ_i is centred on nucleus A we write the integral

$$U_{ii} = \int \chi_i \left(-\frac{h^2}{8\pi^2 m_e} \nabla^2 - \frac{1}{4\pi\epsilon_0} \frac{Z_A}{R_A} \right) \chi_i \, d\tau$$

and determine U from atomic spectral data. The remaining terms in the sum are called *penetration integrals* and are written V_{AB}.

If $i \neq j$ and the basis functions are on different atomic centres, then all three-centre contributions are ignored. The remaining two-centre terms involving atoms A and B are written $\beta^0_{AB} S_{ij}$, where S is the overlap integral and β a 'bonding' parameter. The bonding parameter is non-zero only for bonded pairs.

Collecting terms and simplifying we find that a CNDO HF–LCAO Hamiltonian has elements

$$h^F_{ii} = U_{ii} + \left(P_{AA} - \tfrac{1}{2} P_{ii} \right) \gamma_{AA} + \sum_{B \neq A} \left(P_{BB} \gamma_{AB} - V_{AB} \right)$$

$$h^F_{ij} = \beta^0_{AB} S_{ij} - \tfrac{1}{2} P_{ij} \gamma_{AB}$$

(18.11)

A and B label atoms, i and j label the basis functions and P_{AA} is the sum of the diagonal charge density matrix for those basis functions centred on atom A.

The original parameter scheme was called CNDO/1 (J. A. Pople and G. A. Segal [87]). Electron repulsion integrals were calculated exactly, on the assumption that the basis functions were s-type STOs, and all overlap integrals were calculated exactly. The bonding parameters β^0_{AB} were chosen by comparison with (crude) *ab initio* calculations on relevant small molecules, and a simple additivity scheme was employed

$$\beta^0_{AB} = \beta^0_A + \beta^0_B$$

18.8 CNDO/2

It turned out that CNDO/1 calculations gave poor predictions of molecular geometries, and this failing was analysed as due to the approximations made for U_{ii} and the penetration term V_{AB}. These problems were corrected in CNDO/2. V_{AB}

is no longer calculated exactly; rather it is taken as $-Z_B\gamma_{AB}$. The atomic terms become

$$U_{ii} = -\tfrac{1}{2}(I_i + E_i) - (Z_A - \tfrac{1}{2})\gamma_{AA} \qquad (18.12)$$

where I_i and E_i are valence state ionization energies and electron affinities. The HF–LCAO matrix elements are

$$h_{ii}^{F} = -\tfrac{1}{2}(I_i + E_i) + ((P_{AA} - Z_A) - \tfrac{1}{2}(P_{ii} - 1))\gamma_{AA} + \sum_{B \neq A}(P_{BB}\gamma_{AB} - V_{AB})$$

$$h_{ij}^{F} = \beta_{AB}^{0}S_{ij} - \tfrac{1}{2}P_{ij}\gamma_{AB} \qquad (18.13)$$

18.9 CNDO/S

One failing of the π-electron models was their inability to treat n$\rightarrow\pi^*$ electronic transitions. The CNDO models gave a great breakthrough because they afforded for the first time a systematic treatment of both $\pi\rightarrow\pi^*$ and n$\rightarrow\pi^*$ transitions in conjugated molecules. The treatment essentially followed the lines of CIS discussed in Chapter 17, but the first results were in poor agreement with experiment in that they gave excitation energies that were much too large. The most significant improvement to CNDO for use in spectroscopy was given by the work of J. del Bene and H. H. Jaffé [88]. Their major contribution was to modify the bonding integrals; think of a pair of bonded carbon atoms, with one pair of 2p orbitals pointing along the bond (σ) and the remaining two pairs perpendicular to the bond (π). Del Bene and Jaffé introduced a new parameter κ such that the π-type interaction was reduced. They essentially replaced the term $\beta_{AB}^{0}S_{ij}$ with

$$\beta_{AB}^{\sigma}S_{ij} = \tfrac{1}{2}(\beta_A^0 + \beta_B^0)S_{ij}$$

$$\beta_{AB}^{\pi}S_{ij} = \tfrac{1}{2}\kappa(\beta_A^0 + \beta_B^0)S_{ij} \qquad (18.14)$$

They recommend a value $\kappa = 0.585$, which gives best agreement with experiment for many conjugated molecules.

18.10 Intermediate Neglect of Differential Overlap

The CNDO models make an unnecessary and draconian simplification to the two-electron integrals

$$\iint \chi_i(\mathbf{r}_1)\chi_j(\mathbf{r}_1)\frac{1}{r_{12}}\chi_k(\mathbf{r}_2)\chi_l(\mathbf{r}_2)\, \mathrm{d}\tau_1 \mathrm{d}\tau_2 = \delta_{ij}\delta_{kl}\gamma_{AB}$$

That is to say, even if the orbitals χ_i and χ_j are on the same centre but $i \neq j$, then the entire integral is taken to be zero. Direct calculation shows that the integrals can be far from zero. In fact, it is just one-centre integrals of this kind that give the correct ordering between atomic spectroscopic states. CNDO cannot distinguish between the ^3P, ^1D and ^1S states arising from orbital configuration

$$C: (1s)^2(2s)^2(2p)^2$$

J. A. Pople, D. Beveridge and P. A. Dobosh [89] introduced the intermediate neglect of differential overlap (INDO) scheme, which retained monocentric repulsion integrals. Some of these integrals are taken as parameters to be determined by comparison with experiment rather than by direct calculation.

18.11 Neglect of Diatomic Differential Overlap

The next level of approximation along these lines is the neglect of diatomic differential overlap (NDDO) model, developed by Pople et al. (see for example [90]). All two-electron integrals

$$\int\int \chi_i(\mathbf{r}_1)\chi_j(\mathbf{r}_1)\frac{1}{r_{12}}\chi_k(\mathbf{r}_2)\chi_l(\mathbf{r}_2)\,d\tau_1 d\tau_2$$

are retained when χ_i and χ_j are on the same centre and both χ_k and χ_l are on a single different centre. NDDO never really caught on, it was overtaken by events as follows.

18.12 The Modified INDO Family

The CNDO/2, INDO and NDDO methods are now little used apart from providing initial estimates of the electron density for use in *ab initio* HF–LCAO iterations. Their significance is that they were the first family of semi-empirical models that retained increasing numbers of smaller repulsion integrals in a systematic way. Even NDDO neglects the majority of two-electron integrals for real molecules, and so the models made HF–LCAO calculations on large systems feasible. They did not give very accurate results, perhaps because they were parameterized by comparing with low accuracy *ab initio* calculations. They also needed reliable geometries as input, because geometry optimization was still in its infancy. Pople and Beveridge's book contains a listing of a FORTRAN CNDO/INDO code, and it is interesting to note that there is no mention of geometry optimization.

The MINDO models (Modified INDO) developed by M. J. S. Dewar had a quite different philosophy; their objective was to develop a genuine reliable model for experimentalists. They realized that parameters had to be available for a wide range of atoms and the models had to deliver results to chemical accuracy in reasonable amounts of computer time. Attention was paid to reliable geometry optimization, so enabling the program to accept relatively crude initial geometries for unknown structures.

18.12.1 MINDO/3

The Synopsis for the landmark 1975 MINDO/3 paper by R. C. Bingham *et al.* [91] is terse:

The problems involved in attempts to develop quantitative treatments of organic chemistry are discussed. An improved version (MINDO/3) of the MINDO semiempirical SCF-MO treatment is described. Results obtained for a large number of molecules are summarised

MINDO/3 uses an s, p minimal basis set of STOs and the elements of the HF–LCAO matrix are

$$h_{ii}^F = U_{ii} + \sum_{j \text{ on } A} \left(P_{ij}\gamma_{ij} - \tfrac{1}{2}P_{ij}\lambda_{ij}\right) + \sum_{B \neq A} (P_{BB} - Z_B)\gamma_{AB}$$

$$h_{ij}^F = \begin{cases} -\tfrac{1}{2}P_{ij}\lambda_{ij} & \text{if } \chi_i \text{ and } \chi_j \text{ on } A \\ h_{ij}^{core} - \tfrac{1}{2}P_{ij}\gamma_{AB} & \text{otherwise} \end{cases} \qquad (18.15)$$

I have written the atomic Coulomb and exchange integrals

$$\gamma_{ij} = \frac{e^2}{4\pi\epsilon_0} \int \chi_i(\mathbf{r}_1)\chi_i(\mathbf{r}_1)\frac{1}{r_{12}}\chi_j(\mathbf{r}_2)\chi_j(\mathbf{r}_2)\,d\tau_1 d\tau_2$$

$$\lambda_{ij} = \frac{e^2}{4\pi\epsilon_0} \int \chi_i(\mathbf{r}_1)\chi_j(\mathbf{r}_1)\frac{1}{r_{12}}\chi_i(\mathbf{r}_2)\chi_j(\mathbf{r}_2)\,d\tau_1 d\tau_2$$

for simplicity of notation. The parameters for MINDO/3 were obtained in an entirely different way from the CNDO/INDO/NDDO family; many quantities such as the STO exponents were allowed to vary during the fitting procedure. The bonding parameter β_{AB}^0 was allowed to vary, and experimental data such as enthalpies of formation and accurate molecular geometries were also used to get the best fit.

An interesting feature was the treatment of core–core repulsions (the core in this case being identified with the nucleus plus any inner-shell atomic electrons). The simple Coulomb term

$$U_{AB} = \frac{e^2}{4\pi\epsilon_0}\frac{Z_A Z_B}{R_{AB}}$$

(where Z_A and Z_B are the 'effective' nuclear charges) was modified for various complicated reasons to make it a function of the electron repulsion integrals

$$U_{AB} = Z_A Z_B \left(\gamma_{AB} + \left(\frac{e^2}{4\pi\epsilon_0} \frac{1}{R_{AB}} - \gamma_{AB} \right) \exp\left(-\frac{\alpha_{AB} R_{AB}}{m} \right) \right) \qquad (18.16)$$

Here α_{AB} is a dimensionless constant and it depends on the natures of the atoms A and B. For O–H and N–H bonds a slightly different scheme was adopted.

$$U_{AH} = Z_A Z_H \left(\gamma_{AH} + \left(\frac{e^2}{4\pi\epsilon_0} \frac{1}{R_{AH}} - \gamma_{AB} \right) \alpha_{AH} \exp\left(-R_{AH}/m \right) \right) \qquad (18.17)$$

18.13 Modified Neglect of Differential Overlap

MINDO/3 proved very successful but it had a number of limitations; enthalpies of formation of conjugated molecules were generally too positive, bond angles were not well predicted, and so on. M. J. S. Dewar and W. Thiel [92] introduced the modified neglect of differential overlap (MNDO) model, which they based on NDDO whilst retaining the philosophy of MINDO/3. The core–core repulsions were further modified

$$U_{AB} = \frac{e^2}{4\pi\epsilon_0} Z_A Z_B \int s_A(\mathbf{r}_1) s_A(\mathbf{r}_1) \frac{1}{r_{12}} s_B(\mathbf{r}_2) s_B(\mathbf{r}_2) \, d\tau_1 d\tau_2$$
$$\times \left(1 + \exp\left(-\frac{\alpha_A R_{AB}}{m} \right) + \exp\left(-\frac{\alpha_B R_{AB}}{m} \right) \right) \qquad (18.18)$$

with a different formula for OH and NH bonds.

18.14 Austin Model 1

Next came Austin model 1 (AM1), due to M. J. S. Dewar *et al.* [93]. AM1 was designed to eliminate the problems from MNDO caused by a tendency to overestimate repulsions between atoms separated by the sum of their van der Waals radii. The strategy adopted was to modify the core–core terms by multiplication of the Coulomb term with sums of Gaussian functions. In the original AM1 paper there are four terms in the Gaussian expansion. Each Gaussian is characterized by its position along the A–B vector and by its width. This significantly increased the number of parameters for each atom.

18.15 PM3

PM3 is the third parameterization of MNDO, and the PM3 model contains essentially all the same terms as AM1. The parameters for PM3 were derived by J. J. P. Stewart

[94] in a more systematic way than for AM1, many of which were derived by 'chemical intuition'. As a consequence, some of the parameters are quite different from those of MNDO but the two models seem to predict physical properties to the same degree of accuracy.

18.16 SAM1

The final 1993 offering in Michael Dewar's name was Semi-Ab-Initio Model 1 [95]. In SAM1, two-electron integrals are calculated using a standard STO-3G basis set (and hence the appearance of *ab initio* in the title). The resulting integrals were then scaled, and the Gaussian terms in the core–core repulsions were retained in order to fine-tune the calculations.

18.17 ZINDO/1 and ZINDO/S

Michael Zerner and co-workers developed these models for transition metal compounds [96]. ZINDO/1 is a variant of INDO and ZINDO/S is an INDO-like method used to calculate electronic spectroscopic properties of such molecules.

18.18 Effective Core Potentials

The ground state of a silver atom has an electron configuration

$$\text{Ag: } (1s)^2(2s)^2(2p)^6(3s)^2(3p)^6(3d)^{10}(4s)^2(4p)^6(4d)^{10}(5s)^1$$

and much of its chemistry can be explained in terms of the outer 5s electron. The remaining 46 electrons seem to form an inert core and this suggests that we might try to model the silver atom as a nucleus of charge $Z = 47$ and an inner shell of 46 electrons. This idea was tried in the very early days, and a common effective core potential (ECP) was

$$U_{\text{core}} = -\frac{n_V}{4\pi\epsilon_0 r} + A\frac{\exp(-2kr)}{r} \tag{18.19}$$

where n_V is the number of valence electrons and A and k are constants that have to be determined by fitting an atomic property. Use of such a potential means that we have eliminated most of the two-electron integrals.

The idea of dividing electrons into groups is quite common in chemistry; in CNDO theory we treat the valence electrons only, without explicitly treating the core electrons. In ZDO π-electron models we treat the π electrons explicitly and assume that the effect of the σ electrons can be taken into account in the parameter choice. There is a serious problem that I can explain by reference to silver. The 5s orbital has to be orthogonal to all the inner orbitals, even though we don't use them (and don't necessarily want to calculate them).

Many *ab initio* packages allow the use of *effective core potentials* (ECPs), which replace the atomic cores in valence-only *ab initio* calculations; traditionally they were represented as linear combinations of functions of the type

$$r^{-n} \exp\left(-\alpha r^2\right)$$

with coefficients and exponents determined by fitting the potential generated from accurate HF–LCAO wavefunctions. In recent years it has become fashionable to represent the core potentials by one- and two-term Gaussians obtained directly from the appropriate atomic eigenvalue equation.

19 Electron Correlation

I have mentioned electron correlation at several points in the text, and I gave an operational definition of correlation energy as the difference between the exact HF energy and the true experimental energy. There is a great deal of small print; in general we cannot obtain exact HF wavefunctions for molecules, only LCAO approximations. The Schrödinger treatment takes no account of the theory of relativity, whilst we know from simple atomic spectra that relativistic effects are non-negligible. We have to be careful to treat zero-point energy in a consistent manner when dealing with vibrations, and so on.

19.1 Electron Density Functions

I have put a great deal of emphasis on the electron density. The wavefunction for a molecule with n electrons will depend on the $3n$ spatial electron coordinates $\mathbf{r}_1, \mathbf{r}_2, \ldots, \mathbf{r}_n$ together with the n spin variables s_i (α or β). Many authors combine space and spin variables into a composite $\mathbf{x} = \mathbf{r}s$ and so we would write a wavefunction

$$\Psi(\mathbf{x}_1, \mathbf{x}_2, \ldots, \mathbf{x}_n)$$

According to the Born interpretation,

$$\Psi^*(\mathbf{x}_1, \mathbf{x}_2, \ldots, \mathbf{x}_n)\Psi(\mathbf{x}_1, \mathbf{x}_2, \ldots, \mathbf{x}_n)\, d\tau_1 ds_1 d\tau_2 ds_2 \cdots d\tau_n ds_n$$

gives the probability of finding simultaneously electron 1 in spatial volume element $d\tau_1$ with spin between s_1 and $s_1 + ds_1$, electron 2 in spatial volume element $d\tau_2$ with spin between s_2 and $s_2 + ds_2, \ldots$, electron n in spatial volume element $d\tau_n$ with spin between s_n and $s_n + ds_n$.

As we noted earlier, this expression contains too much information and we ask instead about the probability that any given element will be occupied by one electron with the other electrons anywhere. This is often called the *one-electron density function*

$$\rho_1(\mathbf{x}_1) = n \int \Psi^*(\mathbf{x}_1, \mathbf{x}_2, \ldots, \mathbf{x}_n)\Psi(\mathbf{x}_1, \mathbf{x}_2, \ldots, \mathbf{x}_n)\, d\tau_2 ds_2 \cdots d\tau_n ds_n \qquad (19.1)$$

The x_1 on the left refers to 'point 1' at which the density is to be evaluated rather than the coordinates of electron 1, the indistinguishability of the electrons being accounted for by the factor n.

If we want to know the probability of finding an electron of either spin in the spatial element $d\tau_1$, then we integrate over ds_1 to give the charge density discussed in Chapter 16 and measured by crystallographers

$$P_1(\mathbf{r}_1) = n \int \Psi^*(\mathbf{x}_1, \mathbf{x}_2, \ldots, \mathbf{x}_n)\Psi(\mathbf{x}_1, \mathbf{x}_2, \ldots, \mathbf{x}_n)\, ds_1 d\tau_2 ds_2 \cdots d\tau_n ds_n \qquad (19.2)$$

It is written either P or P_1; I have used P in earlier chapters. It also proves desirable to introduce probabilities for finding different configurations of any number of particles. Thus, the *two-electron* (or *pair*) *density function*

$$\rho_2(\mathbf{x}_1, \mathbf{x}_2) = n(n-1) \int \Psi^*(\mathbf{x}_1, \mathbf{x}_2, \ldots, \mathbf{x}_n)\Psi(\mathbf{x}_1, \mathbf{x}_2, \ldots, \mathbf{x}_n)\, d\tau_3 ds_3 \cdots d\tau_n ds_n$$

$$(19.3)$$

determines the probability of two electrons being found simultaneously at points $\mathbf{x}_1$ and $\mathbf{x}_2$, spins included, whilst

$$P_2(\mathbf{x}_1, \mathbf{x}_2) = n(n-1) \int \Psi^*(\mathbf{x}_1, \mathbf{x}_2, \ldots, \mathbf{x}_n)\Psi(\mathbf{x}_1, \mathbf{x}_2, \ldots, \mathbf{x}_n)\, ds_1 ds_2 d\tau_3 ds_3 \cdots d\tau_n ds_n$$

$$(19.4)$$

determines the probability of finding them at points $\mathbf{r}_1$ and $\mathbf{r}_2$ in ordinary space, irrespective of spin.

Many common one-electron properties depend only on P_1 and since the Schrödinger equation only contains pair interactions we need not consider distributions higher than the pair functions. For a state of definite spin, as distinct from a mixed state, the one-electron density function has the form

$$\rho_1(\mathbf{x}_1) = P_1^\alpha(\mathbf{r}_1)\alpha^2(s_1) + P_1^\beta(\mathbf{r}_1)\beta^2(s_1) \qquad (19.5)$$

where the Ps are spatial functions. There are no cross terms involving both α and β. In orbital models, the Ps are just sums over the squares of the occupied orbitals. The two-electron density function is also given by

$$\rho_2(\mathbf{x}_1, \mathbf{x}_2) = P_2^{\alpha\alpha}(\mathbf{r}_1, \mathbf{r}_2)\alpha^2(s_1)\alpha^2(s_2) + P_2^{\alpha\beta}(\mathbf{r}_1, \mathbf{r}_2)\alpha^2(s_1)\beta^2(s_2)$$
$$+ P_2^{\beta\alpha}(\mathbf{r}_1, \mathbf{r}_2)\beta^2(s_1)\alpha^2(s_2) + P_2^{\beta\beta}(\mathbf{r}_1, \mathbf{r}_2)\beta^2(s_1)\beta^2(s_2) \qquad (19.6)$$

19.1.1 Fermi correlation

The two-electron functions tell us how the two electrons are correlated. In the one-determinant HF model, we find that the Ps are related as follows

$$P_2^{\alpha\alpha}(\mathbf{r}_1, \mathbf{r}_2) = P_1^{\alpha}(\mathbf{r}_1)P_1^{\alpha}(\mathbf{r}_2) - X(\mathbf{r}_1, \mathbf{r}_2)$$
$$P_2^{\alpha\beta}(\mathbf{r}_1, \mathbf{r}_2) = P_1^{\alpha}(\mathbf{r}_1)P_1^{\beta}(\mathbf{r}_2)$$

(19.7)

where X is a function whose form need not concern us, except to note that it exactly cancels the first term when $\mathbf{r}_1 = \mathbf{r}_2$. From these results we get a clear picture of the electron correlation shown by standard closed-shell HF theory. The form of $P_2^{\alpha\beta}$ shows that there is no correlation between electrons of opposite spin, since the simultaneous probability is just the product of the individual ones. This is a defect of HF theory. Electrons of like spin are clearly correlated and they are never found at the same point in space, and HF theory is satisfactory here. This type of correlation arises from antisymmetry and applies to all fermions. It is usually called *Fermi correlation*.

19.2 Configuration Interaction

I have mentioned configuration interaction (CI) at various points in the book, in the dihydrogen discussion of Chapter 15 and in the CIS treatment of excited states in Chapter 17. The idea of modern CI is to start with a reference wavefunction that could be a HF–LCAO closed shell wavefunction, and systematically replace the occupied spinorbitals with virtual ones. So if A, B, ... represent occupied orbitals and X, Y, ... represent virtual ones we would seek to write

$$\Psi^{CI} = \Psi^{HF} + \sum_{A,X} c_A^X \Psi_A^X + \sum_{A,B,X,Y} c_{AB}^{XY} \Psi_{AB}^{XY} + \cdots$$

(19.8)

The expansion coefficients can be found by solving the variational problem, in the usual way. This involves finding the eigenvalues and eigenvectors of the Hamiltonian matrix; usually one is only interested in a few of the electronic states, and methods are available for finding just a small number of eigenvalues and eigenvectors of very large matrices.

In a complete CI calculation we would include every possible excited configuration formed by promoting electrons from the occupied spinorbitals to the virtual ones; for a closed shell singlet state molecule with m electron pairs and n basis functions there are

$$\frac{n!(n+1)!}{m!(m+1)!(n-m)!(n-m+1)!}$$

possible Slater determinants. If we consider a HF/6–311G** wavefunction for benzene, we have $m = 21$ and $n = 144$, giving approximately 5×10^{50} possibilities. We therefore must truncate the CI expansion.

In the CIS model we just take singly excited states. If the reference wavefunction is a closed-shell HF one, then the single excitations don't mix with the ground state and CIS tends to be used as a low-level method for studying excited states.

The next logical step is to include the double excitations. If this is done together with the single excitations, then we have the CISD (*CI singles and doubles*) model. If the double excitations alone are included, then we have CID (*CI doubles*). Such truncations can lead to a problem called *size consistency* that I can illustrate by considering a very simple problem, that of two neon atoms. Table 19.1 shows a CISD/6–311G* calculation for dineon at 5000 pm separation, together with the free atom energy.

Table 19.1   Dineon calculations, 6–311G* basis set

Method	Dineon energy, E_h at 5000 pm	Atom energy, E_h
HF–LCAO	−257.0451061	−128.52255305
CISD	−257.4466147	−128.7283956

The HF–LCAO energy of a pair of neon atoms at large separation is exactly twice the free atom value, but this is not the case for the CISD calculation. If we have an ensemble of n particles and their energy is n times the energy of a single particle, then we say that the theory *scales correctly* (or that the method is *size consistent*). Full CI calculations scale correctly, but truncated CI expansions do not.

After double excitations, quadruple excitations are next in order of importance. If singles, doubles, triples and quadruples are included, then the acronym becomes CISDTQ.

19.3 The Coupled Cluster Method

The coupled cluster (CC) method was first used by physicists studying nuclear structure. R. J. Bartlett's review [97] is fairly recent. The fundamental equation relates a HF wavefunction Ψ^{HF} to the best possible wavefunction Ψ by

$$\Psi = \exp(\hat{T})\Psi^{HF} \tag{19.9}$$

The exponential operator is defined by a Taylor-series expansion

$$\exp(\hat{T}) = \sum_{k=0}^{\infty} \frac{\hat{T}^k}{k!} \tag{19.10}$$

and the *cluster operator* is defined as

$$\hat{T} = \hat{T}_1 + \hat{T}_2 + \cdots + \hat{T}_n \tag{19.11}$$

where n is the number of electrons in the molecule. The operators have the effect of replacing occupied spin orbitals in Ψ^{HF} with virtual spin orbitals. $\hat{T}_1$ performs all singly excited substitutions; $\hat{T}_2$ performs all doubly excited configurations and so on until all n electrons have been promoted from filled to virtual spinorbitals. The effect of the exponential operator on the HF wavefunction is to express it as a linear combination that contains Ψ^{HF} and all possible excitations of electrons from occupied to virtual spinorbitals. We write

$$\hat{T}_1 \Psi^{HF} = \sum_{A,X} c_A^X \Psi_A^X$$

$$\hat{T}_2 \Psi^{HF} = \sum_{A,B,X,Y} c_{AB}^{XY} \Psi_{AB}^{XY}$$

and so on, and the aim of the theory is to determine the numerical coefficients. It is usual to approximate the $\hat{T}$ operator by including only some of the terms, and it is generally accepted that the most important contribution is $\hat{T}_2$. This gives the coupled cluster doubles (CCD) method. Since we take

$$\exp\left(\hat{T}\right) \approx \exp\left(\hat{T}_2\right) = 1 + \hat{T}_2 + \frac{1}{2!}\hat{T}_2^2 + \cdots \tag{19.12}$$

the resulting wavefunction contains the HF wavefunction together with double quadruple, hextuple, etc. substitutions. The CCD method is size consistent.

The next step is to include the T_1 operator and so take

$$\exp\left(\hat{T}\right) \approx \exp\left(\hat{T}_1 + \hat{T}_2\right) = 1 + (\hat{T}_1 + \hat{T}_2) + \frac{1}{2!}(\hat{T}_1 + \hat{T}_2)^2 + \cdots \tag{19.13}$$

which gives CCSD. Inclusion of T_3 gives CCSDT in an obvious notation.

Pople and co-workers [98] have developed a quadratic CI method, which lies between CI and CC. This QCI method exists in size-consistent forms QCISD and QCISDT.

19.4 Møller–Plesset Perturbation Theory

The HF–LCAO model is thought to be satisfactory for treating many molecular properties near minima on the molecular potential energy surface. It cannot generally treat covalent bond-breaking and bond-making, but forms a useful starting-point for the more advanced theories. The HF model averages over electron repulsions; each electron experiences an average field due to the remaining electrons plus the field due to the nuclei. In real life there are 'instantaneous' electron repulsions that go as

$$\frac{e^2}{4\pi\epsilon_0} \frac{1}{r_{ij}}$$

A great advantage of the HF–LCAO model is that the equations can be solved to arbitrary precision, once a basis set has been defined. We therefore have a 'zero-order' problem with an exact solution, together with a perturbation. We discussed perturbation theory in Chapter 14 when dealing with the helium atom; the idea is to write the true Hamiltonian as that for a soluble, zero-order problem plus a perturbation

$$\hat{H} = \hat{H}^{(0)} + \lambda \hat{H}^{(1)}$$

The difference from Chapter 14 is that we now use a HF–LCAO wavefunction as the zero-order wavefunction and treat the difference between the true electron–electron repulsions and the averaged ones from HF theory as the perturbation. This gives us the Møller–Plesset perturbation theory. The method had been in the literature since 1934, but its potential for molecular calculations was not appreciated until the 1970s. It is of immense historical interest because it gave us the first realistic route for a treatment of small-molecule correlation energy, and you ought to read the original paper. The synopsis is reproduced below (see Møller and Plesset [99]:

> *A Perturbation Theory is developed for treating a system of n electrons in which the Hartree Fock solution appears as the zero-order approximation. It is shown by this development that the first order correction for the energy and the charge density of the system is zero. The expression for the second order correction for the energy greatly simplifies because of the special property of the zero order solution. It is pointed out that the development of the higher order approximation involves only calculations based on a definite one-body problem.*

The HF model averages over electron repulsions, and the HF pseudo-one-electron operator for each electron has the form (from Chapters 14 and 16)

$$\hat{h}^{\mathrm{F}}(\mathbf{r}_i) = \hat{h}^{(1)}(\mathbf{r}_i) + \hat{J}(\mathbf{r}_i) - \tfrac{1}{2}\hat{K}(\mathbf{r}_i)$$

The unperturbed Hamiltonian is taken as the sum of the HF operators for each of the n electrons

$$\hat{H}^{(0)} = \sum_{i=1}^{n} \hat{h}^{\mathrm{F}}(\mathbf{r}_i)$$

$$= \sum_{i=1}^{n} \left(\hat{h}^{(1)}(\mathbf{r}_i) + \hat{J}(\mathbf{r}_i) - \tfrac{1}{2}\hat{K}(\mathbf{r}_i) \right) \tag{19.14}$$

whilst the true Hamiltonian makes no reference to averaging

$$\hat{H} = \sum_{i=1}^{n} \hat{h}^{(1)}(\mathbf{r}_i) + \frac{e^2}{4\pi\epsilon_0} \sum_{i=1}^{n-1} \sum_{j=i+1}^{n} \frac{1}{r_{ij}}$$

The first few key equations of the perturbation expansion, taken from Chapter 14 but simplified for our present discussion, are shown below. The electronic state of interest is the ground state, denoted $\Psi^{(0)}$, with energy $\varepsilon^{(0)}$. I have therefore dropped the (i) subscripts, since we are dealing with just the one electronic state. Promoting electrons from the occupied to the virtual HF spin orbitals gives the excited states, and we consider single, double, triple, etc. excitations in the usual fashion of CI

$$\hat{H}^{(0)}\Psi^{(0)} = \varepsilon^{(0)}\Psi^{(0)}$$
$$(\hat{H}^{(0)} - \varepsilon^{(0)})\Psi^{(1)} = (\varepsilon^{(1)} - \hat{H}^{(1)})\Psi^{(0)} \tag{19.15}$$
$$(\hat{H}^{(0)} - \varepsilon^{(0)})\Psi^{(2)} = \varepsilon^{(2)}\Psi^{(0)} + \varepsilon^{(1)}\Psi^{(1)} - \hat{H}^{(1)}\Psi^{(1)}$$

Since the zero-order Hamiltonian is a sum of HF operators, the zero-order energy is a sum of orbital energies

$$\varepsilon^{(0)} = \sum_i \varepsilon_i \tag{19.16}$$

where the sum runs over the occupied spinorbitals. The first-order energy is

$$\varepsilon^{(1)} = \int \Psi^{(0)}\hat{H}^{(1)}\Psi^{(0)}\,d\tau \tag{19.17}$$

and adding $\varepsilon^{(1)}$ to $\varepsilon^{(0)}$ gives the full HF energy ε^{HF}. We therefore need to progress to second-order theory in order to give any treatment of electron correlation. The levels of theory are denoted MP2, MP3, ..., MPn, where n is the order of perturbation theory.

The second-order energy is given by

$$\varepsilon^{(2)} = \sum_j \frac{|\int \Psi^{(0)}\hat{H}^{(1)}\Psi_j\,d\tau|^2}{\varepsilon^{(0)} - \varepsilon_j} \tag{19.18}$$

and the first-order correction to the wavefunction is

$$\Psi^{(1)} = \sum_j \frac{\int \Psi^{(0)}\hat{H}^{(1)}\Psi_j\,d\tau}{\varepsilon^{(0)} - \varepsilon_j}\Psi_j \tag{19.19}$$

where the sum runs over all excited states, written Ψ_j, and energies ε_j. The ground state $\Psi^{(0)}$ is an HF wavefunction and so the integral vanishes for all singly excited states because of the Brillouin theorem. It is also zero when the excited state differs from $\Psi^{(0)}$ by more than two spin orbitals, by the Slater–Condon–Shortley rules. Hence we only need consider doubly excited states in order to find the second-order energy. The $\varepsilon^{(2)}$ numerator is non-negative since it is the square of a modulus. The denominator is negative because $\Psi^{(0)}$ refers to the ground state and Ψ_j to the excited states. The second-order energy correction is therefore always negative or zero. Higher orders of perturbation theory may give corrections of either sign.

The MP1 energy therefore is identical to the HF energy and MP2 is the simplest practicable perturbation treatment for electron correlation. It includes only the effect of double excitations.

By similar considerations, the third, fourth and higher orders of perturbation theory can be determined. The terms rapidly become algebraically complicated and the higher orders are increasingly costly to apply. For n basis functions, HF–LCAO theory scales as n^4, MP2 as n^5, MP3 as n^6 and MP4 as n^7. These should be seen as theoretical upper bounds, since sophisticated use of symmetry and integral cut-offs mean that practical calculations need considerably less resource.

A great simplifying feature of MP2 is that a full four-index transformation is not necessary, all we have to do is to semi-transform the two-electron integrals and this leads to an immense saving in time compared with conventional CI treatments. MP3 also includes only double excitations, whilst MP4 includes a description of single excitations together with some triples and quadruples. The triple contributions in MP4 are the most expensive. In the MP4(SDQ) variation we just include the least expensive singles, doubles and quadruple excitations.

Analytical energy gradients have been determined for MP2, MP3 and MP4, which makes for very effective geometry searching. I can illustrate some points of interest by considering two examples. First, the geometry optimization of ethene shown in Table 19.2.

I ran all calculations using GAUSSIAN98W on my office PC taking the same starting point on the molecular potential energy surface. The HF/STO-3G calculation would these days be regarded as woefully inadequate, and professionals would tend to have faith in the HF/6–311G** calculation. The two 6–311G** HF energies are a little different because the geometries are different. The timings show that there is little to be lost in performing the HF calculation with a decent basis set compared with STO-3G.

There are two sources of error in the calculations: the choice of a finite basis set, and truncation of the MPn series at MP2. According to the literature, the basis set error is the larger of the two. One has to be careful to compare like with like and the results of the calculations should be compared with geometric parameters at the

Table 19.2 Geometry optimization for ethene

	HF/STO-3G	HF/6–311G**	MP2/6–311G**	MP2/cc-pVTZ
ε_{HF}/E_h	−77.0789547	−78.0547244	−78.0539661	−78.0638977
ε_{MP2}/E_h			−78.3442916	−78.3993063
ε_{corr}/E_h			0.2903255	0.3354086
R_{CC} (pm)	130.6	131.7	133.7	133.2
R_{CH} (pm)	108.2	107.6	108.5	108.0
HCH (deg)	115.7	116.7	117.2	117.3
cpu time (s)	28	41	70	881

bottom of the potential energy well. Experimental values very often refer to the lowest vibrational energy level. The estimate of the correlation energy depends markedly on the quality of the basis set. Electron correlation studies demand basis sets that are capable of high accuracy, and a number of suitable basis sets are available in the literature. One popular family is the *correlation consistent* basis sets of Dunning, which go by acronyms such as cc-pV6Z. This contains seven s-type, six p-type, four d-type, two g-type and one h-type primitives.

We usually find that MP2 calculations give a small but significant correction to the corresponding HF–LCAO geometries. It is probably more worthwhile using MPn models to investigate effects that depend crucially on electron correlation; bond-breaking and bond-making phenomena spring to mind. Figure 19.1 shows representative potential energy curves for dihydrogen, following the discussion of Chapter 15.

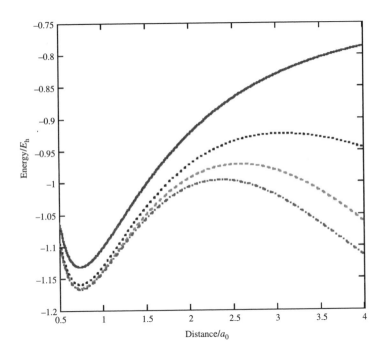

Figure 19.1 Dihydrogen electron correlation

A 6–311G** basis set was used. The top curve is the HF–LCAO calculation, and shows the usual bad dissociation limit. The remaining three curves are (in descending order) MP2, MP3 and MP4(SDQ). The first point to note is that MPn calculations are size consistent. The second point is that none of the curves appears to be approaching the correct dissociation limit of $-1E_h$. It is reported in the literature that MPn calculations do not work particularly well for geometries far from equilibrium. The MPn series has not been proved to converge to the true energy, but calculations using very high orders (MP48) have suggested that convergence will eventually occur.

Finally, MP*n* calculations need an HF–LCAO wavefunction to act as the zero-order approximation. It follows that MP*n* can only be used for those molecules and electronic states where it is possible to find such a wavefunction. This rules out most excited states.

19.5 Multiconfiguration SCF

One problem with traditional CI calculations is that excitations are to the virtual orbitals; the occupied orbitals usually are determined variationally by the HF–LCAO procedure whilst the virtual orbitals give poor representations of excited states. One solution to the problem is to write the ground state wavefunction as a linear combination of Slater determinants and then use the variation procedure to optimize both the linear expansion coefficients and the HF–LCAO coefficients. That is, we write

$$\Psi = \Psi^{HF} + \sum c_j \Psi_j$$

where Ψ^{HF} is the Hartree–Fock wavefunction and Ψ_j an excited state, in the usual sense of CI calculations. In certain cases the ratio of some of the c_j will be fixed by symmetry. The MCSCF model is regarded as highly specialized, with a small number of adherents. A literature search revealed an average of 50 papers per annum in primary journals over the last 10 years.

20 Density Functional Theory and the Kohn–Sham LCAO Equations

The 1998 Nobel Prize for chemistry was awarded jointly to John A. Pople and Walter Kohn, and the texts of their addresses on the occasion of the awards are reported in *Reviews of Modern Physics* **71** (1999) pp. 1253 and 1267. John Pople's name should be familiar to you if you have assiduously studied this text. Walter Kohn opened his address with the following remarks:

> *The citation for my share of the 1998 Nobel Prize in Chemistry refers to the 'Development of Density Functional Theory'. The initial work on Density Functional Theory (DFT) was reported in two publications: the first with Pierre Hohenberg (P Hohenberg and W Kohn, Phys Rev **136** (1964) B864) and the next with Lu J Sham (W Kohn and L J Sham, Phys Rev **140** (1965) A1133). This was almost 40 years after E Schrödinger (1926) published his first epoch-making paper marking the beginning of wave mechanics. The Thomas–Fermi theory, the most rudimentary form of DFT, was put forward shortly afterwards (E Fermi, Att1 Accad Naz Lincei, Cl Sci Fis Mat Nat Rend **6** (1927) 602, L H Thomas, Proc Camb Phil Soc **23** (1927) 542) and received only modest attention.*

It is worth spending a little time on the historical aspects before launching into modern density functional theory (DFT), and the *exchange potential* plays a key role in our discussion. If we consider a closed shell system with electron configuration

$$(\psi_1)^2 (\psi_2)^2 \cdots (\psi_M)^2$$

then in Hartree's original theory we would write the total wavefunction as a simple orbital product for which the electronic energy is

$$\varepsilon_{el} = 2 \sum_{R=1}^{M} \int \psi_R(\mathbf{r}_1) \hat{h}(\mathbf{r}_1) \psi_R(\mathbf{r}_1)\, d\tau_1$$

$$+ \sum_{R=1}^{M} \sum_{S=1}^{M} 2 \iint \psi_R(\mathbf{r}_1) \psi_R(\mathbf{r}_1) \hat{g}(\mathbf{r}_1, \mathbf{r}_2) \psi_S(\mathbf{r}_2) \psi_S(\mathbf{r}_2)\, d\tau_1 d\tau_2$$

In Hartree–Fock (HF) theory we write the wavefunction as a Slater determinant and this gives an additional exchange term in the electronic energy

$$\varepsilon_{el} = 2\sum_{R=1}^{M} \int \psi_R(\mathbf{r}_1)\hat{h}(\mathbf{r}_1)\psi_R(\mathbf{r}_1)\,d\tau_1$$

$$+ \sum_{R=1}^{M}\sum_{S=1}^{M} 2 \iint \psi_R(\mathbf{r}_1)\psi_R(\mathbf{r}_1)\hat{g}(\mathbf{r}_1,\mathbf{r}_2)\psi_S(\mathbf{r}_2)\psi_S(\mathbf{r}_2)\,d\tau_1 d\tau_2$$

$$- \sum_{R=1}^{M}\sum_{S=1}^{M} \iint \psi_R(\mathbf{r}_1)\psi_S(\mathbf{r}_1)\hat{g}(\mathbf{r}_1,\mathbf{r}_2)\psi_R(\mathbf{r}_2)\psi_S(\mathbf{r}_2)\,d\tau_1 d\tau_2$$

These expressions are general to Hartree and HF theory in that they don't depend on the LCAO approximation. You should also be familiar with the LCAO versions, where the energy expressions are written in terms of the 'charges and bond orders' matrix $\mathbf{P}$ together with the matrices $\mathbf{h}^{(1)}$, $\mathbf{J}$ and $\mathbf{K}$

$$\varepsilon_{el,H} = \text{tr}(\mathbf{Ph}_1) + \tfrac{1}{2}\text{tr}(\mathbf{PJ})$$

for the Hartree model and

$$\varepsilon_{el,HF} = \text{tr}(\mathbf{Ph}_1) + \tfrac{1}{2}\text{tr}(\mathbf{PJ}) - \tfrac{1}{4}\text{tr}(\mathbf{PK})$$

for the HF version. The $\mathbf{J}$ and $\mathbf{K}$ matrices depend on $\mathbf{P}$ in a complicated way, but the Hartree and HF electronic energies are completely determined from knowledge of the electron density $\mathbf{P}$.

20.1 The Thomas–Fermi and $X\alpha$ Models

There is nothing sinister in the exchange term; it arises because of the fermion nature of electrons. Nevertheless, it caused considerable confusion among early workers in the field of molecular structure theory and much effort was spent in finding effective model potentials that could mimic electron exchange. In the meantime, solid-state physics had been developing along quite different lines. I discussed the earliest models of metallic conductors in Chapter 12. The Pauli model is the simplest one to take account of the quantum mechanical nature of the electrons; the electrons exist in a three-dimensional infinite potential well, the wavefunction obeys the Pauli Principle and at 0 K the electrons occupy all orbitals having energy less than or equal to energy ε_F (which of course defines the Fermi energy). The number N of conduction electrons can be related to ε_F and we find

$$\varepsilon_F = \frac{h^2}{8m_e}\left(\frac{3N}{\pi L^3}\right)^{2/3} \tag{20.1}$$

Now N/L^3 is the number density of conduction electrons and so Pauli's model gives a simple relationship between the Fermi energy and the number density of electrons. Physicists very often use the symbol n for the number density; it can vary at points in space and so we write $n(\mathbf{r})$ but in this simple model, the number density is constant throughout the dimensions of the metallic box.

$$n = \frac{N}{L^3}$$
$$= \frac{8\pi}{3h^2}(2m_e)^{3/2}(\varepsilon_F)^{3/2} \tag{20.2}$$

We now switch on an *external potential* $U_{ext}(\mathbf{r})$ that is slowly varying over the dimensions of the metallic conductor, making the number density inhomogeneous. This could be (for example) due to the set of metallic cations in such a conductor, or due to the nuclei in a molecule. A little analysis suggests that the number density at position $\mathbf{r}$ should be written

$$n(\mathbf{r}) = \frac{8\pi}{3h^2}(2m_e)^{3/2}(\varepsilon_F - U_{ext}(\mathbf{r}))^{3/2} \tag{20.3}$$

This *Thomas–Fermi relation* relates the number density at points in space to the potential at points in space. The number density of electrons at a point in space is just the charge density $P_1(\mathbf{r})$ discussed in several previous chapters and so we can write it in more familiar chemical language

$$P_1(\mathbf{r}) = \frac{8\pi}{3h^2}(2m_e)^{3/2}(\varepsilon_F - U_{ext}(\mathbf{r}))^{3/2} \tag{20.4}$$

Thomas and Fermi suggested that such a statistical treatment would be appropriate for molecular systems where the number of electrons is 'large' and in the case of a molecule we formally identify the external potential as the electrostatic potential generated by the nuclei. The Thomas–Fermi approach replaces the problem of calculating an N-electron wavefunction by that of calculating the electron density in three-dimensional position space.

Paul Dirac [100] studied the effects of exchange interactions on the Thomas–Fermi model, and discovered that these could be modelled by an extra term

$$U_X(\mathbf{r}) = C(P_1(\mathbf{r}))^{1/3} \tag{20.5}$$

where C is a constant. R. Gáspár is also credited with this result, which J. C. Slater rediscovered in 1951 [101] but with a slightly different numerical coefficient of $\frac{2}{3}C$. The disagreement between Dirac's and Slater's numerical coefficient seems to have been first resolved by Gáspár [102] and authors began to write the *local exchange potential* as

$$U_{X\alpha}(\mathbf{r}) = C\alpha(P_1(\mathbf{r}))^{1/3} \tag{20.6}$$

where α could take values between $\frac{2}{3}$ and 1 and this is the so-called $X\alpha$ *model*. The use of the symbol α here is not to be confused with the use of the same symbol for a spin eigenfunction.

Slater then had the ingenious idea of writing the atomic HF eigenvalue equation

$$\hat{h}^F(\mathbf{r})\psi(\mathbf{r}) = \varepsilon\psi(\mathbf{r})$$
$$(\hat{h}^{(1)}(\mathbf{r}) + \hat{J}(\mathbf{r}) - \tfrac{1}{2}\hat{K}(\mathbf{r}))\psi(\mathbf{r}) = \varepsilon\psi(\mathbf{r})$$

as

$$(\hat{h}^{(1)}(\mathbf{r}) + \hat{J}(\mathbf{r}) + \hat{U}_{X\alpha}(\mathbf{r}))\psi(\mathbf{r}) = \varepsilon\psi(\mathbf{r}) \qquad (20.7)$$

and such calculations are usually referred to as *atomic $X\alpha$–HF*. The resulting $X\alpha$ orbitals differ from conventional HF orbitals in one major way, namely that Koopmans' theorem is no longer valid for every orbital and so the orbital energies cannot generally be used to estimate ionization energies. Koopmans' theorem now applies only to the highest occupied orbital. A key difference between standard HF theory and density functional calculations is the way we conceive the *occupation number ν* of each orbital. In molecular HF theory, the occupation number is 2, 1 or 0 depending on whether a given spatial orbital is fully occupied by two electrons (one of either spin), singly occupied, or a virtual orbital. For a system comprising very many electrons we focus on the statistical occupation of each orbital and the occupation number becomes a continuous variable having a value between 0 and 2. The relationship between the electronic energy ε_{el} and the occupation number of orbital i is

$$\frac{\partial \varepsilon_{el}}{\partial \nu_i} = \varepsilon_i \qquad (20.8)$$

so that the ionization energy from a particular orbital ψ_i is given by

$$\varepsilon_{el}(\nu_i = 0) - \varepsilon_{el}(\nu_i = 1) = \int_1^0 \varepsilon_i d\nu_i$$

Note, however, that the energy is determined by the charge density $P(\mathbf{r})$. When the calculations are performed, the resulting orbitals closely resemble those from standard HF theory and people use them in much the same way.

20.2 The Hohenberg–Kohn Theorems

The Thomas–Fermi and $X\alpha$ approaches were constructed as approximations to the quantum mechanical problem of calculating the electronic properties of a system of interest. The density functional theory of Hohenberg and Kohn, to be discussed, is in principle an exact theory. The first keynote paper contains two fundamental theorems.

The first theorem is what mathematicians call an *existence theorem*; it proves that the ground state energy of any electronic system is determined by a *functional* of the electron density. The theorem means that in principle we only need to know the electron density in three-dimensional space and not the full wavefunction in order to calculate any ground-state property including the energy.

The term *functional* is mathematical jargon; a function is a mapping from one set of numbers to another (see Figure 20.1), whilst a functional is a mapping from a set of functions to a set of numbers.

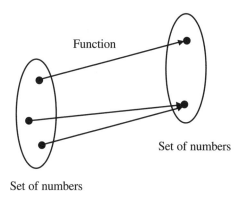

Function

Set of numbers

Set of numbers

Figure 20.1   A function

The proof of the theorem is quite simple, but will not be reproduced here. The paper is so important that you should see at least the Abstract. Hohenberg and Kohn [103] use square brackets [. . .] to denote functionals:

This paper deals with the ground state of an interacting electron gas in an external potential $v(\mathbf{r})$. It is proved that there exists a universal functional of the density, $F[n(\mathbf{r})]$, independent of $v(\mathbf{r})$, such that the expression $E \equiv \int v(\mathbf{r}) n(\mathbf{r}) d\mathbf{r} + F[n(\mathbf{r})]$ has as its minimum value the correct ground state energy associated with $v(\mathbf{r})$. The functional $F[n(\mathbf{r})]$ is then discussed for two situations: (1) $n(\mathbf{r}) = n_0 + \tilde{n}(\mathbf{r})$, $\tilde{n}/n_0 \gg 1$, and (2) $n(\mathbf{r}) = \varphi(\mathbf{r}/r_0)$ with φ arbitrary and $r_0 \to \infty$. In both cases F can be expressed entirely in terms of the correlation energy and linear and higher order electronic polarizabilities of a uniform electron gas. This approach also sheds some light on generalized Thomas–Fermi methods and their limitations. Some new extensions of these methods are presented.

The second theorem gives a variation principle for the density functionals; it states (in chemical language) that

$$\varepsilon_{\mathrm{el}}[P(\mathbf{r})] \geq \varepsilon_{\mathrm{el}}[P_0(\mathbf{r})] \tag{20.9}$$

where P_0 is the true density for the system and P any other density obeying

$$\int P(\mathbf{r})\, \mathrm{d}\tau = \int P_0(\mathbf{r})\, \mathrm{d}\tau = N \tag{20.10}$$

where N is the number of electrons. The difficulty is that the Hohenberg–Kohn theorems give us no clue as to the nature of the density functional, nor how to set about finding it.

20.3 The Kohn–Sham (KS–LCAO) Equations

The Kohn–Sham paper [104] gives us a practical solution to the problem, based on HF–LCAO theory; again, it is such an important paper that you should have sight of the Abstract that in any case is self-explanatory:

From a theory of Hohenberg and Kohn, approximation methods for treating an inhomogeneous system of interacting electrons are developed. These methods are exact for systems of slowly varying or high density. For the ground state, they lead to self-consistent equations analogous to the Hartree and Hartree–Fock equations, respectively. In these equations the exchange and correlation portions of the chemical potential of a uniform electron gas appears as additional effective potentials. (The exchange portion of our effective potential differs from the due to Slater by a factor of 2/3). Electronic systems at finite temperatures and in magnetic fields are also treated by similar methods. An appendix deals with a further correction for systems with short-wavelength density oscillations.

The Kohn–Sham equations are modifications of the standard HF equations and we write

$$(\hat{h}^{(1)}(\mathbf{r}) + \hat{J}(\mathbf{r}) + \hat{U}_{XC}(\mathbf{r}))\psi(\mathbf{r}) = \varepsilon\psi(\mathbf{r}) \qquad (20.11)$$

where $U_{XC}(\mathbf{r})$ is the local exchange–correlation term that accounts for the exchange phenomenon and the dynamic correlation in the motions of the individual electrons.

We speak of the KS–LCAO procedure. It is usual to split $U_{XC}(\mathbf{r})$ into an exchange term and a correlation term and treat each separately

$$(\hat{h}^{(1)}(\mathbf{r}) + \hat{J}(\mathbf{r}) + \hat{U}_X(\mathbf{r}) + \hat{U}_C(\mathbf{r}))\psi(\mathbf{r}) = \varepsilon\psi(\mathbf{r}) \qquad (20.12)$$

The electronic energy is usually written in DFT applications as

$$\varepsilon_{el}[P] = \varepsilon_1[P] + \varepsilon_J[P] + \varepsilon_X[P] + \varepsilon_C[P] \qquad (20.13)$$

where the square brackets denote a functional of the one-electron density $P(\mathbf{r})$. The first term on the right-hand side gives the one-electron energy, the second term is the Coulomb contribution, the third term the exchange and the fourth term gives the correlation energy. We proceed along the usual HF–LCAO route; we choose a basis set and then all that is needed in principle is knowledge of the functional forms of $U_X(\mathbf{r})$ and $U_C(\mathbf{r})$. It is then a simple matter in principle to modify an existing standard HF–LCAO computer code to include these additional terms. Either component can be of two distinct types: *local functionals* that depend only on the electron density at a point in space, and *gradient-corrected functionals* that depend both on the electron density and its gradient at that point.

In their most general form, the exchange and correlation functionals will depend on the density of the α and the β electrons, and the gradients of these densities. Each energy term ε_C and ε_X will therefore be given by a volume integral

$$\varepsilon_X = \int f_X(P^\alpha, P^\beta, \operatorname{grad} P^\alpha, \operatorname{grad} P^\beta) \, d\tau$$

$$\varepsilon_C = \int f_C(P^\alpha, P^\beta, \operatorname{grad} P^\alpha, \operatorname{grad} P^\beta) \, d\tau$$

(20.14)

where f is an energy density.

20.4 Numerical Integration (Quadrature)

Given the functionals and the electron density, we have to integrate over the space of the molecule to obtain the energy. These extra integrals cannot usually be done analytically and so a numerical integration has to be done to each HF–LCAO cycle. What we do is to replace the integral by a sum over quadrature points, e.g.

$$\varepsilon_{X/C} = \sum_A \sum_i w_{Ai} f(P_1^\alpha, P_1^\beta, \operatorname{grad}(P_1^\alpha), \operatorname{grad}(P_1^\alpha); \mathbf{r}_{Ai})$$

(20.15)

where the first summation is over the atoms and the second is over the numerical quadrature grid points. The w_{Ai} are the quadrature weights and the grid points are given by the sum of the position of nucleus A and a suitable one-centre grid

$$\mathbf{r}_{Ai} = \mathbf{R}_A + \mathbf{r}_i$$

The keynote paper for numerical integration is that due to A. D. Becke [105], and once again the paper is so important that you should see the Abstract:

We propose a simple scheme for decomposition of molecular functions into single-center components. The problem of three-dimensional integration in molecular systems thus reduces to a sum of one-center, atomic-like integrations, which are treated using standard numerical techniques in spherical polar coordinates. The resulting method is tested on representative diatomic and polyatomic systems for which we obtain five- or six-figure accuracy using a few thousand integration points per atom.

Becke's scheme rigorously separates the molecular integral into atomic contributions that may be treated by standard single-centre techniques. Numerical integration grids usually are specified as a specific number of radial shells around each atom, each of which contains a set number of integration points. For example, a GAUSSIAN98 use a grid designated (75 302) which has 75 radial shells per atom, each containing 302 points giving a total of 22 650 integration points.

Over the years, a large number of exchange and correlation functionals have been proposed. Software packages such as GAUSSIAN98 and HyperChem7 offer a selection.

20.5 Practical Details

There are two versions of DFT, one for closed shell systems analogous to the Roothaan closed shell treatment, and one for the open shell case analogous to the UHF technique and we implement the Kohn–Sham LCAO (KS–LCAO) equations by including the relevant exchange/correlation term(s) instead of the traditional HF exchange term in the HF–LCAO equations. That is, for the UHF case we write

$$(\hat{h}^{(1)}(\mathbf{r}) + \hat{J}(\mathbf{r}) + \hat{U}_X^\alpha(\mathbf{r}) + \hat{U}_C^\alpha(\mathbf{r}))\psi^\alpha(\mathbf{r}) = \varepsilon^\alpha \psi^\alpha(\mathbf{r})$$

for the α-spin electrons and a similar expression for the β-spin. The elements of the α-spin X/C matrices are related to the energy density f and the basis functions χ by

$$(U_{X/C}^\alpha)_{i,j} = \int \left\{ \frac{\partial f}{\partial P^\alpha} \chi_i \chi_j + \left[2 \frac{\partial f}{\partial \gamma^{\alpha\alpha}} \mathrm{grad}\, P^\alpha + \frac{\partial f}{\partial \gamma^{\alpha\beta}} \mathrm{grad}\, P^\beta \right] \cdot \mathrm{grad}(\chi_i \chi_j) \right\} d\tau$$

$$(20.16)$$

where I have written

$$\gamma^{\alpha\beta} = \mathrm{grad}(P^\alpha)\mathrm{grad}(P^\beta)$$

The β-spin electrons have a similar matrix.

So, for example, the Slater–Xα exchange density is

$$f(\mathbf{r}) = -\frac{9}{4}\alpha^* \left(\frac{3}{4\pi}\right)^{1/3} ((P^\alpha(\mathbf{r}))^{4/3} + (P^\beta(\mathbf{r}))^{4/3})$$

$$(20.17)$$

where the α^* is the $X\alpha$ parameter, not to be confused with the spin variable. The derivatives are simple

$$\frac{\partial f}{\partial P_1^\alpha} = -3\alpha^* \left(\frac{3}{4\pi}\right)^{1/3} (P^\alpha)^{1/3} \tag{20.18}$$

with a similar expression for the β-spin term.

By itself, the Slater–$X\alpha$ local term is not adequate for describing molecular systems. Becke [106] formulated the following gradient-corrected exchange functional

$$f = (P^\alpha)^{4/3} g(x^\alpha) + (P^\beta)^{4/3} g(x^\beta)$$

where

$$g(x) = -\frac{3}{2}\left(\frac{3}{4\pi}\right)^{1/3} - \frac{bx^2}{1 + 6bx \sinh^{-1} x} \tag{20.19}$$

$$x^\alpha = \frac{\gamma_{aa}}{(P^\alpha)^{4/3}} \quad \text{and} \quad x^\beta = \frac{\gamma_{\beta\beta}}{(P^\beta)^{4/3}}$$

The derivatives are again straightforward.

In a similar way there are local and gradient-corrected correlation functionals. One of the oldest correlation functionals is that due to C. Lee, W. Yang and R. G. Parr [107]

$$f = -\frac{4a}{1 + d(P)^{-1/3}}\frac{P^\alpha P^\beta}{P} - 2^{11/3}\frac{3}{10}(3\pi^2)^{2/3}ab\omega(P)P^\alpha P^\beta((P^\alpha)^{8/3} + (P^\alpha)^{8/3})$$

$$+ \frac{\partial f}{\partial \gamma_{\alpha\alpha}}\gamma_{\alpha\alpha} + \frac{\partial f}{\partial \gamma_{\alpha\beta}}\gamma_{\alpha\beta} + \frac{\partial f}{\partial \gamma_{\beta\beta}}\gamma_{\beta\beta} \tag{20.20}$$

where

$$\omega(P) = \frac{\exp(-cP^{-1/3})}{1 + dP^{-1/3}} \tag{20.20}$$

where a, b, c and d are constants.

A modern reference that contains a wealth of detail for implementing DFT calculations is that of Johnson *et al.* [108].

20.6 Custom and Hybrid Functionals

Hybrid functionals are those that contain mixtures of the HF exchange with DFT exchange correlation, whilst *ab initio* codes usually give the user a choice of constructing their own linear combination.

20.7 An Example

We worked through a typical HF–LCAO calculation on L-phenylanine using the 6–31G* basis set in Chapter 17. The following output relates to a corresponding KS–LCAO single point calculation. First the route:

```
---------------------------------------------------
# B3LYP/6–31G* SCF = Direct #T
---------------------------------------------------
1/38 = 1/1;
2/17 = 6, 18 = 5/2;
3/5 = 1, 6 = 6, 7 = 1, 11 = 2, 25 = 1, 30 = 1/1, 2, 3;
4//1;
5/5 = 2, 32 = 1, 38 = 4, 42 = −5/2;
6/7 = 3, 28 = 1/1;
99/5 = 1, 9 = 1/99;
---------------------------------------------------
L-phenylanine DFT single point
---------------------------------------------------
```

I chose the Becke three-parameter exchange and the Lee–Yang–Parr correlation functionals. The usual iterative procedure is followed; a major difference from HF–LCAO theory is that the numerical integration has to be done at the end of each cycle, although this is not explicit in the next section of output.

```
Standard basis: 6–31G(d) (6D, 7F)
    There are 202 symmetry adapted basis functions of A symmetry.
    Crude estimate of integral set expansion from redundant integrals = 1.000.
    Integral buffers will be 2 62 144 words long.
    Raffenetti 2 integral format.
    Two-electron integral symmetry is turned on.
        202 basis functions     380 primitive gaussians
        44 alpha electrons      44 beta electrons
            nuclear repulsion energy     695.8409407052 Hartrees.
    One-electron integrals computed using PRISM.
    NBasis = 202 RedAO = T NBF = 202
    NBsUse = 202 1.00D-04 NBFU = 202
    Projected INDO Guess.
    Warning! Cutoffs for single-point calculations used.
    Requested convergence on RMS density matrix = 1.00D-04 within 64 cycles.
    Requested convergence on MAX density matrix = 1.00D-02.
    Requested convergence on     energy = 5.00D-05.
    SCF Done: E(RBLYP) = −554.684599482     A.U. after 7 cycles
```

Convg = 0.2136D-04 −V/T = 2.0080
S**2 = 0.0000

Next come the KS–LCAO orbital energies and LCAO coefficients (not shown here):

```
Alpha occ. eigenvalues  --   −19.21906  −19.15371  −14.31807  −10.32931  −10.23197
Alpha occ. eigenvalues  --   −10.18919  −10.18384  −10.17994  −10.17848  −10.17813
Alpha occ. eigenvalues  --   −10.17637  −10.14697   −1.15467   −1.03083   −0.90142
Alpha occ. eigenvalues  --    −0.84164   −0.78208   −0.73302   −0.72415   −0.68266
Alpha occ. eigenvalues  --    −0.62064   −0.59768   −0.57072   −0.55001   −0.53327
Alpha occ. eigenvalues  --    −0.49791   −0.48769   −0.46859   −0.46130   −0.44216
Alpha occ. eigenvalues  --    −0.42887   −0.42198   −0.40246   −0.39305   −0.37604
Alpha occ. eigenvalues  --    −0.36527   −0.34305   −0.33539   −0.32529   −0.29565
Alpha occ. eigenvalues  --    −0.26874   −0.24712   −0.23422   −0.22807
```

The LCAO orbitals tend to be used just like ordinary HF–LCAO orbitals, and authors make use of the standard Mulliken population analysis indices:

Total atomic charges:

		1
1	N	−0.692262
2	H	0.308081
3	C	−0.094081
4	H	0.310106
5	H	0.185908
6	C	0.561568
7	C	−0.338514
8	O	−0.460986
9	H	0.172216
10	H	0.150270
11	C	0.139207
12	O	−0.532463
13	C	−0.178535
14	C	−0.228343
15	H	0.413517
16	H	0.116707
17	H	0.248340
18	C	−0.135999
19	C	−0.169497
20	H	0.122653
21	H	0.110562
22	C	−0.126440
23	H	0.117986

Sum of Mulliken charges = 0.00000

Comparison with Chapter 17 shows that the Mulliken indices are quite different; there is nothing particularly sinister in this, we simply have to be careful to compare like with like when using them. Finally, the properties and timing:

```
Electronic spatial extent (au): ⟨R**2⟩ = 2204.5013
Charge = 0.0000 electrons
Dipole moment (Debye):
     X =        0.0147      Y =       1.3960      Z =       0.1895    Tot =      1.4088
Quadrupole moment (Debye-Ang):
    XX =      -72.7761     YY =      -63.9087     ZZ =      -72.8707
    XY =        1.2446     XZ =       -3.1819     YZ =       -3.2701
Octapole moment (Debye-Ang**2):
   XXX =        7.3252    YYY =       31.8738    ZZZ =       -0.0206    XYY =       1.1233
   XXY =        0.0154    XXZ =       -4.9841    XZZ =       10.8590    YZZ =       2.4692
   YYZ =        9.6757    XYZ =       -7.1274
Hexadecapole moment (Debye-Ang**3):
  XXXX =    -1977.5412   YYYY =     -555.0004   ZZZZ =    -178.6904   XXXY =    -13.4681
  XXXZ =      -51.1200   YYYX =       46.6035   YYYZ =      -2.9691   ZZZX =       0.4550
  ZZZY =        0.5732   XXYY =     -457.6858   XXZZ =    -406.2508   YYZZ =    -135.1196
  XXYZ =      -45.4948   YYXZ =       19.4003   ZZXY =       3.9922
```

Job cpu time: 0 days 0 hours 8 minutes 6.0 seconds.
File lengths (MBytes): RWF = 27 Int = 0 D2E = 0 Chk = 10 Scr = 1

The electric moments are similar to the HF values, and the increased cpu time comes about because of the numerical integrations.

20.8 Applications

DFT is a relatively new branch of chemical modelling, and the number of papers has grown enormously over the last 10 years. In many cases these papers are used to support experimental work, in others to obtain alternative theoretical information compared with that obtained with the more traditional HF–LCAO models. Michael Springborg gives a critical review of the literature every two years in his contributions to the Royal Society of Chemistry Specialist Periodical Reports Series 'Chemical Modelling'. In standard HF–LCAO theory, we can systematically improve the accuracy of our calculations by increasing the sophistication of the basis set and this option is still available in DFT theory. There is no systematic way that the form of the DFT exchange and correlation terms can be improved other than by refining the basic model, the free-electron gas. Michael's comments in Volume 1 are well worth reading [109]:

> There is, however, a fundamental difference . . . The HF approximation represents
> a first approximation to the exact solution of the exact many-body Schrödinger

equation so that one may in principle systematically improve the situation. On the other hand the Kohn–Sham equations are currently approximated and it is not obvious whether more accurate solutions yield better agreement with experiment. However, for all but the smallest systems, one can solve the Kohn–Sham equations more accurately than one can solve the Schrödinger equation. Taking everything together this suggests that both approaches have advantages and disadvantages and that the best way of developing the understanding of the properties of materials is to continue to apply both types of method and not to abandon any of them.

21 Miscellany

There are a finite number of pages in any textbook and I have obviously had to leave a number of loose ends and miss out a number of topics, such as time dependence, relativistic effects, enumeration, atoms-in-molecules and scattering that go to make up modern molecular modelling. I just haven't had the space to deal with them, and I want to finish the text by picking out four topics that interest me. I hope they will also interest you.

If you want to keep up with the literature without all the trouble of actually making the literature searches yourself, you might like to read the Royal Society of Chemistry (RSC) Specialist Periodical Reports (SPRs) on *Chemical Modelling; Applications and Theory*. These appear every two years, and experts ('Reporters', in RSC-speak) write the individual chapters. Set your web browser to http://www.rsc.org and follow the links to our SPR.

21.1 Modelling Polymers

Many modern materials are *polymers*, compounds in which chains or networks of small repeating units called *monomers* form giant molecules. The essential requirement that a small molecule should qualify as a monomer is that it should be at least bifunctional (for example hydroxyethanoic acid (I), vinyl chloride (II) adipic acid (III) and hexamethylenediamine (IV)) in order that the monomers can link together to form a chain (Figure 21.1).

Thus (I) can condense with another identical hydroxy acid molecule through the two reactive groups and the polymerization reaction in this case is a simple condensation. The double bond in (II) is bifunctional and activation by a free radical leads to polymer formation. If the monomer units are all the same, we speak of a *homopolymer* (see Figure 21.2).

Condensation of the two different monomers (III) and (IV) yields 6:6 nylon, which is an example of a *copolymer*. You have probably come across the man-made polymers polyethene, nylon and polytetrafluoroethene (PTFE); they are often named after their chemical source, but a wide variety of trade names are in common use (for example, Teflon® is PTFE). The prefix *poly* is attached to the name of a monomer in

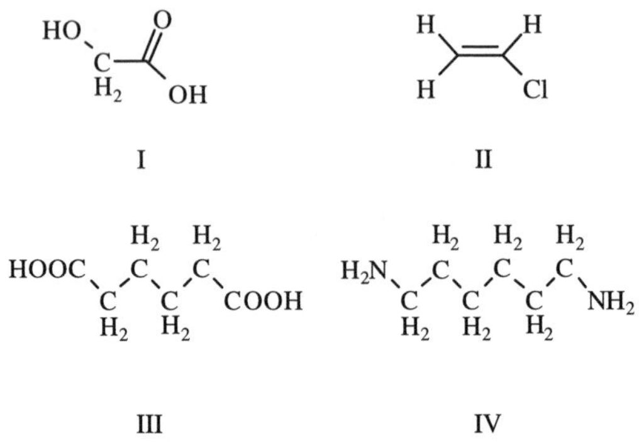

Figure 21.1　Four monomers

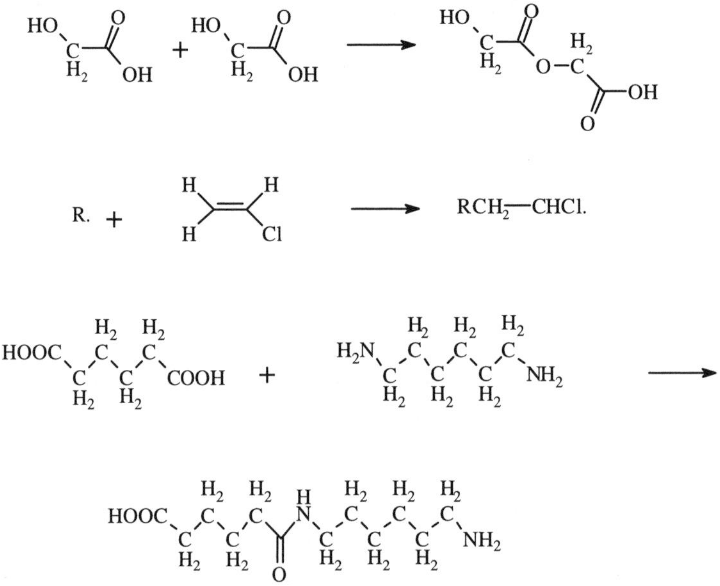

Figure 21.2　Polymer forming unit reactions

addition polymers. When the monomer has a multi-worded name, then this is enclosed in parentheses and prefixed as in poly(vinyl chloride).

There are also many naturally occurring polymers, such as natural rubber, shown in Figure 21.3. Michael Faraday established its constitution to be $(C_5H_8)_n$; it is a polymer of isoprene with a perfectly regular chain. Every fourth carbon atom in the chain carries a methyl group and the presence of the double bond in each monomer unit determines the chemical reactivity and its ability to react with sulphur in the vulcanization process (which forms cross-chains by reaction across the double bonds of different chains). The structure Gutta–Percha, the other natural polymer of isoprene, differs significantly from that of natural rubber.

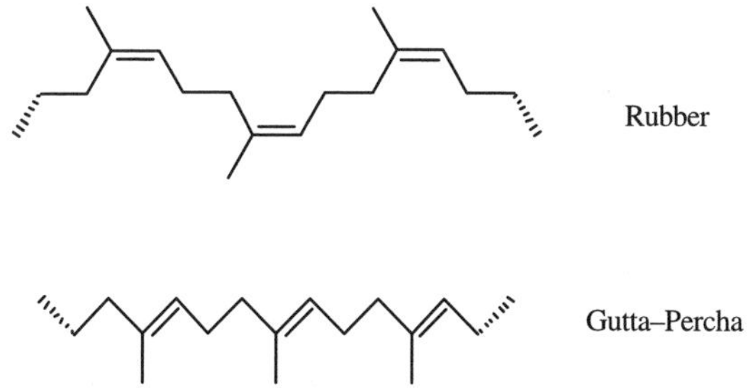

Rubber

Gutta–Percha

Figure 21.3   Rubber

One feature that distinguishes a polymer from a monomer is that it is not possible to assign an exact molar mass to a polymer sample. A given polymer chain obviously has a chemical formula and molar mass, but the length of a polymer chain is determined by random events and the product of a polymerization reaction is a mixture of chains of different lengths.

There are two experimentally determined measures of the molar mass. Colligative properties, such as the osmotic pressure and depression of freezing point, give the number average molar mass defined as

$$\langle M \rangle_n = \frac{\sum_i N_i M_i}{\sum_i N_i}$$

where N_i is the number of species with molar mass M_i. Light scattering measurements determine a different kind of average called the weight average

$$\langle M \rangle_w = \frac{\sum_i N_i M_i^2}{\sum_i N_i M_i}$$

21.2 The End-to-End Distance

The simplest place to start the discussion of polymer structure and modelling is with a long hydrocarbon chain, illustrated in Figure 21.4. One's first reaction is to draw the chain as I have done and assume that the fully stretched conformation will be the lowest energy one (as it probably will) and therefore the only one of interest.

In the early days of conformational studies, it was assumed that there was completely free rotation about every single bond (that is to say, no barrier to rotation). We now know better and for each typical subunit there will be three possible local minima, which can be interchanged by rotation about the C—C bond. There will be 3^{n-1} conformations for a chain of n carbon atoms and all will be very similar in

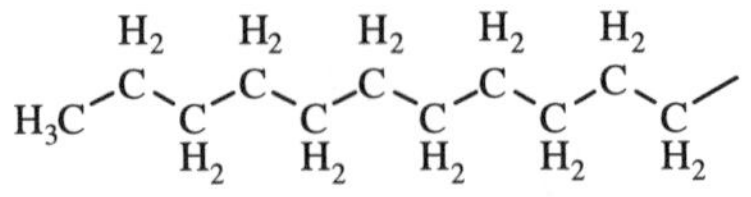

$$H_3C \overset{\overset{\displaystyle H_2}{C}}{\underset{}{}} \quad \overset{\overset{\displaystyle H_2}{C}}{\underset{\displaystyle H_2}{C}} \quad \overset{\overset{\displaystyle H_2}{C}}{\underset{\displaystyle H_2}{C}} \quad \overset{\overset{\displaystyle H_2}{C}}{\underset{\displaystyle H_2}{C}} \quad \overset{\overset{\displaystyle H_2}{C}}{\underset{\displaystyle H_2}{C}} \quad \overset{\overset{\displaystyle H_2}{C}}{\underset{\displaystyle H_2}{C}}$$

Figure 21.4   Fragment of long hydrocarbon chain

energy. In spite of the fact that the lowest energy state will be the fully extended conformation, there will be a spread amongst all the possible conformations at room temperature, determined by their Boltzmann factors. The end-to-end distance affords a rough-and-ready measure of the complexity of the conformation of a long chain, and the early theories of polymer structure focused on this quantity.

21.3 Early Models of Polymer Structure

The simple models ignore hydrogens and other side groups and just treat the heavy atoms, and workers in the field usually follow the notation shown in Figure 21.5. The heavy atoms are labelled $A_0, A_1, \ldots, A_n$, and monomer A_0 is taken as the coordinate origin. It is assumed that n is large, and the polymer will not generally be planar. The relative position vectors are as shown; $\mathbf{r}_1$ points from A_0 to A_1 and so on, and it is clear from elementary vector analysis that the position vector $\mathbf{r}$ of the end atom (A_n) is

$$\mathbf{r} = \sum_{i=1}^{n} \mathbf{r}_i \tag{21.1}$$

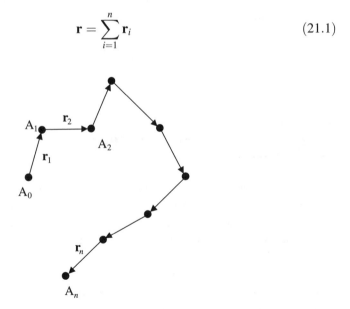

Figure 21.5   Notation for simple chain of $n + 1$ heavy atoms

The scalar r (the magnitude of $\mathbf{r}$) is the *end-to-end distance* and it is given by

$$r^2 = \mathbf{r} \cdot \mathbf{r} = \sum_{i=1}^{n} \sum_{j=1}^{n} \mathbf{r}_i \cdot \mathbf{r}_j \tag{21.2}$$

This can also be written in terms of the individual bond lengths (that is, the r_i)

$$r^2 = \sum_{i=1}^{n} r_i^2 + 2 \sum_{i=1}^{n-1} \sum_{j=i+1}^{n} \mathbf{r}_i \cdot \mathbf{r}_j \tag{21.3}$$

The number of conformations available to a long chain molecule is immense, and it is futile to try to study each and every one of them in isolation. What we do instead is to adopt the methods of statistical thermodynamics and investigate appropriate averages. We focus on the end atom A_n and ask about the probability that this atom will occupy a certain volume element $d\tau$ (which would be written $dxdydz$ in Cartesian coordinates or $r^2 \sin\theta \, d\theta \, d\phi \, dr$ in the more appropriate spherical polar coordinates) located at the arbitrary position $\mathbf{r}$. If we write this probability as $W(\mathbf{r})d\tau$, then the simplest theories draw analogies with the Maxwell–Boltzmann distribution of velocities and the theory of random errors to suggest that

- $W(\mathbf{r})$ should only depend on the scalar distance r, not on the vector $\mathbf{r}$
- $W(r)$ should have a Gaussian distribution.

The appropriate expression turns out to be

$$W(r) = \left(\frac{3}{2\pi \langle r^2 \rangle} \right)^{3/2} \exp\left(-\frac{3}{2\langle r^2 \rangle} r^2 \right) \tag{21.4}$$

where I have used the familiar convention that $\langle r^2 \rangle$ means the average of r^2 over all possible conformations. If I take two spheres of radii r and $r + dr$ centred on the coordinate origin (heavy atom A_0), then the probability of finding the end atom A_n somewhere within these two spheres is

$$4\pi r^2 W(r) dr$$

The Gaussian expression for $W(r)$ can be derived formally, assuming an infinite chain length.

With these ideas in mind, we return to the expression for the end-to-end distance

$$r^2 = \sum_{i=1}^{n} r_i^2 + 2 \sum_{i=1}^{n-1} \sum_{j=i+1}^{n} \mathbf{r}_i \cdot \mathbf{r}_j \tag{21.5}$$

If each bond between the heavy atoms in the polymer has the same length l and we take account of the statistical averaging, then we can write

$$\langle r^2 \rangle = nl^2 + 2 \sum_{i=1}^{n-1} \sum_{j=i+1}^{n} \langle \mathbf{r}_i \cdot \mathbf{r}_j \rangle \tag{21.6}$$

Over the years, many researchers have made contributions to this field and I will describe two very simple models.

21.3.1 The freely jointed chain

This model consists of n heavy atom bonds of equal length l and the angles at the bond junctions may take all values with equal probability. Rotations about the bonds are completely free in all directions, and there is no correlation between the directions of neighbouring bonds. Thus

$$\langle \mathbf{r}_i \cdot \mathbf{r}_j \rangle = 0$$

and so

$$\langle r^2 \rangle = nl^2$$

Workers in the field compare the results of their theories by calculating the *characteristic ratio*

$$C_n = \frac{\langle r^2 \rangle}{nl^2} \tag{21.7}$$

which is unity for the freely jointed chain.

21.3.2 The freely rotating chain

In this model we take n bonds of equal length l joined at fixed bond angles, as shown in Figure 21.6. Free rotation is permitted about any bond, so that every possible dihedral angle is equally likely. In modern parlance, the barrier to rotation is set to zero.

The projection of bond $i + 1$ on bond i is $l \cos \theta$ and the projection of bond $i + 1$ in a direction perpendicular to bond i averages to zero under the assumption of free rotation. This gives

$$\langle \mathbf{r}_{i+1} \cdot \mathbf{r}_i \rangle = l^2 \cos \theta$$

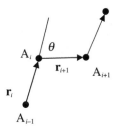

Figure 21.6 Freely rotating chain

The projection of bond $i+2$ on bond i is $(l\cos\theta)\cos\theta = l\cos^2\theta$, and the general formula is that the projection of bond $i+k$ on bond i is $l\cos^k\theta$. We therefore find

$$\langle r^2 \rangle = nl^2 + 2l^2 \sum_{i=1}^{n-1} \sum_{j=1+1}^{n} (\cos\theta)^{j-i}$$

This can be summed to give a characteristic ratio

$$C_n = \frac{1+\cos\theta}{1-\cos\theta} - \frac{2\cos\theta}{n} \frac{1-\cos^n\theta}{(1-\cos\theta)^2} \tag{21.8}$$

and so C_n for a freely rotating chain varies roughly as $1/n$. For an infinite chain of tetrahedrally bonded atoms $C_\infty = 2$. In fact the experimental ratio usually comes out to be about 7 so these simple models leave much to be desired. Janos J. Ladik reported on recent advances in the field in Volume 1 of the *Chemical Modelling Specialist Periodical Reports*.

21.4 Accurate Thermodynamic Properties; The G1, G2 and G3 Models

Quantum chemical calculations of thermodynamic data have developed beyond the level of simply reproducing experimental values, and can now make accurate predictions for molecules whose experimental data are unknown. The target is usually set as $\pm 2\,\text{kcal mol}^{-1}$ for energy quantities.

21.4.1 G1 theory

The *Gaussian-1 (G1) method* was introduced by J. A. Pople *et al.* in 1989 [110] in order to systematically investigate the shortcomings in the levels of theory known at that time. It has been recognized for many years that bond dissociation energies are poorly predicted by standard HF–LCAO calculations. The development of MPn perturbation theory marked a turning point in the treatment of molecular correlation, and made the treatment of correlation energy at last possible for molecules of moderate size (but at a high cost in computer resource). The MPn methodology usually is implemented up to MP4 level, but the convergence of the perturbation series leaves something to be desired. For this reason a further quadratic CI correction was thought desirable.

The effect of polarization functions is also important, and even f orbitals make a contribution to the total energies of first-row atoms.

There are eight distinct stages in a G1 calculation for molecules containing first- and second-row atoms; in the GAUSSIAN suite the steps are followed automatically once the G1 keyword is selected.

1. An initial structure is obtained at the HF/6–31G* level of theory.

2. The equilibrium structure is revised to the MP2/6–31G* level of theory. All electrons are included in the correlation treatment; there are no frozen cores.

3. The geometry from step 2 is now used in a number of single-point calculations starting with MP4/6–311G**. This energy is improved in four distinct stages and these four improvements are assumed to be additive.

4. As a first correction we add diffuse s and p basis functions at the MP4 level of theory. These are known to be important for molecules with lone pairs. The correction is obtained by comparing MP4/6–311 + G** and MP4/6–311G** energies.

5. As a second correction we take account of polarization functions on non-hydrogens. This correction is found by comparing MP4/6–311G(2df) and MP4/6–311G** energies.

6. The third correction allows for the inadequacies of the MP4 treatment. We make an expensive QCISD(T)/6–311G** calculation.

7. The fourth correction is to add an empirical term devised to give agreement with experiment for the hydrogen atom and dihydrogen. It is referred to as a *higher level* correction.

8. Finally, harmonic frequencies are obtained at the HF/6–31G* level of theory and scaled uniformly by a well-accepted factor of 0.8929.

Total atomization energies for a set of 31 molecules were found to agree with experimental thermochemical data to an accuracy of better than $\pm 2\,kcal\,mol^{-1}$.

Similar agreement was obtained for ionization energies, electron and proton affinities.

21.4.2 G2 theory

G1 theory was originally tested against experimental values for a range of simple first- and second-row molecules. It was observed that G1 theory did badly with ionic molecules, with triplet state molecules and with hypervalent molecules. *Gaussian-2 (G2) theory* was introduced by L. A. Curtiss *et al.* in 1991[111], and it eliminates some of the difficulties by making three modifications:

1. G2 theory eliminates the assumption of additivity of the diffuse sp and the 2df basis functions used in G1 theory. This change gives a significant improvement for ionic species and some anions.

2. It adds a third d function to the non-hydrogen atoms and a second p function to the hydrogens. The third d function is especially important for some hypervalent molecules such as SO_2, whilst the second p function significantly improves the atomization energy of some hydrogen-containing molecules

3. The higher-level correction is determined by a least-squares fit for 55 molecules rather than just the hydrogen atom and dihydrogen. This also contributes to an improvement in calculated energies.

A comparison was made for 79 well-established molecules, including 43 that were not included in the original G1 paper. The final total energies are essentially at the QCISD(T)/6–311 + G(3df,2p) level of theory. It was subsequently found that significant savings in computer resource could be obtained at little cost in accuracy by reducing the order of the MP4 calculation to MP3 (giving G2(MP3) theory) or to MP2 (giving G2(MP2) theory).

21.4.3 G3 theory

A recent 1998 reassessment by L. A. Curtiss *et al.* [112] of G2 theory used 302 energies which included 148 enthalpies of formation, 88 ionization energies, 58 electron affinities and 8 proton affinities for larger and more diverse molecules. This revealed some interesting discrepancies, for example the enthalpy of formation of CF_4 is too positive by $7.1 \, kcal \, mol^{-1}$, whilst that of SiF_4 is too negative by $5.5 \, kcal \, mol^{-1}$. The deviations were also much larger for unsaturated systems than for saturated ones. These considerations led the authors to propose *Gaussian-3 (G3) theory*, which follows along the same lines as the earlier G1 and G2 theories in that it is a well-defined sequence of *ab initio* steps to arrive at the total energy of a given molecule. G3 differs from G2 theory in several major ways.

1. An initial HF/6–31G(d) equilibrium structure is obtained using the RHF or UHF treatment (as in G2 theory).

2. HF/6–31G(d) structure is used to find vibration frequencies that are then scaled by 0.8929. The zero-point energy is calculated (as in G2 theory).

3. Equilibrium geometry is refined at the MP2/6–31G(d) level, including all electrons. This geometry is then used in single-point calculations (as in G2 theory).

4. The first higher level correction is a complete MP4/6–31G(d) single point, which is then modified to allow for diffuse functions, higher polarization functions, higher correlation terms using QCI, larger basis sets and the effect of non-additivity.

5. The MP4/6–31G(d) energy and the four corrections are combined along with a spin-orbit correction.

6. A higher level correction (HLC) is added to take account of remaining deficiencies. This is

$$\begin{aligned} \text{HLC} &= -An_\beta - B(n_\alpha - n_\beta) \text{ for molecules} \\ \text{HLC} &= -Cn_\beta - D(n_\alpha - n_\beta) \text{ for atoms} \end{aligned} \tag{21.9}$$

where n_α is the number of α-spin electrons, etc., and A through D are constants chosen to give the smallest average deviation from experiment for the G2 test set.

7. The total energy at 0 K is obtained by adding the zero-point energy from step 2.

The overall agreement with experiment for 299 energies is $1.02 \text{ kcal mol}^{-1}$ compared with $1.48 \text{ kcal mol}^{-1}$ for G2 theory. Use of MP3 rather than MP4 gives a saving in computer resource but the agreement with experiment becomes $1.22 \text{ kcal mol}^{-1}$. A further variation uses B3LYP/6–31G(d) geometries instead of those from MP2.

21.5 Transition States

Geometry optimization plays a key role in modern molecular modelling. *Ab initio* packages invariably contain powerful options for geometry optimization that can locate stationary points on a molecular potential energy surface starting from naïve back-of-an-envelope molecular sketches. Whether that stationary point is a global minimum is of course open to question.

Most current methods are based on a Taylor series expansion of the energy about a reference point $\mathbf{X}_0$; the energy at point $\mathbf{X}$ is given in terms of the gradient $\mathbf{g}$ and the Hessian $\mathbf{H}$ by

$$\varepsilon = \varepsilon_0 + (\mathbf{X} - \mathbf{X}_0)^{\mathrm{T}}\mathbf{g} + \tfrac{1}{2}(\mathbf{X} - \mathbf{X}_0)^{\mathrm{T}}\mathbf{H}(\mathbf{X} - \mathbf{X}_0) + \cdots$$

The Newton–Raphson method truncates the expression at the quadratic term, and tells us that the best step to take in order to reach a stationary point is

$$\mathbf{X} = \mathbf{X}_0 - \mathbf{H}^{-1}\mathbf{g}$$

The gradient and the Hessian have to be calculated at the initial point, and an iterative calculation is normally done. A great deal of effort has gone into determining explicit algorithms for the gradient and the Hessian at many levels of theory.

Minima correspond to points on the molecular potential energy surface where the eigenvalues of the Hessian matrix are all positive. Transition states are characterized as stationary points having just one negative eigenvalue. The requirement for a single negative eigenvalue means that one has to be much more careful with the step taken. For example, it is always possible to take a steepest descent step in a minimum search, which will lower the energy or leave it unchanged. Such a step is not appropriate for transition state searching; it is therefore harder to find a transition state than a minimum.

The key concept in our discussion is the Hessian matrix $\mathbf{H}$. This is a real symmetric matrix and the eigenvalues and eigenvectors can be found by standard numerical methods. Simons *et al.* [113] showed that each Newton–Raphson step is directed in the negative direction of the gradient for each eigenvector that has a positive Hessian eigenvalue, and along the positive direction of the gradient for each eigenvector that has a negative eigenvalue.

For a transition state search, if you are in a region of the molecular potential energy surface where the Hessian does indeed have one negative eigenvalue, then the Newton–Raphson step is appropriate. If you have landed on some region of the surface where the Hessian does not have the desired structure, then you must somehow get out of the region and back to a region where the Hessian has the correct structure.

For minima, qualitative theories of chemical structure are a valuable aid in choosing starting geometries, whilst for transition states one only has a vague notion that the saddle point must lie somewhere between the reactants and products.

One of the earliest algorithms that could take corrective action when the wrong region of the molecular potential energy surface was chosen, was that due to D. Poppinger [114]. This author suggested that the lowest eigenvalue of the Hessian should be followed uphill. The technique has come to be known as *eigenvector following*. The problem with this algorithm is that the remaining directions are left in isolation whilst the fault is corrected.

If we think of a chemical reaction (reactants) as a minimum on the molecular potential energy surface, then there is no unique way of moving uphill on the surface since all directions go uphill. The *linear synchronous transit* (LST) algorithm searches for a maximum along a path between reactants and products. It frequently yields a structure with two or more negative Hessian eigenvalues, and this is not a transition state. The *quadratic synchronous transit* (QST) method searches along a parabola for a minimum in all directions perpendicular to the parabola.

21.6 Dealing with the Solvent

Chemical reactions generally take place in condensed media. I have had very little to say so far about the presence or absence of a solvent, and many of the applications discussed have referred to single, isolated molecules at 0 K and in free space. The benchmark molecular dynamics (MD) and Monte Carlo (MC) applications discussed in earlier chapters dealt with arrays of particles with particularly simple potentials, and the quantum mechanical models have made no mention of a solvent.

The obvious way to take account of the solvent in a MM calculation is to physically add solvent molecules and then optimize the molecular geometry, but such calculations tend to be particularly compute- intensive and the necessary computer power simply was not available to the pioneers in the field. The same comment obviously holds for quantum mechanical studies.

In a medium of relative permittivity ϵ_r, the mutual potential energy of two point charges Q_A and Q_B is reduced by a factor of ϵ_r. Typical values are shown in Table 21.1. Relative permittivities are temperature dependent, and they are usually recorded as a power series in the temperature

$$\epsilon_r = a + b\left(\frac{T}{K}\right) + c\left(\frac{T}{K}\right)^2 + \cdots \tag{21.10}$$

Early workers in the MM field attempted to allow for the presence of a solvent by modifying any electrostatic contribution to their force field. The problem was that the

Table 21.1   Representative relative permittivities

Substance	ϵ_r
Free space	1
Air	1.0006
Glass	6
Water	81

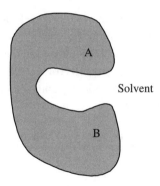

Figure 21.7 Atoms A and B in a molecule (grey) surrounded by a solvent (white)

chosen factor was at first sight quite arbitrary, and bore little relation to the values given in Table 21.1. For example, a value of 2.5 was often used for liquid water. All kinds of convincing arguments were used to justify such choices, for example the two atoms A and B in the molecule of Figure 21.7 will not have much exposure to a solvent and so a pure solvent value of ϵ_r is not appropriate.

21.7 Langevin Dynamics

Molecules in solution undergo collisions with other molecules and the solvent, and they can also be thought to experience frictional forces as they move through the solvent. The Langevin technique allows for both these effects. The collisions are modelled by the addition of a random force **R**, and the frictional effects are modelled by adding a frictional drag that is directly proportional to the velocity of a given particle. Solvent molecules are not explicitly included.

We saw in Chapter 9 that the key equation for a MD simulation is Newton's second law

$$m\frac{d^2\mathbf{r}}{dt^2} = \mathbf{F}$$

In Langevin dynamics, the force is modified according to the ideas discussed above

$$m\frac{d^2\mathbf{r}}{dt^2} = \mathbf{F} - m\gamma\frac{d\mathbf{r}}{dt} + \mathbf{R} \qquad (21.11)$$

The frictional term introduces energy and the random force removes it as kinetic energy. The quantity γ is the collision frequency, and $1/\gamma$ is often called the *velocity relaxation time*; it can be thought of as the time taken for a particle to forget its initial velocity. There is an equation due to Einstein that relates γ to the diffusion constant D of the solvent

$$\gamma = \frac{k_B T}{mD}$$

Integration of the new equation of motion proceeds along the lines discussed in Chapter 9. A straightforward algorithm has been given by D. L. Ermak and H. Buckholtz [115]. The equations of motion are integrated over a time interval Δt that is sufficiently short so that the interparticle forces remain approximately constant. The algorithm for advancing the position r_A and velocity v_A of particle A is then similar to those discussed in Chapter 9; we have

$$
\mathbf{r}_A(t + \Delta t) = \mathbf{r}_A(t) + c_1 \left(\frac{d\mathbf{r}_A}{dt}\right)_t \Delta t + c_2 \left(\frac{d^2\mathbf{r}_A}{dt^2}\right)_t (\Delta t)^2 + \Delta \mathbf{r}_A^G
$$
$$
\mathbf{v}_A(t + \Delta t) = c_0 \left(\frac{d\mathbf{r}_A}{dt}\right)_t + c_1 \left(\frac{d^2\mathbf{r}_A}{dt^2}\right)_t \Delta t + \Delta \mathbf{v}_A^G
$$

(21.12)

Here $\mathbf{r}_A(t)$, $(d\mathbf{r}_A/dt)_t$ and $(d^2\mathbf{r}_A/dt^2)_t$ are the instantaneous position, velocity and acceleration vector of particle A. The acceleration is calculated from the force. $\Delta \mathbf{r}^G$ and $\Delta \mathbf{v}^G$ are random vectors chosen from a Gaussian distribution with zero mean and standard deviations

$$
\sigma_r^2 = \Delta t^2 \left(\frac{k_B T}{m_A}\right) \frac{1}{\gamma \Delta t} \left(2 - \frac{1}{\gamma \Delta t}(3 - 4\exp(-\gamma \Delta t) + \exp(-2\gamma \Delta t))\right)
$$
$$
\sigma_v^2 = \left(\frac{k_B T}{m_A}\right)(1 - \exp(-\gamma \Delta t))
$$

(21.13)

The numerical coefficients are given by

$$
c_0 = \exp(-\gamma \Delta t)
$$
$$
c_1 = \frac{1 - c_0}{\gamma \Delta t}
$$
$$
c_2 = \frac{1 - c_1}{\gamma \Delta t}
$$

At low values of the friction coefficient, the dynamic aspects dominate. If the interparticle forces are taken to vary linearly with time between each time step, then the equations of motion can be rewritten in a form that is said to produce a more accurate simulation;

$$
\mathbf{r}_A(t + \Delta t) = \mathbf{r}_A(t) + c_1 \left(\frac{d\mathbf{r}_A}{dt}\right)_t \Delta t + c_2 \left(\frac{d^2\mathbf{r}_A}{dt^2}\right)_t (\Delta t)^2 + \Delta \mathbf{r}_A^G
$$
$$
\mathbf{v}_A(t + \Delta t) = c_0 \left(\frac{d\mathbf{r}_A}{dt}\right)_t + (c_1 - c_2)\left(\frac{d^2\mathbf{r}_A}{dt^2}\right)_t \Delta t + c_2 \left(\frac{d^2\mathbf{r}_A}{dt^2}\right)_{t+\Delta t} \Delta t + \Delta \mathbf{v}_A^G
$$

(21.14)

and as $\gamma \rightarrow 0$ we recover the *velocity Verlet* algorithm discussed in Chapter 9. For large values of γ, the random collisions dominate and the motion becomes diffusion-like.

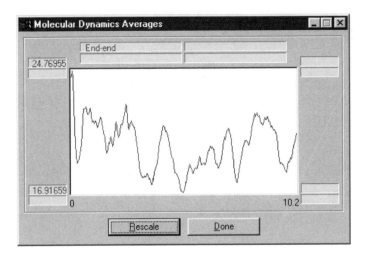

Figure 21.8 Langevin dynamics on C20 hydrocarbon

The same considerations apply to Langevin dynamics as to standard molecular dynamics; there are three stages to a simulation, the heating phase, the data collection stage and the optional cooling stage. For the sake of illustration, Figure 21.8 shows the end-to-end distance in the hydrocarbon $C_{20}H_{42}$ discussed in an earlier chapter, over a 10 ps experiment with a $4\,ps^{-1}$ friction coefficient. The end-to-end distance varied much less than the gas-phase experiment, ranging from 2477 pm to 1692 pm, with a mean of 1993 pm.

21.8 The Solvent Box

In Langevin dynamics the solvent is simulated; no solvent molecules are explicitly included in the calculation. Such calculations are comparable with standard MD calculations in their consumption of computer resource.

I mentioned the interest in water as solvent in Chapters 9 and 10, and it is worth reading the Abstract to Jorgensen's classic 1983 paper [116]:

Classical Monte Carlo simulations have been carried out for liquid water in the NPT ensemble at 25 °C and 1 atm using six of the simpler intermolecular potential functions for the water dimer: Bernal–Fowler (BF), SPC, ST2, TIPS2, TIP3P and TIP4P. Comparisons are made with experimental thermodynamic and structural data including the recent neutron diffraction data of Thiessen and Narten. The computed densities and potential energies are in reasonable accord with experiment except for the original BF model, which yields an 18% overestimate of the density and poor structural results. The TIPS2 and TIP4P potentials yield oxygen–oxygen partial structure functions in good agreement with the neutron

Figure 21.9   Phenylanine in a periodic solvent box

diffraction results. The accord with the experimental OH and HH partial structure functions is poorer; however, the computed results for these functions are similar for all the potential functions. Consequently, the discrepancy may be due to the correction terms needed in processing the neutron data or to an effect uniformly neglected in the computations. Comparisons are also made for self-diffusion coefficients obtained from molecular dynamics simulations. Overall, the SPC, SDT2, TIPS2 and TIP4P models give reasonable structural and thermodynamic descriptions of liquid water and they should be useful in simulations of aqueous solutions. The simplicity of the SPC, TIPS2, and TIP4P is also attractive from a computational standpoint.

Modern MM packages such as HyperChem usually include the option of a periodic solvent box, often the TIP3P box of 216 water molecules. The example in Figure 21.9 is from an MM optimization using phenylanine in water. The standard box is a cube of side 1870 pm, and it is conventional to choose the appropriate box with sides in multiples of this.

21.9 ONIOM or Hybrid Models

One recent approach to the simulation of chemistry in solution is to use a combination of quantum mechanical models for the solute and less accurate models such as an MM solvent box for the solvent. It all started with a key paper in the *Journal of Molecular Biology*, which by and large is not read by chemists. Once again, I'll let the two authors A. Warshel and M. Levitt [117] give you their ideas in their own words:

A general method for detailed study of enzymic reactions is presented. The method considers the complete enzyme–substrate complex together with the surrounding solvent and evaluates all the different quantum mechanical and classical energy factors that can affect the reaction pathway. These factors include the quantum mechanical energies associated with bond cleavage and charge redistribution of the substrate and the classical energies of steric and electrostatic interactions between the substrate and the enzyme. The electrostatic polarization of the enzyme atoms and the orientation of the dipoles of the surrounding water molecules is simulated by a microscopic dielectric model. The solvation energy resulting from this polarization is considerable and must be included in any realistic calculation of chemical reactions involving anything more than an isolated atom in vacuo. Without it, acidic groups can never become ionized and the charge distribution on the substrate will not be reasonable. The same dielectric model can also be used to study the reaction of the substrate in solution. In this way, the reaction is solution can be compared with the enzymic reaction . . . .

What the authors did was to combine an MM potential for the solvent with an early (MINDO/2) quantum mechanical model for the solute. By 1998 such hybrid methods had become sufficiently important to justify an American Chemical Society Symposium (see J. Gao and M.A. Thompson [118]).

If we consider phenylanine in the water solvent box, Figure 21.9, the idea is to treat the solute phenylanine by a 'rigorous' quantum mechanical method (and the choice is semi-empirical, *ab initio*, DFT) and the solute by a less rigorous and so less costly method. The obvious choice for the solvent is MM but we could also choose a semi-empirical treatment for the solvent and a DFT treatment for the solute.

Most implementations use a two-layer approach, but multilayer approaches are becoming popular. Such ONIOM hybrid calculations are now routinely used for solvent modelling.

Appendix
A Mathematical *Aide-Mémoire*

There are several mathematical topics, such as vectors, vector calculus, determinants, matrices and linear operators, that appear throughout this book. You should have them at your fingertips if you want to fully understand the subject; it is no use speaking glibly about the scalar and cross products of two vectors, gradients and hessians, eigenvalues and eigenvectors, Hamiltonian operators and so on when you haven't the slightest idea what the terms mean. I assume that you have come across most of these topics elsewhere, and so the Appendix is essentially an *aide-mémoire*.

Angular momentum dominates the theory of atomic structure (and if you know about commuting linear operators, you will understand why), so I have included a section on this topic.

A.1 Scalars and Vectors

I should emphasize that physical quantities such as mass, temperature and time have two component parts: a measure (i.e. how many) and a unit (e.g. a kelvin or a second). It is usual these days to express all physical quantities in the system of units referred to as the *Système International*, SI for short. The International Unions of Pure and Applied Physics, and of Pure and Applied Chemistry both recommend SI units. These are based on the metre, kilogram, second and the ampere as the fundamental units of length, mass, time and electric current. There are three additional units in SI: the kelvin, mole and candela that are units of thermodynamic temperature, amount of substance and luminous intensity, respectively.

A *scalar s* is a quantity such as mass, temperature or time that can be represented by a single value. The *modulus* (or *magnitude*, or *size*) of s, denoted $|s|$, is the value of s irrespective of its sign. So if T is a temperature and $T = -273$ K, then the modulus of T is 273 K. If you know about thermodynamics, you will realize that temperatures cannot be lower than 0 K, but that's another story.

A vector quantity is a quantity such as displacement, force or velocity that has both a magnitude and a direction in space. I am going to write vectors in bold, e.g. **a**, **b**, **E**. Other authors follow different conventions, and you might find it easier to write vectors by underlining them $\underline{a}$, $\underline{b}$, $\underline{E}$ rather than trying to write them in bold with a biro. My continental European colleagues often write vectors with an arrow above the symbol $\vec{a}, \vec{b}, \vec{E}$.

The size or *modulus* of vector **v** is denoted by v or $|\mathbf{v}|$. It is important to distinguish vector quantities from scalar ones; they are certainly not the same thing. Develop the habit of writing vectors and their moduli in a distinctive style. Once again, you should recognize that vectors representing physical quantities have a measure and an SI unit. A *unit vector* is one with unit modulus (apart from the SI unit), and a unit vector in the direction of vector **v** is $\mathbf{v}/v$. Some authors write unit vectors with a special 'hat', typically $\hat{\mathbf{v}}$. I am not going to follow this convention, because I want to reserve the hat symbol for a *linear operator*, of which more shortly.

A.2 Vector Algebra

Vector notation makes many equations look simpler than they would otherwise be, and so I will make extensive use of vectors through the book. I need to remind you of certain elementary properties.

A.2.1 Vector addition and scalar multiplication

It is common practice when drawing vectors to show their direction by including an arrowhead, as illustrated in Figure A.1. The vector 2**v** is a vector in the same direction as **v** but with twice the modulus. The vector $-2\mathbf{v}$ has twice the modulus of **v** but it points in

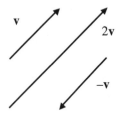

Figure A.1   Scalar multiplication of a vector

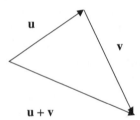

Figure A.2   Addition of two vectors

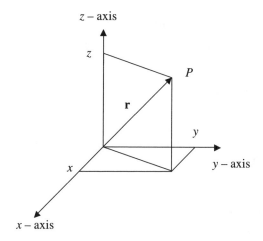

Figure A.3   Cartesian components of a vector

the opposite direction. For many kinds of vector, it doesn't matter where on the page we draw them. Two vectors are equal if they have the same direction and magnitude.

Two vectors **u** and **v** are added together according to the *parallelogram rule*; we draw **u** and then add **v** onto **u** according to Figure A.2. Vectors obey the law that $\mathbf{u} + \mathbf{v} = \mathbf{v} + \mathbf{u}$ and we could have equally drawn **v** first and then added **u** to get the sum of the two.

A.2.2 Cartesian coordinates

The point P in Figure A.3 has Cartesian coordinates (x, y, z). The Cartesian unit vectors $\mathbf{e}_x$, $\mathbf{e}_y$ and $\mathbf{e}_z$ are directed parallel to the x-, y- and z-axes, respectively. They are sometimes written **i**, **j** and **k** rather than $\mathbf{e}_x$, $\mathbf{e}_y$ and $\mathbf{e}_z$. The *position vector* **r** of point P can be expressed by drawing an arrow from the origin to P, and in view of the laws for vector addition and scalar multiplication we have

$$\mathbf{r} = x\mathbf{e}_x + y\mathbf{e}_y + z\mathbf{e}_z \tag{A.1}$$

I have indicated the *components* x, y, z in the figure.

A.2.3 Cartesian components of a vector

Vectors **u** and **v** can be specified by their Cartesian components

$$\mathbf{u} = u_x\mathbf{e}_x + u_y\mathbf{e}_y + u_z\mathbf{e}_z$$
$$\mathbf{v} = v_x\mathbf{e}_x + v_y\mathbf{e}_y + v_z\mathbf{e}_z$$

so that the vector $k\mathbf{u}$ (where k is a scalar) has components

$$k\mathbf{u} = ku_x\mathbf{e}_x + ku_y\mathbf{e}_y + ku_z\mathbf{e}_z \tag{A.2}$$

whilst the vector sum $\mathbf{u} + \mathbf{v}$ has components

$$\mathbf{u} + \mathbf{v} = (u_x + v_x)\mathbf{e}_x + (u_y + v_y)\mathbf{e}_y + (u_z + v_z)\mathbf{e}_z \tag{A.3}$$

A.2.4 Vector products

Vectors can be multiplied by scalars, but some care is needed when considering products of two vectors.

The scalar (or dot) product

The scalar (or dot) product of two vectors $\mathbf{u}$ and $\mathbf{v}$ is

$$\mathbf{u} \cdot \mathbf{v} = |\mathbf{u}||\mathbf{v}| \cos \theta \tag{A.4}$$

where θ is the angle between $\mathbf{u}$ and $\mathbf{v}$, and $|\mathbf{u}|$, $|\mathbf{v}|$ are the moduli of the vectors. If $\mathbf{u} \cdot \mathbf{v} = 0$ and neither $\mathbf{u}$ nor $\mathbf{v}$ is a zero vector, then we say that $\mathbf{u}$ and $\mathbf{v}$ are *orthogonal* (in other words, they are at right angles to each other).

Scalar products obey the rules

$$\mathbf{u} \cdot \mathbf{v} = \mathbf{v} \cdot \mathbf{u}$$

$$\mathbf{u} \cdot (\mathbf{v} + \mathbf{w}) = \mathbf{u} \cdot \mathbf{v} + \mathbf{u} \cdot \mathbf{w}$$

and the Cartesian unit vectors satisfy

$$\mathbf{e}_x \cdot \mathbf{e}_x = \mathbf{e}_y \cdot \mathbf{e}_y = \mathbf{e}_z \cdot \mathbf{e}_z = 1$$

$$\mathbf{e}_x \cdot \mathbf{e}_y = \mathbf{e}_x \cdot \mathbf{e}_z = \mathbf{e}_y \cdot \mathbf{e}_z = 0$$

It follows that the scalar product of $\mathbf{u}$ and $\mathbf{v}$ can be written

$$\mathbf{u} \cdot \mathbf{v} = u_x v_x + u_y v_y + u_z v_z \tag{A.5}$$

and the modulus of vector $\mathbf{v}$ is

$$|\mathbf{v}| = (\mathbf{v} \cdot \mathbf{v})^{1/2} = (v_x^2 + v_y^2 + v_z^2)^{1/2}$$

It also follows that the angle θ between the vectors $\mathbf{u}$ and $\mathbf{v}$ is given by

$$\cos \theta = \frac{u_x v_x + u_y v_y + u_z v_z}{(u_x^2 + u_y^2 + u_z^2)^{1/2}(v_x^2 + v_y^2 + v_z^2)^{1/2}}$$

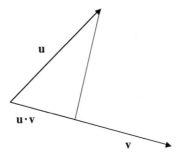

Figure A.4 Projection of **u** in the direction of **v**

The scalar product of **u** and **v** gives the projection of **u** in the direction of **v** (Figure A.4), which by symmetry is the same as the projection of **v** in the direction of **u**.

The vector (or cross) product

In addition to the scalar product $\mathbf{u} \cdot \mathbf{v}$, there is a *vector* (or *cross*) product $\mathbf{u} \times \mathbf{v}$

$$\mathbf{u} \times \mathbf{v} = |\mathbf{u}||\mathbf{v}| \sin \theta \mathbf{n} \tag{A.6}$$

where θ is the angle between **u** and **v** and **n** is a unit vector normal to the plane containing **u** and **v**. The direction of $\mathbf{u} \times \mathbf{v}$ is given by the direction a screw would advance if rotated from the direction of **u** to the direction of **v**. This is shown in Figure A.5.

Vector products obey the rules

$$\mathbf{u} \times \mathbf{v} = -\mathbf{v} \times \mathbf{u}$$
$$\mathbf{u} \times (\mathbf{v} + \mathbf{w}) = \mathbf{u} \times \mathbf{v} + \mathbf{u} \times \mathbf{w}$$

and the Cartesian unit vectors satisfy

$$\mathbf{e}_x \times \mathbf{e}_x = \mathbf{e}_y \times \mathbf{e}_y = \mathbf{e}_z \times \mathbf{e}_z = \mathbf{0}$$

$$\mathbf{e}_x \cdot \mathbf{e}_y = \mathbf{e}_z$$

$$\mathbf{e}_y \cdot \mathbf{e}_z = \mathbf{e}_x$$

$$\mathbf{e}_z \cdot \mathbf{e}_x = \mathbf{e}_y$$

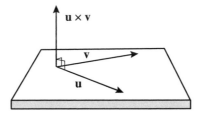

Figure A.5 Vector product of **u** and **v**

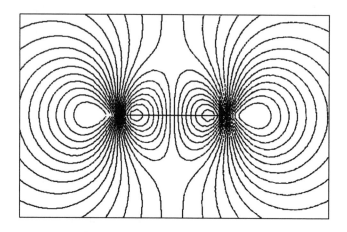

Figure A.6   Contour lines for a scalar field

It follows that the vector product of **u** and **v** can be written

$$\mathbf{u} \times \mathbf{v} = (u_y v_z - u_z v_y)\mathbf{e}_x + (u_z v_x - u_x v_z)\mathbf{e}_y + (u_x v_y - u_y v_x)\mathbf{e}_z \qquad \text{(A.7)}$$

A.3 Scalar and Vector Fields

Mathematically, a field is a function that describes a physical property at points in space. In a *scalar field*, this physical property is completely described by a single value for each point (e.g. temperature, electron density). A scalar field can be represented pictorially by contours, which are lines or surfaces that link together points with the same value of the field *in a given plane*. An example is shown in Figure A.6 (taken from a quantum mechanical study of the electron distribution in a diatomic molecule; the contour lines join points of equal electron density in a plane drawn so as to contain the two nuclei, which on close inspection can be identified as the circles joined by a 'bond').

For quantitative work we need to know the values of the field at each contour; this is sometimes achieved by labelling the contours or by a system of colour coding. For *vector fields* (e.g. gravitational force, electrostatic field intensity), both a direction and a magnitude are required for each point.

A.4 Vector Calculus

In this section I want to explain how fields can be differentiated. I will then discuss two aspects of scalar fields: their *gradient* (which describes their rate of change in

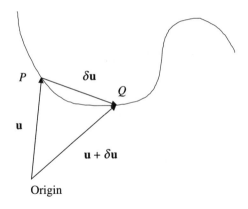

Figure A.7 A curve in space

space) and their *integrals* over surfaces and volumes. There are other advanced concepts that you might have come across, for example the divergence and curl of a vector field. I don't intend to draw on these ideas in this book.

A.4.1 Differentiation of fields

Suppose that the vector field $\mathbf{u}(t)$ is a continuous function of the scalar variable t. As t varies, so does $\mathbf{u}$, and if $\mathbf{u}$ denotes the position vector of a point P, then P moves along a continuous curve in space.

For most of this book we will identify the variable t as time and we might be interested in studying the trajectory of a particle along the above curve in space. Suppose then that a particle moves from point P to point Q as in Figure A.7, in time δt.

Mathematicians tell us that we will profit from a study of what happens to the ratio $\delta\mathbf{u}/\delta t$ as δt gets smaller and smaller, just as we do in ordinary differential calculus. The first differential $d\mathbf{u}/dt$ is defined as the limit of this ratio as the interval δt becomes progressively smaller.

$$\frac{d\mathbf{u}}{dt} = \lim \frac{\delta\mathbf{u}}{\delta t}$$
$$= \lim \left(\frac{\delta u_x}{\delta t}\mathbf{e}_x + \frac{\delta u_y}{\delta t}\mathbf{e}_y + \frac{\delta u_z}{\delta t}\mathbf{e}_z \right)$$
$$= \frac{du_x}{dt}\mathbf{e}_x + \frac{du_y}{dt}\mathbf{e}_y + \frac{du_z}{dt}\mathbf{e}_z \tag{A.8}$$

It is a vector directed along the tangent at P (left to right in the figure). The derivative of a vector is the vector sum of the derivatives of its components. The usual rules for differentiation apply

$$\frac{d}{dt}(\mathbf{u} + \mathbf{v}) = \frac{d\mathbf{u}}{dt} + \frac{d\mathbf{v}}{dt}$$

$$\frac{d}{dt}(k\mathbf{u}) = k\frac{d\mathbf{u}}{dt}$$

$$\frac{d}{dt}(f\mathbf{u}) = \frac{df}{dt}\mathbf{u} + f\frac{d\mathbf{u}}{dt}$$

where k is a scalar and f a scalar field.

A.4.2 The gradient

One of the most important properties of a scalar field is the rate at which it changes in space. A vector called the *gradient vector* describes this. Suppose that f is a scalar field, and we wish to investigate how f changes between the points $\mathbf{r}$ and $\mathbf{r} + d\mathbf{r}$. Here

$$d\mathbf{r} = \mathbf{e}_x dx + \mathbf{e}_y dy + \mathbf{e}_z dz$$

and the change in f is

$$df = f(\mathbf{r} + d\mathbf{r}) - f(\mathbf{r})$$

We know from the rules of partial differentiation that

$$df = \left(\frac{\partial f}{\partial x}\right)dx + \left(\frac{\partial f}{\partial y}\right)dy + \left(\frac{\partial f}{dz}\right)dz$$

and so we identify df as a certain scalar product

$$df = \left(\frac{\partial f}{\partial x}\mathbf{e}_x + \frac{\partial f}{\partial y}\mathbf{e}_y + \frac{\partial f}{\partial z}\mathbf{e}_z\right) \cdot (\mathbf{e}_x dx + \mathbf{e}_y dy + \mathbf{e}_z dz)$$

The first vector on the right-hand side is called the *gradient of f*, and it is written grad f (in this book)

$$\text{grad} f = \frac{\partial f}{\partial x}\mathbf{e}_x + \frac{\partial f}{\partial y}\mathbf{e}_y + \frac{\partial f}{dz}\mathbf{e}_z \tag{A.9}$$

An alternative notation, used in other texts, involves the use of the so-called gradient operator ∇ (pronounced *del*)

$$\nabla = \frac{\partial}{\partial x}\mathbf{e}_x + \frac{\partial}{\partial y}\mathbf{e}_y + \frac{\partial}{\partial z}\mathbf{e}_z$$

In this case the gradient of f is written ∇f.

The vector grad f is a vector field whose direction at any point is the direction in which f is increasing most rapidly and whose magnitude is the rate of change of f in that direction. The spatial rate of change of the scalar field f in the direction of an arbitrary unit vector $\mathbf{e}$ is given by the scalar product $\mathbf{e} \cdot \text{grad} f$.

A.4.3 Volume integrals of scalar fields

In order to evaluate quantities such as the mass or electric charge contained within a region of space, it is necessary to evaluate certain *volume integrals*. I can best illustrate the concept with the simple example shown in Figure A.8; suppose that the electric charge density ρ inside a cubic box whose faces are the planes $x = 0$, $x = 1$; $y = 0$, $y = 1$ and $z = 0$, $z = 1$ is given by $\rho = \rho_0(x + y + z)$ where ρ_0 is a constant. We divide each axis into differential elements, dxs along the x-axis, dys along the y-axis and dzs along the z-axis, giving a number of infinitesimal differential volume elements each of volume dxdydz. The charge enclosed by each differential volume element dxdydz is ρdxdydz and so the total Q enclosed by the box is

$$Q = \int_0^1 \int_0^1 \int_0^1 \rho_0(x + y + z) \, \mathrm{d}x\mathrm{d}y\mathrm{d}z$$

We form three single integrals. First of all, we draw within the region a column having cross section dz, and constant x and y. In order to add these contributions to Q, we integrate between the limits of $z = 0$ and $z = 1$ to give

$$\int_0^1 \rho_0(x + y + z) \, \mathrm{d}z = \rho_0(x + y + \tfrac{1}{2})$$

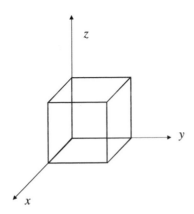

Figure A.8 A cubic volume

Next we form the y integral by drawing a slice parallel to the yz plane and including the column. The limits of the y integration are $y=0$ and $y=1$

$$\int_0^1 \rho_0(x+y+\tfrac{1}{2})\,dy = \rho_0(x+1)$$

Finally, we complete the volume by adding all the slabs, and the limits of the x integration are $x=0$ to $x=1$

$$\int_0^1 \rho_0(x+1)\,dx = \tfrac{3}{2}\rho_0$$

I have used the symbol $d\tau$ to denote a volume element; other commonly used symbols are dV, $d\mathbf{r}$ and $d\vec{r}$.

A.4.4 Line integrals

Figure A.9 refers to a particle at point P moving along a curve in space under the influence of a force $\mathbf{F}$. I have taken the force to be centred at the origin; for example, the particle at P might be a point charge moving under the influence of another point charge at the origin. The work done, w, in moving the short distance $\delta\mathbf{r}$ is the projection of $\mathbf{F}$ along the displacement $\mathbf{F}\cdot\delta\mathbf{r}$. In order to calculate the total work done as the particle moves from initial point A to final point B, we divide the curve into small segments $\delta\mathbf{r}_1, \delta\mathbf{r}_2, \ldots, \delta\mathbf{r}_N$. We then have

$$w \approx \sum_{i=1}^{N} \mathbf{F}(\mathbf{r}_i)\cdot\delta\mathbf{r}_i$$

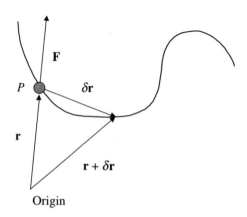

Figure A.9 Work done under the influence of a force

We let the number of points become infinitely large, and the summation approaches a limit called the *line integral*

$$\int_C \mathbf{F} \cdot \mathbf{dr} = \lim \sum_{i=1}^{N} \mathbf{F}(\mathbf{r}_i) \cdot \delta \mathbf{r}_i \qquad (A.10)$$

If the points A and B coincide, then the line integral refers to a *closed curve* and we often write the line integral with a circle round the integral sign $\oint_C \mathbf{F} \cdot \mathbf{dr}$. If the line integral of a certain vector field $\mathbf{F}$ is zero around any arbitrary closed path, then the vector field is called a *conservative field*. It can be shown that every conservative field can be written as the gradient of a suitable scalar field

$$\mathbf{F} = -\text{grad} \, \phi \qquad (A.11)$$

The importance of scalar fields such as ϕ is that changes in ϕ depend only on the starting and finishing points, and not on the path chosen to get between these points.

A.5 Determinants

The set of simultaneous linear equations

$$a_{11}x_1 + a_{12}x_2 = b_1$$
$$a_{21}x_1 + a_{22}x_2 = b_2$$

has solution

$$x_1 = \frac{b_1 a_{22} - b_2 a_{12}}{a_{11} a_{22} - a_{12} a_{21}}; \qquad x_2 = \frac{b_2 a_{11} - b_2 a_{21}}{a_{11} a_{22} - a_{12} a_{21}}$$

If we define a *determinant of order* 2 by the symbol

$$\begin{vmatrix} a & b \\ c & d \end{vmatrix} \equiv (ad - bc) \qquad (A.12)$$

then these solutions may be written as

$$x_1 = \frac{\begin{vmatrix} b_1 & a_{12} \\ b_2 & a_{22} \end{vmatrix}}{\begin{vmatrix} a_{11} & a_{12} \\ a_{21} & a_{22} \end{vmatrix}}; \qquad x_2 = \frac{\begin{vmatrix} a_{11} & b_1 \\ a_{21} & b_b \end{vmatrix}}{\begin{vmatrix} a_{11} & a_{12} \\ a_{21} & a_{22} \end{vmatrix}}$$

We speak of the elements, rows, columns and diagonal of the determinant in an obvious manner. In a similar fashion, solution of the three simultaneous linear equations

$$a_{11}x_1 + a_{12}x_2 + a_{13}x_3 = b_1$$
$$a_{21}x_1 + a_{22}x_2 + a_{23}x_3 = b_2$$
$$a_{31}x_1 + a_{32}x_2 + a_{33}x_3 = b_3$$

involves a denominator

$$a_{11}a_{22}a_{33} - a_{11}a_{23}a_{32} + a_{21}a_{32}a_{13} - a_{21}a_{12}a_{33} + a_{31}a_{12}a_{23} - a_{31}a_{22}a_{13}$$

and so we define a determinant of order three by

$$\begin{vmatrix} a_{11} & a_{12} & a_{13} \\ a_{21} & a_{22} & a_{23} \\ a_{31} & a_{32} & a_{33} \end{vmatrix} = a_{11}a_{22}a_{33} - a_{11}a_{23}a_{32} + a_{21}a_{32}a_{13} - a_{21}a_{12}a_{33} + a_{31}a_{12}a_{23} - a_{31}a_{22}a_{13}$$

$$(A.13)$$

By inspection, this can be written in terms of certain determinants of order 2

$$\begin{vmatrix} a_{11} & a_{12} & a_{13} \\ a_{21} & a_{22} & a_{23} \\ a_{31} & a_{32} & a_{33} \end{vmatrix} = a_{11}\begin{vmatrix} a_{22} & a_{23} \\ a_{32} & a_{33} \end{vmatrix} - a_{21}\begin{vmatrix} a_{12} & a_{13} \\ a_{32} & a_{33} \end{vmatrix} + a_{31}\begin{vmatrix} a_{12} & a_{13} \\ a_{22} & a_{23} \end{vmatrix}$$

The three determinants of order 2 are called the *minors* of a_{11}, a_{21} and a_{31}, respectively; they are the determinants produced by striking out the first column and successive rows of the determinant of order 3. A little analysis shows that this determinant may be expanded down any of its columns or along any of its rows by suitably combining products of elements and their minors.

A.5.1 Properties of determinants

1. The value of a determinant is unchanged by interchanging the elements of any corresponding rows and columns.

2. The sign of a determinant is reversed by interchanging any two of its rows or columns.

3. The value of a determinant is zero if any two of its rows (or columns) are identical.

A.6 Matrices

A matrix is a set of $m \times n$ quantities arranged in a rectangular array of m rows and n columns, for example

$$\mathbf{A} = \begin{pmatrix} a_{11} & a_{12} & \cdots & a_{1n} \\ a_{21} & a_{22} & \cdots & a_{2n} \\ \cdots & \cdots & \cdots & \cdots \\ a_{m1} & a_{m2} & \cdots & a_{mn} \end{pmatrix}$$

Throughout this book I will denote matrices by bold letters, just like vectors. The matrix above is said to be of *order m by n* (denoted $m \times n$). The null matrix $\mathbf{0}$ is one whose elements are all zero. If $m = n$, then the matrix is said to be square of order n, and a square matrix whose only non-zero elements are the diagonal elements is said to be a *diagonal matrix*. Thus,

$$\begin{pmatrix} a_{11} & 0 & 0 \\ 0 & a_{22} & 0 \\ 0 & 0 & a_{33} \end{pmatrix}$$

is a diagonal matrix. The unit matrix ($\mathbf{1}$ or $\mathbf{I}$) of order n is a diagonal matrix whose elements are all equal to 1.

A *row vector* is a matrix of order $1 \times n$ and a *column vector* is a matrix of order $n \times 1$. For example $(a_1, a_2, \ldots, a_n)$ is a row vector and $\begin{pmatrix} a_1 \\ a_2 \\ \cdots \\ a_n \end{pmatrix}$ is a column vector.

A.6.1 The transpose of a matrix

Interchanging rows and columns of a determinant leaves its value unchanged, but interchanging the rows and columns of a matrix $\mathbf{A}$ produces a new matrix called the *transpose* $\mathbf{A}^{\mathrm{T}}$. Thus, for example, if

$$\mathbf{A} = \begin{pmatrix} a_{11} & a_{12} \\ a_{21} & a_{22} \\ a_{31} & a_{32} \end{pmatrix}, \text{ then } \mathbf{A}^{\mathrm{T}} = \begin{pmatrix} a_{11} & a_{21} & a_{31} \\ a_{12} & a_{22} & a_{32} \end{pmatrix}$$

In the case that $\mathbf{A} = \mathbf{A}^{\mathrm{T}}$, then we say that the matrix $\mathbf{A}$ is *symmetric*. A symmetric matrix is square and has $a_{ij} = a_{ji}$.

A.6.2 The trace of a square matrix

In the square matrix

$$
\mathbf{A} = \begin{pmatrix} a_{11} & a_{12} & a_{13} \\ a_{21} & a_{22} & a_{13} \\ a_{31} & a_{32} & a_{33} \end{pmatrix}
$$

the elements a_{11}, a_{22} and a_{33} are called the *diagonal elements*. The sum of the diagonal elements is called the *trace* of the matrix

$$
\mathrm{tr}\,\mathbf{A} = a_{11} + a_{22} + a_{33} \tag{A.14}
$$

and the following rules are obeyed

$$
\begin{aligned}
\mathrm{tr}(k\mathbf{A}) &= k\,\mathrm{tr}\mathbf{A} \\
\mathrm{tr}(\mathbf{AB}) &= \mathrm{tr}(\mathbf{BA}) \\
\mathrm{tr}(\mathbf{A} + \mathbf{B}) &= \mathrm{tr}\mathbf{A} + \mathrm{tr}\mathbf{B}
\end{aligned} \tag{A.15}
$$

A.6.3 Algebra of matrices

We need to be aware of a few simple matrix properties as follows.

1. If $\mathbf{A}$ and $\mathbf{B}$ are two matrices of the same order with elements a_{ij} and b_{ij}, then their sum $\mathbf{S} = \mathbf{A} + \mathbf{B}$ is defined as the matrix whose elements are $c_{ij} = a_{ij} + b_{ij}$.

2. Two matrices $\mathbf{A}$ and $\mathbf{B}$ with elements a_{ij} and b_{ij} are equal only if they are of the same order, and all their corresponding elements are equal, $a_{ij} = b_{ij}$.

3. The result of multiplying a matrix $\mathbf{A}$ whose elements are a_{ij} by a scalar k is a matrix whose elements are ka_{ij}.

4. The definition of *matrix multiplication* is such that two matrices $\mathbf{A}$ and $\mathbf{B}$ can only be multiplied together to form their product $\mathbf{AB}$ when the number of columns of $\mathbf{A}$ is equal to the number of rows of $\mathbf{B}$. Suppose $\mathbf{A}$ is a matrix of order $(m \times p)$, and $\mathbf{B}$ is a matrix of order $(p \times n)$. Their product $\mathbf{C} = \mathbf{AB}$ is a matrix of order $(m \times n)$ with elements

$$
c_{ij} = \sum_{k=1}^{p} a_{ik}b_{kj} \tag{A.16}
$$

Thus, for example, if

$$\mathbf{A} = \begin{pmatrix} 1 & 3 \\ 2 & 4 \\ 3 & 6 \end{pmatrix} \quad \text{and} \quad \mathbf{B} = \begin{pmatrix} 1 & 2 & 3 \\ 4 & 5 & 6 \end{pmatrix}$$

then

$$\mathbf{AB} = \begin{pmatrix} 13 & 17 & 21 \\ 18 & 24 & 30 \\ 27 & 36 & 45 \end{pmatrix} \quad \text{and} \quad \mathbf{BA} = \begin{pmatrix} 14 & 29 \\ 32 & 68 \end{pmatrix}$$

so we see from this simple example that $\mathbf{AB}$ is not necessarily equal to $\mathbf{BA}$. If $\mathbf{AB} = \mathbf{BA}$, then we say that the matrices *commute*.

A.6.4 The inverse matrix

Let $\mathbf{A}$ be the square matrix

$$\mathbf{A} = \begin{pmatrix} a_{11} & a_{12} & \cdots & a_{1n} \\ a_{21} & a_{22} & \cdots & a_{2n} \\ \cdots & \cdots & \cdots \\ a_{n1} & a_{n2} & \cdots & a_{nn} \end{pmatrix}$$

If a suitable matrix $\mathbf{X}$ can be found such that $\mathbf{AX} = \mathbf{1}$, then we refer to $\mathbf{X}$ as the *inverse* of $\mathbf{A}$, and write it $\mathbf{A}^{-1}$. We say that $\mathbf{A}$ is *invertible* or *non-singular*. Not all matrices are invertible; the rule is that square matrices with a non-zero determinant such as $\begin{pmatrix} 1 & -1 \\ 1 & 1 \end{pmatrix}$ are invertible, whilst those with zero determinant such as $\begin{pmatrix} 1 & 1 \\ 1 & 1 \end{pmatrix}$ are not. Matrices with zero determinant are often called *singular*.

A.6.5 Matrix eigenvalues and eigenvectors

Consider once again an $(n \times n)$ matrix $\mathbf{A}$. If we form the product of $\mathbf{A}$ with a suitable but arbitrary column vector $\mathbf{u}$, then sometimes it will happen that the product $\mathbf{Au}$ is a linear multiple λ of $\mathbf{u}$, $\mathbf{Au} = \lambda \mathbf{u}$. In this case we say that $\mathbf{u}$ is an *eigenvector* (or *eigenfunction*) of $\mathbf{A}$ with *eigenvalue* λ. There are exactly n eigenvalues and eigenvectors. Thus, for example, if

$$\mathbf{A} = \begin{pmatrix} 0 & 1 & 0 \\ 1 & 0 & 1 \\ 0 & 1 & 0 \end{pmatrix}$$

then we find

$$\lambda_1 = \sqrt{2} \quad \text{and} \quad \mathbf{u}_1^T = \left(\frac{1}{2} \quad \frac{\sqrt{2}}{2} \quad \frac{1}{2} \right)$$

$$\lambda_2 = 0 \quad \text{and} \quad \mathbf{u}_2^T = \left(-\frac{1}{2} \quad 0 \quad \frac{1}{2} \right)$$

$$\lambda_3 = -\sqrt{2} \quad \text{and} \quad \mathbf{u}_3^T = \left(\frac{1}{2} \quad -\frac{\sqrt{2}}{2} \quad \frac{1}{2} \right)$$

If two or more eigenvalues are the same, then we say that they are *degenerate*.

A.7 Angular Momentum

Figure A.8 shows a particle moving along a curve in space. If the particle has mass m, then its *linear momentum* $\mathbf{p}$ is defined as

$$\mathbf{p} = m \frac{d\mathbf{r}}{dt} \tag{A.17}$$

Newton's second law of motion relates force to the rate of change of linear momentum

$$\mathbf{F} = \frac{d\mathbf{p}}{dt} \tag{A.18}$$

In the absence of an applied force, the linear momentum is constant. When the mass is constant, Equation (A.18) can be written

$$\mathbf{F} = m \frac{d^2\mathbf{r}}{dt^2}$$

The *angular momentum* $\mathbf{l}$ is defined as a vector cross product

$$\mathbf{l} = \mathbf{r} \times \mathbf{p} \tag{A.19}$$

Care has to be exercised when dealing with angular momentum, because $\mathbf{l}$ depends on the point chosen to be the coordinate origin. If we were to choose a new coordinate origin such that the position vector of the particle were

$$\mathbf{r}' = \mathbf{r} + \mathbf{R}$$

where $\mathbf{R}$ is a constant vector, then we have

$$\mathbf{l}' = \mathbf{r}' \times \mathbf{p}$$
$$= \mathbf{l} + \mathbf{R} \times \mathbf{p}$$

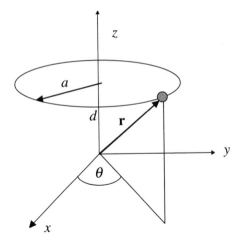

Figure A.10   Circular motion around the z-axis

On differentiation of both sides of Equation (A.19) with respect to time we find

$$\frac{d\mathbf{l}}{dt} = \frac{d}{dt}\mathbf{r} \times \mathbf{p}$$

$$\frac{d\mathbf{l}}{dt} = \frac{d\mathbf{r}}{dt} \times \mathbf{p} + \mathbf{r} \times \frac{d\mathbf{p}}{dt}$$

$$\frac{d\mathbf{l}}{dt} = \mathbf{r} \times \mathbf{F}$$

The vector cross product of $\mathbf{r}$ and the applied force $\mathbf{F}$, $\mathbf{r} \times \mathbf{F}$, is known as the *torque* and so I have proved that in the absence of an applied torque, the angular momentum of a particle remains constant in time.

Consider now a particle of mass m executing circular motion around the z-axis, as shown in Figure A.10

$$\mathbf{r} = x\mathbf{e}_x + y\mathbf{e}_y + z\mathbf{e}_z$$
$$= a \sin\theta\mathbf{e}_x + a \cos\theta\mathbf{e}_y + d\mathbf{e}_z$$

Differentiating we have

$$\frac{d\mathbf{r}}{dt} = -a\frac{d\theta}{dt}\cos\theta\mathbf{e}_x + a\frac{d\theta}{dt}\sin\theta\mathbf{e}_y$$

$$= \frac{d\theta}{dt}(-y\mathbf{e}_x + x\mathbf{e}_y)$$

$$= \frac{d\theta}{dt}\mathbf{e}_z \times \mathbf{r}$$

It also follows that

$$\mathbf{I} = \mathbf{r} \times \mathbf{p}$$

$$= m\mathbf{r} \times \frac{d\mathbf{r}}{dt}$$

$$= m\frac{d\theta}{dt}\mathbf{r} \times (-y\mathbf{e}_x + x\mathbf{e}_y)$$

$$= m\frac{d\theta}{dt}(-zx\mathbf{e}_x + zy\mathbf{e}_y + (x^2 + y^2)\mathbf{e}_z)$$

The vector $\mathbf{I}$ is not directed along the z-axis because of the choice of 'origin' along the z-axis but for rotation with $z = 0$ we have

$$\mathbf{I} = m\frac{d\theta}{dt}(x^2 + y^2)\mathbf{e}_z$$

$$= m\frac{d\theta}{dt}a^2\mathbf{e}_z \tag{A.20}$$

The quantity ma^2 turns out to have a special significance for circular motion, and we call it the *moment of inertia* about the z-axis.

A.8 Linear Operators

I now need to remind you about linear operators, mathematical entities that form a key concept in the language of quantum mechanics. An operator $\hat{A}$ can be thought of as a mathematical entity that turns one function into another. It is conventional to write them with the hat sign in order to distinguish them from other mathematical constructs. To keep the discussion simple, I will just deal in this section with operators that work on functions of a real single variable that I will write consistently as x. Thus, for example, we might have

$$\hat{A}f(x) = (x^2 + 1)f(x) \tag{A.21}$$

which means that the action of $\hat{A}$ is to replace $f(x)$ by $(x^2 + 1)f(x)$. So, if $f(x) = \exp(x)$, then we would have $f(0.1) = 1.105$ and $\hat{A}f(0.1) = (0.1^2 + 1) \times 1.105 = 1.116$. I should emphasize that this is not an equation that must somehow be solved for a value of x, it is a mapping between two sets of functions often represented as a Venn diagram (Figure A.11).

Now consider the operator $\hat{B}$ which operates on a differentiable function $f(x)$ to give the first derivative $df(x)/dx$. We can write this as

$$\hat{B}f(x) = \frac{d}{dx}f(x) \tag{A.22}$$

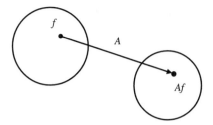

Figure A.11 Venn diagram

So $\hat{B}$ acting on $\exp(x)$ gives $\exp(x)$. It is clear that the action of these two operators on the same function generally gives a different result.

An operator is *linear* if it satisfies the following two properties

$$\hat{A}(f(x) + g(x)) = \hat{A}f(x) + \hat{A}g(x)$$
$$\hat{A}(cf(x)) = c\hat{A}f(x)$$

where c is a scalar and both of the two operators discussed above are linear.

Consider now the product of two operators $\hat{A}$ and $\hat{B}$ which we write as $\hat{A}\hat{B}$. We define this product as follows

$$(\hat{A}\hat{B})f(x) = \hat{A}(\hat{B}f(x)) \tag{A.23}$$

That is to say, we calculate the result $\hat{B}f(x)$ first and then operate with $\hat{A}$ on the result. Using the two operators discussed so far we have

$$\hat{A}\hat{B}f(x) = (x^2 + 1)\frac{df(x)}{dx}$$

whilst

$$\hat{B}\hat{A}f(x) = 2xf(x) + (x^2 + 1)\frac{df(x)}{dx}$$

They certainly don't give the same result and it therefore matters which way round we apply the operators. If we now define a third operator $\hat{C}$ whose effect on a function is to multiply by x such that

$$\hat{C}f(x) = xf(x)$$

then we see that

$$\hat{B}\hat{A}f(x) = 2\hat{C}f(x) + \hat{A}\hat{B}f(x)$$

This is true for all differentiable functions $f(x)$ and so it forms an *operator identity*. We would write it as

$$\hat{B}\hat{A} = 2\hat{C} + \hat{A}\hat{B}$$

In the special case where two operators satisfy the equality

$$\hat{B}\hat{A} = \hat{A}\hat{B} \tag{A.24}$$

we say that they *commute* (you will recognize much of the terminology that we used for matrices; once again, that's a different story). Commuting operators play a major role in quantum mechanics, as we will shortly see.

To every linear operator there belongs a special set of functions $u_i(x)$ and scalars a_i such that

$$\hat{A}u_i(x) = a_i u_i(x) \tag{A.25}$$

The scalars can be real or complex. We say that the $u_i(x)$ are the *eigenvectors* (or *eigenfunctions*) of the operator and the a_i are the *eigenvalues*. Depending on the physical circumstances and the operator, there may be a finite set of eigenvalues, a countably infinite set or a continuum.

To finish this section, let me give you an important result from the theory of linear operators (together with a simplified proof).

Theorem A.1 *If two linear operators commute, then it is possible to find a set of simultaneous eigenfunctions.*

Proof (simplified version)   Suppose that two operators have each a single eigenvector and eigenvalue, given by

$$\hat{A}u(x) = au(x)$$
$$\hat{B}v(x) = bv(x)$$

If we operate on the first equation by the second operator we find

$$\hat{B}\hat{A}u(x) = a\hat{B}u(x)$$

but we know that the operators commute and so

$$\hat{A}\hat{B}u(x) = a\hat{B}u(x)$$

or

$$\hat{A}[\hat{B}u(x)] = a[\hat{B}u(x)]$$

The function in square brackets is therefore a linear multiple of the eigenfunction of the operator $\hat{A}$

$$\hat{B}u(x) = cu(x)$$

which shows that the eigenfunctions of the two operators are the same, apart from some arbitrary multiplicative constant.

Linear operators play an important role in quantum mechanics because

1. Every observable such as position, momentum, energy and angular momentum can be represented by a linear operator.

2. The results of measurements of these observables are given by the eigenvalues of the operators.

3. Observables can be measured simultaneously to arbitrary precision only when their operators commute

We usually give some concrete form to the linear operators, especially those representing position and linear momentum. In the Schrödinger representation we substitute as follows:

- for a position operator such as $\hat{x}$ we write x
- for a momentum operator such as $\hat{p}_x$ we substitute

$$-j\frac{h}{2\pi}\frac{\partial}{\partial x}$$

A little operator analysis shows that

$$xp_x - p_x x = j\frac{h}{2\pi} \tag{A.26}$$

and this is consistent with the famous Heisenberg uncertainty principle.

A.9 Angular Momentum Operators

The classical angular momentum vector of a particle with position vector $\mathbf{r}$ and linear momentum $\mathbf{p}$ is

$$\mathbf{l} = \mathbf{r} \times \mathbf{p} \tag{A.27}$$

In classical mechanics we can measure the three components and the square of the length, l_x, l_y, l_z and $l^2 = (l_x{}^2 + l_y{}^2 + l_z{}^2)$ simultaneously with arbitrary precision. Things are far more interesting for quantum mechanical systems.

Expanding the vector product Equation (A.27) we see that the x component of l is

$$l_x = yp_z - zp_y$$

The corresponding operator is

$$\hat{l}_x = \hat{y}\hat{p}_z - \hat{z}\hat{p}_y$$

which becomes, in Schrödinger representation

$$\hat{l}_x = -j\frac{h}{2\pi}\left(\hat{y}\frac{\partial}{\partial z} - \hat{z}\frac{\partial}{\partial y}\right)$$

After some operator algebra it can be shown that the various components of the angular momentum operator do not commute with each other and so cannot be measured simultaneously to arbitrary precision. They are subject to Heisenberg's uncertainty principle. We find in addition that the square of the magnitude and any of the three components can be measured simultaneously to arbitrary precision. We conclude that at best we can simultaneously measure the square of the magnitude and one component, which is by convention called the z-component.

The eigenvalues and eigenfunctions of angular momentum can be found most easily in spherical polar coordinates

$$\hat{l}_z = -j\frac{h}{2\pi}\frac{\partial}{\partial\phi}$$
$$\hat{l}^2 = -\frac{h^2}{4\pi^2}\left(\frac{\partial^2}{\partial\theta^2} + \cot\theta\frac{\partial}{\partial\theta} + \frac{1}{\sin^2\theta}\frac{\partial^2}{\partial\phi^2}\right)$$

(A.28)

This problem is tackled in detail in all the traditional quantum chemistry texts such as Eyring, Walter and Kimball, and we can write

$$\hat{l}^2 Y_{l,m_l}(\theta,\phi) = l(l+1)\frac{h^2}{4\pi^2} Y_{l,m_l}(\theta,\phi)$$
$$\hat{l}_z Y_{l,m_l}(\theta,\phi) = m_l\frac{h}{2\pi} Y_{l,m_l}(\theta,\phi)$$

(A.29)

The eigenfunctions are well known from mathematical physics and are called *spherical harmonics*. The quantum numbers l and m_l are restricted to certain discrete values: $l = 0, 1, 2, 3 \ldots$ and for each value of l; m_l can take integral values ranging from $-l$ to $+l$.

We visualize angular momentum as follows. The vector precesses round the z-axis in such a way that its z-component is constant in time, but that the x- and

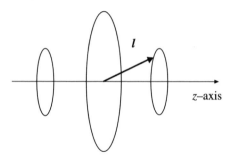

Figure A.12   Angular momentum with $l = 1$

y-components vary. There are $2l + 1$ possible orientations of the vector. I have illustrated this for the case $l = 1$ in Figure A.12, where just one of the three possible vector alignments is shown. For each orientation the vector precesses around the axis keeping the z-component fixed at $m_l(h/2\pi)$. The length of the vector is $\sqrt{l(l+1)}(h/2\pi)$.

References

[1] Hirschfelder, J. O., C. F. Curtiss and R. B. Bird, *Molecular Theory of Gases and Liquids*, John Wiley & Sons, New York, 1954.

[2] Latimer, W. M. and W. H. Rodebush, *J. Am. Chem. Soc.* **42** (1920) 1419.

[3] Linnett, J. W., *Trans. Farad. Soc.* **36** (1940) 1123; **38** (1942) 1.

[4] Andrews, D. H., *Phys. Rev.* **36** (1930) 544.

[5] Snyder, R. G. and J. H. Schachtschneider, *Spectrochim. Acta* **21** (1965) 169.

[6] Mayo, S. L., B. D. Olafson and W. A. Goddard III, *J. Chem. Phys.* **94** (1990) 8897.

[7] Allinger, N. L., in *Advances in Physical Organic Chemistry*, Vol. 13, eds. V. Gold and D. Bethell, Academic Press, London, 1976.

[8] Allinger, N. L., *J. Am. Chem. Soc.* **99** (1977) 8127.

[9] Weiner, S. J., P. A. Kollman, D. A. Case and V. C. Singh, *J. Am. Chem. Soc.* **106** (1984) 765.

[10] Jorgensen, W. L. and J. Tirado-Rives, *J. Am. Chem. Soc.* **110** (1988) 1657.

[11] Johnson, R. A., *Phys. Rev.* **134** (1964) A1329.

[12] Westheimer, F. H., in *Steric Effects in Organic Chemistry*, ed. M. S. Newman, John Wiley & Sons, New York, 1956.

[13] Hendrickson, J. B., *J. Am. Chem. Soc.* **83** (1961) 4537.

[14] Wiberg, K., *J. Am. Chem. Soc.* **87** (1965) 1070.

[15] Fletcher, R. and C. M. Reeves, *Comput. J.* **7** (1964) 149.

[16] Polak, E. and G. Ribière, *Rev. Fr. Inform. Rech. Operation*, **16-R1** (1969) 35.

[17] Fletcher, R. and M. J. D. Powell, *Comput. J.* **6** (1963) 163.

[18] Baker, J. and W. J. Hehre, *J. Comp. Chem.* **12** (1991) 606.

[19] Pulay, P., G. Fogarasi, F. Pang and J. E. Boggs, *J. Am. Chem. Soc.* **101** (1979) 2550.

[20] Peng, C., P. Y. Ayala, H. B. Schlegel and M. J. Frisch, *J. Comp. Chem.* **17** (1996) 49.

[21] Gasteiger, J. and M. Marsili, *Tetrahedron* **36** (1980) 3219.

[22] Pauling, L. and D. M. Yost, *Proc. Natl Acad. Sci. USA* **14** (1932) 414.

[23] Mulliken, R. S., *J. Chem. Phys.* **2** (1934) 782.

[24] Sanderson, R. T., *Science* **144** (1951) 670.

[25] Hinze, J., M. A. Whitehead and H. H. Jaffé, *J. Am. Chem. Soc.* **85** (1963) 148.

[26] Silberstein, L., *Philos. Mag.* **33** (1917) 92.

[27] Miller, K. J. and J. A. Savchik, *J. Am. Chem. Soc.* **101** (1979) 7206.

[28] Rekker, R. E., *The Hydrophobic Fragment Constant*, Elsevier, Amsterdam, 1976.

[29] Leo, A., P. Y. C. Jow, C. Silipo and C. Hansch, *J. Med. Chem.* **18** (1975) 865.

[30] Klopman, G. and L. D. Iroff, *J. Comp. Chem.* **2** (1981) 157.

[31] Bodor, N., Z. Gabany1 and C.-K. Wong, *J. Am. Chem. Soc.* **111** (1989) 3783.

[32] Alder, B. J. and T. E. Wainwright, *J. Chem. Phys.* **27** (1957) 1208.

[33] Alder, B. J. and T. E. Wainwright, *J. Chem. Phys.* **31** (1959) 459.

[34] Rahman, A., *Phys. Rev.* **1136** (1964) A405.

[35] Woodcock, L. V., *Chem. Phys. Lett.* **10** (1971) 257.

[36] Rahman, A. and F. H. Stillinger, *J. Chem. Phys.* **55** (1971) 3336.

[37] Jorgensen, W. L., *J. Am. Chem. Soc.* **103** (1981) 335.

[38] Bernal, J. D. and R. H. Fowler, *J. Chem. Phys.* **1** (1933) 515.

[39] Jorgensen, W. L., J. Chandrasekhar, J. D. Madura, R. W. Impey and M. L. Klein, *J. Chem. Phys.* **79** (1983) 926.

[40] Nose, S., *Mol. Phys.* **52** (1984) 255.

[41] Metropolis, N., A. W. Rosenbluth, M. N. Rosenbluth, A. H. Teller and E. Teller, *J. Chem. Phys.* **21** (1953) 1087.

[42] Wood, W. W. and F. R. Parker, *J. Chem. Phys.* **27** (1957) 720.

[43] Heisenberg, W., *Z. Phys.* **39** (1926) 499.

[44] Atkins, P. W., *Physical Chemistry* 5th edn, Oxford University Press, 1982.

[45] Stern, O. and W. Gerlach, *Z. Phys.* **8** (1921) 110.

[46] Goudsmit, S. and G. E. Uhlenbeck, *Naturwissenschaften* **13** (1925) 953.

[47] Moore, C. E., *Atomic Energy Levels*, Vol. I: *Hydrogen through Vanadium*, Circular of the National Bureau of Standards 467, US Government Printing Office, Washington, DC, 1949; Vol. II: *Chromium through Niobium*, 1952; Vol. III: *Molybdenum through Lanthanum and Hafnium through Actinium*, 1958.

[48] Hylleraas, E., *Z. Phys.* **65** (1930) 209.

[49] Frankowski, K. and C. L. Pekeris, *Phys. Rev.* **146** (1966) 46.

[50] Hartree, D. R., *Proc. Cambridge Philos. Soc.* **24** (1927) 89.

[51] Hartree, D. R., *The Calculation of Atomic Structures*, John Wiley & Sons, New York, 1957.

[52] Fock, V., *Z. Phys.* **61** (1930) 126; **62** (1930) 795.

[53] Zener, C., *Phys. Rev.* **36** (1930) 51.

[54] Slater, J. C., *Phys. Rev.* **36** (1930) 57.

[55] Born, M. and J. R. Oppenheimer, *Ann. Phys.* **84** (1927) 457.

[56] Teller, E., *Z. Phys.* **61** (1930) 458.

[57] Burrau, O., *Kgl Danske Videnskab. Selskab.* **7** (1) 1927.

[58] Bates, D. R., K. Ledsham and A. L. Stewart, *Philos. Trans. R. Soc.* **A246** (1953) 215.

[59] Wind, H., *J. Chem. Phys.* **42** (1965) 2371.

[60] James, H. M., *J. Chem. Phys.* **3** (1935) 7.

[61] Heitler, W. and F. London, *Z. Phys.* **44** (1927) 455.

[62] Suguira, Y., *Z. Phys.* **45** (1937) 484.

[63] James, H. and M. Coolidge, *J. Chem. Phys.* **1** (1933) 825.

[64] Hellmann, H., *Einführung in die Quantenchemie*, Franz Deuticke, 1937, p. 133.

[65] Mulliken, R. S., *J. Chem. Phys.* **23** (1955) 1833.

[66] Roothaan, C. C. J., *Rev. Mod. Phys.* **23** (1951) 161.

[67] Roothaan, C. C. J., *Rev. Mod. Phys.* **32** (1960) 179.

[68] Clementi, E. and D. L. Raimondi, *J. Chem. Phys.* **38** (1963) 2686.

[69] Clementi, E., *J. Chem. Phys.* **40** (1964) 1944.

[70] Boys, S. F., *Proc. R. Soc. Series A*, **200** (1950) 542.

[71] Hehre, W. J., R. F. Stewart and J. A. Pople, *J. Chem. Phys.* **51** (1969) 2657.

[72] Ditchfield, R., W. J. Hehre and J. A. Pople, *J. Chem. Phys.* **54** (1971) 724.

[73] Collins, J. B., P. Von, R. Schleyer, J. S. Binkley and J. A. Pople, *J. Chem. Phys.* **64** (1976) 5142.

[74] Dunning, Jr, T. H., *J. Chem. Phys.* **55** (1975) 716.

[75] Raffeneti, R. C., *Chem. Phys. Lett.* **20** (1973) 335.

[76] Scrocco, E. and J. Tomasi, in *Topics in Current Chemistry*, 42, *New Concepts* II, Springer-Verlag, Berlin 1978.

[77] Brillouin, L., *Actual Sci. Ind.* **71** (1933) 159.

[78] Foresman, J. B., M. Head-Gordon, J. A. Pople and M. J. Frisch, *J. Phys. Chem.* **96** (1992) 135.

[79] Clar, E., *Aromatische Kohlenwasserstoffe*, Springer-Verlag, Berlin, 1952.

[80] Mason, S. F., *J. Chem. Soc.* (1962) 493.

[81] Townes, C. H. and B. P. Dailey, *J. Chem. Phys.* **17** (1949) 782.

[82] Streitwieser, A. A., *Molecular Orbital Theory for Organic Chemists*, John Wiley & Sons, New York, 1961.

[83] Wolfsberg, M. and L. Helmholtz, *J. Chem. Phys.* **20** (1952) 837.

[84] Hoffmann, R., *J. Chem. Phys.* **39** (1963) 1397.

[85] Pople, J. A., *Trans. Farad. Soc.* **49** (1953) 1375.

[86] Pariser, R. and R. G. Parr, *J. Chem. Phys.* **21** (1953) 466.

[87] Pople, J. A. and G. A. Segal, *J. Chem. Phys.* **43** (1965) S129.

[88] Del Bene, J. and H. H. Jaffé, *J. Chem. Phys.* **48** (1968) 1807.

[89] Pople, J. A., D. Beveridge and P. A. Dobosh, *J. Chem. Phys.* **47** (1967) 2026.

[90] Pople, J. A. and D. L. Beveridge, *Approximate Molecular Orbital Theory*, McGraw-Hill, New York, 1970.

[91] Bingham, R. C., M. J. S. Dewar and D. H. Lo, *J. Am. Chem. Soc.* **97** (1975) 1285.

[92] Dewar, M. J. S. and W. Thiel, *J. Am. Chem. Soc.* **99** (1977) 4907.

[93] Dewar, M. J. S., E. G. Zoebisch, E. F. Healey and J. J. P. Stewart, *J. Am. Chem. Soc.* **107** (1985) 3902.

[94] Stewart, J. J. P., *J. Comp. Chem.* **10** (1989) 209, 221.

[95] Dewar, M. J. S., C. Jie and J. Yu, *Tetrahedron* **49** (1993) 5003.

[96] Zerner, M., in *Reviews of Computational Chemistry*, eds. K. B. Lipkowitz and D. B. Boyd, VCH Publishing, New York, 1991.

[97] Bartlett, R. J., *J. Phys. Chem.* **93** (1989) 1697.

[98] Pople, J. A., M. Head-Gordon and K. Raghavachari, *J. Chem. Phys.* **87** (1987) 5968.

[99] Møller, C. and M. S. Plesset, *Phys. Rev.* **46** (1934) 618.

[100] Dirac, P. A. M., *Proc. Cambridge Philos. Soc.* **26** (1930) 376.

[101] Slater, J. C., *Phys. Rev.* **81** (1951) 385.

[102] Gáspár, R., *Acta Phys. Acad. Hung.* **3** (1954) 263.

[103] Hohenberg, P. and W. Kohn, *Phys. Rev.* **136** (1964) B864.

[104] Kohn, W. and L. J. Sham, *Phys. Rev.* **140** (1965) A1133.

[105] Becke, A. D., *J. Chem. Phys.* **88** (1988) 2547.

[106] Becke, A. D., *Phys. Rev.* **A38** (1988) 3098.

[107] Lee, C., W. Yang and R. G. Parr, *Phys. Rev.* **B37** (1988) 785.

[108] Johnson, B. G., P. M. Gill and J. A. Pople, *J. Chem. Phys.* **98** (1993) 5612.

[109] Springborg, M., *Methods of Electronic Structure Calculations: From Molecules to Solids*, John Wiley & Sons, Chichester, 2000

[110] Pople, J. A., M. Head-Gordon, D. J. Fox, K. Raghavachari and L. A. Curtiss, *J. Chem. Phys.* **90** (1989) 5622.

[111] Curtiss, L. A., K. Raghavachari, G. W. Trucks and J. A. Pople, *J. Chem. Phys.* **94** (1991) 7221.

[112] Curtiss, L. A., K. Raghavachari, P. C. Redfern, V. Rassolov and J. A. Pople, *J. Chem. Phys.* **109** (1998) 7764.

[113] Simons, J., P. Jorgensen, H. Taylor and J. Szment, *J. Phys. Chem.* **87** (1983) 2745.

[114] Poppinger, D., *Chem. Phys. Lett.* **35** (1975) 550.

[115] Ermak, D. L. and H. Buckholtz, *J. Comp. Phys.* **35** (1980) 169.

[116] Jorgensen, W. L., J. Chandresekhar and J. D. Madura, *J. Phys. Chem.* **79** (1983) 926.

[117] Warshel, A. and M. Levitt, *J. Mol. Biol.* **103** (1976) 227

[118] Gao, J. and M. A. Thompson, eds., *Combined Quantum Mechanical and Molecular Mechanical Methods*, ACS Symposium Series 712, American Chemical Society, Washington, DC, 1998.

Index